AF355960

Heritage Conservation and Restoration: Surface Characterization, Cleaning and Treatments

Heritage Conservation and Restoration: Surface Characterization, Cleaning and Treatments

Maduka Lankani Weththimuni

Basel • Beijing • Wuhan • Barcelona • Belgrade • Novi Sad • Cluj • Manchester

Editor
Maduka Lankani Weththimuni
Department of Chemistry
University of Pavia
Pavia
Italy

Editorial Office
MDPI AG
Grosspeteranlage 5
4052 Basel, Switzerland

This is a reprint of articles from the Special Issue published online in the open access journal *Coatings* (ISSN 2079-6412) (available at: www.mdpi.com/journal/coatings/special_issues/Heritage_Cleaning).

For citation purposes, cite each article independently as indicated on the article page online and as indicated below:

Lastname, A.A.; Lastname, B.B. Article Title. *Journal Name* **Year**, *Volume Number*, Page Range.

ISBN 978-3-7258-2230-0 (Hbk)
ISBN 978-3-7258-2229-4 (PDF)
doi.org/10.3390/books978-3-7258-2229-4

Contents

About the Editor

Maduka Lankani Weththimuni

Weththimuni Maduka Lankani (Assistant Professor/Ricercatore RTDb) received her B.Sc. Special degree in Chemistry from the University of Sri Jayewardenepura, Sri Lanka, and Ph.D. in Chemistry from the University of Pavia, Italy. She completed her research doctorate (Ph.D.) in Inorganic Chemistry (in the field of Materials Science) in 2013. After that, she worked as a postdoctoral researcher, and she is currently an Assistant Professor/Researcher (RTDb) at the same university, continuing her research and teaching activities there. Her research interests include the synthesis and characterization of nanoparticles, doped and functionalized nanomaterials, nanocomposites, hydrogels, and polymers, and applying them for different purposes.

Preface

Why are heritage conservation and restoration necessary? It is necessary because heritage sites (historical buildings, artefacts, sculptures, paintings, etc.) are symbols of history and are subject to several degradation processes and ageing due to several factors. In particular, decay processes of historical materials related to physical and chemical phenomena occur mainly in the presence of water, which is one of the most deteriorating factors, especially for porous materials (wood and stone). In addition, air pollution, dyes, and microorganism colonisation are frequently related to the harmful degradation of building materials, which contribute to the decay of external historic buildings. Thus, the conservation and restoration of cultural heritage items has become a necessity in a lot of countries. In addition to their intrinsic value, effective conservation and maintenance not only helps in preserving and safeguarding resources but also in revitalising local economies and in bringing about a sense of identity, pride, and belonging to residents. For conservation and restoration purposes, numerous new materials and methods have been developed during the past decade.

Toward this goal, this reprint aims to encourage researchers and to provide them with a platform to read original novel studies together as a book. Eleven original articles were recommended for acceptance and publication. These published articles mainly cover original research articles and reviews on the following areas related to the topic: surface characterisation methods, new tools for characterisation purposes, new materials and methods for surface cleaning, analysing the decay processes, new materials and methods for conservation (novel treatments), novel approaches for restoration, and durability assessments.

We hope that the research achievements presented in the following reprint can contribute to a new platform for the researchers to develop their research with new goals. We would like to express our thanks to all authors who have contributed to this reprint and, in particular, the editorial staff of *Processes* for their support and assistance.

Maduka Lankani Weththimuni
Editor

Editorial

Heritage Conservation and Restoration: Surface Characterization, Cleaning and Treatments

Maduka L. Weththimuni [1,*] and **Maurizio Licchelli** [1,2]

1   Department of Chemistry, University of Pavia, Via Taramelli 12, 27100 Pavia, Italy
2   CISRiC, University of Pavia, Via Ferrata 1, 27100 Pavia, Italy
*   Correspondence: madukalankani.weththimuni@unipv.it

Citation: Weththimuni, M.L.; Licchelli, M. Heritage Conservation and Restoration: Surface Characterization, Cleaning and Treatments. *Coatings* **2023**, *13*, 457. https://doi.org/10.3390/coatings13020457

Received: 9 February 2023
Accepted: 15 February 2023
Published: 17 February 2023

Conservation is not the same thing as restoration. Conservation attempts to preserve an artifact in its current condition. By contrast, the restoration process strives to return an artifact back to its original condition. In other words, conservation lets a viewer see the damage that naturally and inevitably occurs as a result of the aging of the artifact over the course of its life, whereas restoration lets a viewer see what the artifact originally looked like. Why are the conservation and restoration of cultural heritage items relevant or even necessary? Heritage items (historical buildings, artifacts, sculptures, paintings, and books) are symbols of our cultural patrimony, which are subject to aging and degradation processes due to several factors (physical, chemical, biological, and air pollution) [1–4] and, therefore, need to be properly preserved for the benefit of future generations. Generally, decay processes of historical materials related to physical and chemical phenomena, mainly occur in the presence of water, which is one of the most deteriorating factors, especially for porous materials (paper, wood, and stone). Highly porous materials often undergo faster degradation, as water and other decay agents carried by water may easily penetrate below their surface through their wide and highly accessible pores [1,5–7]. For instance, history books and wood artifacts may undergo drastic damage due to this phenomenon. In fact, due to the hydrophilic nature of their components (cellulose and hemicellulose), they can easily adsorb water (both as vapor and liquid), establishing an equilibrium with the air moisture [8]. The presence of water mainly promotes hydrolytic reactions and biodegradation processes (e.g., bacteria, fungi, and xylophages), particularly when the moisture content of those materials exceeds the fiber saturation point [8]. Besides water, environmental agents (e.g., air pollution, solar radiation, and temperature variations) and microorganism colonization are frequently related to the harmful degradation of cultural heritage items [7–10]. For instance, building materials undergo weathering due to different processes (freezing–thawing cycles, salt crystallization, acid rain, and interactions with atmospheric pollutants such as SO_2), when exposed to outdoor conditions, which can produce a variety of types of damage to the original stone substrates, such as surface erosion, exfoliation, loss of material, and, in the worst cases, disaggregation [4,11,12]. Absorption of solar light, particularly of the ultraviolet (UV) component, is responsible for the photodegradation of paper and wood materials owing to different photolytic and/or photo-oxidative reactions [13,14]. As a consequence, these kinds of materials undergo different decay processes when exposed to solar light, ranging from changes in their esthetic appearance to a severe decrease in their mechanical properties [8,15].

Thus, the conservation and restoration of cultural heritage materials have become a necessity in many countries. In addition to their intrinsic value, effective conservation and maintenance not only help in preserving and safeguarding resources, but also in revitalizing local economies and in bringing about a sense of identity, pride, and belonging to residents. Hence, in past decades, scientists have focused their research on discovering appropriate materials and procedures for conservation and restoration purposes.

The removal of any undesired matter (e.g., dust, aged coatings, deposits of pollutants, graffiti, and dirt) is one of the most important and delicate operations that affects the different ways (potentially invasive, aggressive, and completely irreversible) the original material should be approached, and occurs before any conservation or restoration process begins [5,16–18]. This is an essential approach because it helps to remove foreign matter which could promote the degradation of the heritage materials as well as cause changes to their original appearance.

If the cleaning intervention is not correctly performed, it may irreversibly damage the original substrate of the considered heritage items. Therefore, there are many conditions to control when selecting the appropriate cleaning agent. For instance, when the removal of undesired materials involves a water-sensible substrate, no cleaning procedures involving aqueous solutions can be carried out, making it a more challenging task than when using other substrates [17]. Hence, high-performing cleaning systems, which are able to perform controlled and selective cleaning actions, are highly desired [16]. Traditionally, several organic solvents have been used to dissolve or swell aged polymers, varnishes, and other unwanted materials, whose removal from substrate surfaces can also accomplished through mechanical action [17,19,20]. However, this method has declined in use due to several drawbacks: uncontrolled penetration of the organic solvents into the substrates, the solvents reacting with and damaging the original materials, poor selectivity, residues left behind, and safety problems (for the environment as well as for users) [16,17]. Thus, several other methods have been introduced to this field, but have produced unsatisfactory results. For instance, the cleaning methods based on cellulose pulp poultice and gels (the inclusion of the solvents within a cellulose matrix and a polymer network, respectively) also showed some unsatisfactory results, particularly when the residual gel or cellulose (or other components, e.g., surfactants) were used on the treated surface [16,21,22]. Therefore, there is high demand to develop smart cleaning tools that can overcome all these limitations. In recent decades, conservation scientists have investigated alternative products such as physical and chemical gel materials based on natural (e.g., xanthan gum, gellan gum, agar, chitosan, sodium alginate, and glucomannan-borax) and artificial polymers (e.g., polyacrylamide and p(HEMA)/PVP)), as well as conducted research into modifying them to act as a transporting medium for different aqueous cleaning systems (micro-emulsions, mixed solvent systems, micellar solutions, and nanoemulsions), which could be correctly used to clean heritage items and overcome the above-mentioned limitations [16,23–26].

After successfully cleaning the undesired materials from the heritage items, further steps are required for the conservation/restoration process. As mentioned previously, the process completely depends on the substrate and its porosity. Traditionally, using different kinds of materials for the surface treatment has been the common method of consolidation and protection, especially in the case of wood and stone artifacts [7,27–29]. A variety of protecting and consolidating products, based on both organic and inorganic materials, have been developed in recent decades, but their uses were often limited due to different adverse phenomena. For instance, synthetic polymers generally show poor durability and insufficient compatibility, as well as significantly affect the properties of the original substrate. On the other hand, inorganic consolidants (e.g., $Ca(OH)_2$, $Ba(OH)_2$, and $(NH_4)_2C_2O_4$) show good compatibility and good durability properties, but, at the same time, they are characterized by an insufficient penetration depth and often induce surface whitening and poor strengthening of the substrates [4,30,31]. The use of the well-known consolidant tetraethoxysilane (TEOS) also depends on the substrate and has several limitations when applied to carbonate stones [4]. In recent decades, the use of nanomaterials in all fields has been a successful development, providing new materials, concepts, and advantages to the research areas concerning conservation/restoration. For instance, different inorganic nanomaterials have been studied as active components for the conservation of stone and wood artifacts [32–34]. For instance, Baglioni et al., reported that alcoholic dispersion of calcium hydroxide nanoparticles is highly compatible with carbonate-based materials, such as in wall paintings and calcareous stone [32]. The dispersion was commercialized as Nanorestore Plus®, which is a good alternative to the traditional consolidation products [35]. Moreover, other alkaline hydroxides ($Sr(OH)_2$, $Ba(OH)_2$, and $Mg(OH)_2$), silica nanoparticles, and nano-hydroxyapatite (the in situ reaction of calcium hydroxide nanoparticles with diammonium hydrogen phosphate) have been also tested as consolidation agents for different kind of porous stones [1,4,32,36–39]. Meanwhile, conservation scientists have a great interest in developing multifunctional products that are able to provide surface protection, consolidation, and self-cleaning efficiency when applied to the heritage materials. For example, Crupi et al., reported the self-cleaning efficiency of a TiO_2–SiO_2–PDMS coating [40]; Kapridaki et al., analyzed different types of coatings (a SiO_2-crystalline TiO_2 nanocomposite coating and TEOS-nanocalcium oxalate consolidant) for the same purposes [41,42]; La Russa et al., studied nano-TiO_2 coatings with different binder materials [43]; Ag-TiO_2/PDMS and Gd-doped TiO_2/PDMS nanocomposites were analyzed by Ben Chobba et al. [44–47]; and Weththimuni et al., examined a nanocomposite coating based on ZrO_2-doped-ZnO-PDMS as a self-cleaning, multifunctional, and durable coating [48–50]. Nanomaterials were also used in chemical modification, impregnation, and to improve the properties of natural coatings (e.g., shellac). For instance, Cristea et al., examined the improvement of wood coatings by using zinc oxide (ZnO), silica (SiO_2), and titanium dioxide (TiO_2) nanoparticles [51], whereas Weththimuni et al., studied the enhancement of properties displayed by traditional shellac varnish by introducing inorganic nanoparticles (e.g., ZnO, SiO_2, ZrO_2, and montmorillonite) as well as functionalized nanoparticles [52,53]. The same group also reported the enhancement of wood resistance to decay induced by ZnO nanorod-based treatments [8]. Although numerous new materials and methods have been developed over the past decade, conservation/restoration efficiency still depends on several factors, including the nature of the substrate, porosity, the chemical composition of the original material, decay phenomena happening on the original heritage item, and the location and conservation conditions of original artifacts. Hence, all products and methods must always be used with some precaution and in a controlled way.

Therefore, research scientists are still focusing their research to discover more appropriate new materials and their applications to conserve and restore our heritage items for future generations. With this goal in mind, we are assembling a Special Issue of *Coatings* with the aim of encouraging researchers to publish novel studies this topic by providing them with a platform. Subtopics may include (but are not limited to): surface characterization methods, new tools for characterization purposes, analyses of decay processes, new materials and methods for surface cleaning, new materials and methods (novel treatments) for conservation, novel approaches to restoration, and durability assessments. In this Special Issue, original research articles and reviews are welcome.

Funding: This research received no external funding.

Institutional Review Board Statement: Not applicable.

Informed Consent Statement: Not applicable.

Conflicts of Interest: The authors declare no conflict of interest.

References

1. Licchelli, M.; Malagodi, M.; Weththimuni, M.; Zanchi, C. Nanoparticles for conservation of bio-calcarenite stone. *Appl. Phys. A* **2014**, *114*, 673–683. [CrossRef]
2. Warscheid, T.; Braams, J. Biodeterioration of stone: A review. *Int. Biodeterior. Biodegrad.* **2000**, *46*, 343–368. [CrossRef]
3. Jroundi, F.; Gómez-Suaga, P.; Jimenez-Lopez, C.; González-Muñoz, M.T.; Fernandez-Vivas, M.A. Stone-isolated carbonatogenic bacteria as inoculants in bioconsolidation treatments for historical limestone. *Sci. Total Environ.* **2012**, *425*, 89–98. [CrossRef] [PubMed]
4. Weththimuni, M.L.; Licchelli, M.; Malagodi, M.; Rovella, N.; Russa, M.L. Consolidation of bio-calcarenite stone by treatment based on diammonium hydrogenphosphate and calcium hydroxide nanoparticles. *Measurement* **2018**, *127*, 396–405. [CrossRef]
5. Licchelli, M.; Malagodi, M.; Weththimuni, M.; Zanchi, C. Anti-graffiti nanocomposite materials for surface protection of a very porous stone. *Appl. Phys. A* **2014**, *116*, 1525–1539. [CrossRef]

6. Shupe, T.; Lebow, S.; Ring, D. *Causes and Control of Wood Decay, Degradation & Stain, in Degradation & Stain*; Pub. (Louisiana Cooperative Extension Service) -2703; Louisiana State University Agricultural Center: Baton Rouge, LA, USA, 2008.
7. Licchelli, M.; Malagodi, M.; Weththimuni, M.L.; Zanchi, C. Water-repellent properties of fluoroelastomers on a very porous stone: Effect of the application procedure. *Prog. Org. Coat.* **2013**, *76*, 495–503. [CrossRef]
8. Weththimuni, M.L.; Capsoni, D.; Malagodi, M.; Licchelli, M. Improving Wood Resistance to Decay by Nanostructured ZnO-Based Treatments. *J. Nanomater.* **2019**, *2019*, 6715756. [CrossRef]
9. Belfiore, C.M.; Fichera, G.V.; la Russa, M.F.; Pezzino, A.; Ruffolo, S.A. The Baroque architecture of Scicli (south-eastern Sicily): Characterization of degradation materials and testing of protective products. *Period. Mineral.* **2012**, *81*, 19–33.
10. Ricca, M.; Le Pera, E.; Licchelli, M.; Macchia, A.; Malagodi, M.; Randazzo, L.; Rovella, N.; Ruffolo, S.A.; Weththimuni, M.L.; La Russa, M.F. The CRATI Project: New Insights on the Consolidation of Salt Weathered Stone and the Case Study of San Domenico Church in Cosenza (South Calabria, Italy). *Coatings* **2019**, *9*, 330. [CrossRef]
11. Ruffolo, S.A.; La Russa, M.F.; Aloise, P.; Belfiore, C.M.; Macchia, A.; Pezzino, A.; Crisci, G.M. Efficacy of nanolime in restoration procedures of salt weathered limestone rock. *Appl. Phys. A.* **2014**, *114*, 753–758. [CrossRef]
12. Poli, T.; Toniolo, L.; Chiantore, O. The protection of different Italian marbles with two partially flourinated acrylic copolymers. *Appl. Phys. A* **2004**, *79*, 347–351.
13. Sun, Q.; Lu, Y.; Zhang, H.; Yang, D.; Wang, Y.; Xu, J.; Tu, J.; Liu, Y.; Li, J. Improved UV resistance in wood through the hydrothermal growth of highly ordered ZnO nanorod arrays. *J. Mater. Sci.* **2012**, *47*, 4457–4462. [CrossRef]
14. Pandey, K.K. Study of the effect of photo-irradiation on the surface chemistry of wood. *Polym. Degrad. Stab.* **2005**, *90*, 9–20. [CrossRef]
15. Nzokou, P.; Kamdem, D.P.; Temiz, A. Effect of accelerated weathering on discoloration and roughness of finished ash wood surfaces in comparison with red oak and hard maple. *Prog. Org. Coat.* **2011**, *71*, 350–354. [CrossRef]
16. Domingues, J.A.L.; Bonelli, N.; Giorgi, R.; Fratini, E.; Gorel, F.; Baglioni, P. Innovative Hydrogel Based on Semi-Intrepenetrating p(HEMA)/PVP Networks for the Cleaning of Water-Sensitive Cultural Heritage Artifacts. *Langmuir* **2013**, *29*, 2746–2755. [CrossRef]
17. Baglioni, M.; Alterini, M.; Chelazzi, D.; Giorgi, R.; Baglioni, P. Removing Polymeric Coatings With Nanostructured Fluids: Influence of Substrate, Nature of the Film, and Application Methodology. *Front. Mater.* **2019**, *6*, 1–18. [CrossRef]
18. Licchelli, M.; Marzolla, S.J.; Poggi, A.; Zanchi, C. Crosslinked fluorinated polyurethanes for the protection of stone surfaces from graffiti. *J. Cult. Herit.* **2011**, *12*, 34–43. [CrossRef]
19. Baglioni, M.; Giorgi, R.; BERTI, D.; BAGLIONI, P. Smart cleaning of cultural heritage: A new challenge for soft nanoscience. *Nanoscale* **2012**, *4*, 42–53. [CrossRef]
20. Burnstock, A.; White, R. A preliminary assessment of the aging/degradation of ethomeen C-12 residues from solvent gel formulations and their potential for inducing changes in resinous paint media. *Stud. Conserv.* **2000**, *45*, 34–38. [CrossRef]
21. Cremonesi, P. *L'Uso di Tensioattivi e Chelanti nella Pulitura di Opere Policrome*; Il Prato: Padua, Italy, 2004.
22. Burnstock, A.; Kieslich, T. A Study of the Clearance of Solvent Gels Used for Varnish Removal from Paintings. In *ICOM Committee for Conservation, Proceedings of the 11th Triennial Conference, Edinburgh, Scotland, 1–6 September 1996*; James & James: London, UK, 1996; pp. 253–262.
23. Richteringa, W.; Saunders, B.R. Gel architectures and their complexity. *Soft Matter* **2014**, *10*, 3695–3702. [CrossRef]
24. Lee, C.; Volpi, F.; Fiocco, G.; Weththimuni, M.L.; Licchelli, M.; Malagodi, M. Preliminary Cleaning Approach with Alginate and Konjac Glucomannan Polysaccharide Gel for the Surfaces of East Asian and Western String Musical Instruments. *Materials* **2022**, *15*, 1100. [CrossRef] [PubMed]
25. Lee, C.; Di Turo, F.; Vigani, B.; Weththimuni, M.L.; Rossi, S.; Beltram, F.; Pingue, P.; Licchelli, M.; Malagodi, M.; Fiocco, G.; et al. Biopolymer Gels as a Cleaning System for Differently Featured Wooden Surfaces. *Polymers* **2023**, *15*, 36. [CrossRef] [PubMed]
26. Thakur, S.; Arotiba, O.A. Synthesis, swelling and adsorption studies of a pH responsive sodium alginate–poly(acrylic acid) super absorbent hydrogel. *Polym. Bull.* **2018**, *75*, 4587–4606. [CrossRef]
27. Weththimuni, M.L.; Canevari, C.; Legnani, A.; Licchelli, M.; Malagodi, M.; Ricca, M.; Zeffiro, A. Experimental Characterization of Oil-Colophony Varnishes: A Preliminary Study. *Int. J. Conserv. Sci.* **2016**, *7*, 813–826.
28. Spinella, A.; Malagodi, M.; Saladino, M.L.; Weththimuni, M.L.; Caponetti, E.; Licchelli, M. A Step Forward in Disclosing the Secret of Stradivari's Varnish by NMR Spectroscopy. *J. Polym. Sci. Part A Polym. Chem.* **2017**, *55*, 3949–3954. [CrossRef]
29. Licchelli, M.; Malagodi1, M.; Somaini, M.; Weththimuni, M.; Zanchi, C. Surface treatments of wood by chemically modified shellac. *Surf. Eng.* **2013**, *29*, 121–127. [CrossRef]
30. Hansen, E.; Doehne, E.; Fidler, J.; Larson, J.; Martin, B.; Matteini, M.; Rodriguez-Navarro, C.; Pardo, E.S.; Price, C.; De Tagle, A.; et al. A review of selected inorganic consolidants and protective treatments for porous calcareous materials. *Stud. Conserv.* **2003**, *48*, 13–25. [CrossRef]
31. Weththimuni, M.; Crivelli, F.; Galimberti, C.; Malagodi, M.; Licchelli, M. Evaluation of commercial consolidating agents on very porous biocalcarenite. *Int. J. Conserv. Sci.* **2020**, *11*, 251–260.
32. Dei, L.; Salvadori, B. Nanotechnology in cultural heritage conservation: Nanometric slaked lime saves architectonic and artistic surfaces from decay. *J. Cult. Herit.* **2006**, *7*, 110–115. [CrossRef]
33. D'Arienzo, L.; Scarfato, P.; Incarnato, L. New polymeric nanocomposites for improving the protective and consolidating efficiency of tuff stone. *J. Cult. Herit.* **2008**, *9*, 253–260. [CrossRef]

34. La Russa, M.F.; Ruffolo, S.A.; Rovella, N.; Belfiore, C.M.; Palermo, A.M.; Guzzi, M.T.; Crisci, G.M. Multifunctional TiO$_2$ coatings for cultural heritage. *Prog. Org. Coat.* **2012**, *74*, 186–191. [CrossRef]
35. Baglioni, P.; Chelazzi, D.; Giorgi, R. *Nanotechnologies in the Conservation of Cultural Heritage: A Compendium of Materials and Techniques*; Springer: Dordrecht, The Netherlands, 2014; ISBN 9401793026.
36. Chelazzi, D.; Poggi, G.; Jaidar, Y.; Toccafondi, N.; Giorgi, R.; Baglioni, P. Hydroxide nanoparticles for cultural heritage: Consolidation and protection of wall paintings and carbonate materials. *J. Colloid Interface Sci.* **2013**, *392*, 42–49. [CrossRef]
37. Giorgi, R.; Ambrosi, M.; Toccafondi, N.; Baglioni, P. Nanoparticles for cultural heritage conservation: Calcium and barium hydroxide nanoparticles for wall painting consolidation. *Chem. Eur. J.* **2010**, *16*, 9374–9382. [CrossRef]
38. Ciliberto, E.; Condorelli, G.G.; La Delfa, S.; Viscuso, E. Nanoparticles of Sr(OH)$_2$: Synthesis in homogeneous phase at low temperature and application for cultural heritage artefacts. *Appl. Phys. A.* **2008**, *92*, 137–141. [CrossRef]
39. Sassoni, E.; Franzoni, E.; Pigino, B.; Scherer, G.W.; Naidu, S. Consolidation of calcareous and siliceous sandstones by hydroxyapatite: Comparison with a TEOS-based consolidant. *J. Cult. Herit.* **2013**, *14S*, 103–108. [CrossRef]
40. Crupi, V.; Fazio, B.; Gessini, A.; Kis, Z.; La Russa, M.F.; Majolino, D.; Masciovecchio, C.; Ricca, M.; Rossi, B.; Ruffolo, S.A.; et al. TiO$_2$–SiO$_2$–PDMS nanocomposite coating with self-cleaning effect for stone material: Finding the optimal amount of TiO$_2$. *Constr. Build. Matter.* **2018**, *166*, 464–471. [CrossRef]
41. Kapridaki, C.; Pinho, L.; Mosquera, M.J.; Maravelaki-kalaitzaki, P. Producing photoactive, transparent and hydrophobic SiO$_2$-crystalline TiO$_2$ nanocomposites at ambient conditions with application as self-cleaning coatings. *Appl. Catal. B Environ.* **2014**, *156–157*, 416–427. [CrossRef]
42. Kapridaki, C.; Verganelaki, A.; Dimitriadou, P.; Maravelaki-Kalaitzaki, P. Conservation of Monuments by a Three-Layered Compatible Treatment of TEOS-Nano-Calcium Oxalate Consolidant and TEOS-PDMS-TiO$_2$ Hydrophobic/Photoactive Hybrid Nanomaterials. *Materials* **2018**, *11*, 684. [CrossRef]
43. la Russa, M.F.; Rovella, N.; de Buergo, M.A.; Belfiore, C.M.; Pezzino, A.; Crisci, G.M.; Ruffolo, S.A. Nano-TiO$_2$ coatings for cultural heritage protection: The role of the binder on hydrophobic and self-cleaning efficacy. *Prog. Org. Coat.* **2016**, *91*, 1–8. [CrossRef]
44. Chobba, M.B.; Weththimuni, M.L.; Messaoud, M.; Urzi, C.; Bouaziz, J.; de Leo, F.; Licchelli, M. Ag-TiO$_2$/PDMS nanocomposite protective coatings: Synthesis, characterization, and use as a self-cleaning and antimicrobial agent. *Prog. Org. Coat.* **2021**, *158*, 106342. [CrossRef]
45. Ben Chobba, M.; Weththimuni, M.L.; Messaoud, M.; Sacchi, D.; Bouaziz, J.; Leo, F.D.; Urzi, C.; Licchelli, M. Multifunctional and Durable Coatings for Stone Protection Based on Gd-Doped Nanocomposites. *Sustainability* **2021**, *13*, 11033.
46. Ben Chobba, M.; Weththimuni, M.L.; Messaoud, M.; Bouaziz, J.; Salhi, R.; De Leo, F.; Urzì, C.; Licchelli, M. Silver-Doped TiO$_2$-PDMS Nanocomposite as a Possible Coating for the Preservation of Serena Stone: Searching for Optimal Application Conditions. *Heritage* **2022**, *5*, 3411–3426. [CrossRef]
47. Ben Chobba, M.; Weththimuni, M.L.; Messaoud, M.; Bouaziz, J.; Licchelli, M. Enhanced Gd Doped TiO$_2$ NPs-PDMS Nanocomposites as Protective Coatings for Bio-Calcarenite Stone: Preliminarily Analysis. In *Design and Modeling of Mechanical Systems—V.*; Walha, L., Jarraya, A., Djemal, F., Chouchane, M., Aifaoui, N., Chaari, F., Abdennadher, M., Benamara, A., Haddar, M., Eds.; Springer International Publishing: Cham, Switzerland, 2022; pp. 885–893.
48. Weththimuni, M.L.; Ben Chobba, M.; Tredici, I.; Licchelli, M. Polydimethylsiloxane (PDMS)/ZrO2-Doped ZnO Nanocomposites as Protective Coatings for Stone Materials, TC4 MetroArchaeo 2020-IMEKO TC4 International Conference on Metrology for Archaeology and Cultural Heritage. 2020, pp. 527–531. Available online: https://www.imeko.org/publications/tc4-Archaeo-20 20/IMEKO-TC4-MetroArchaeo2020-100 (accessed on 24 October 2020).
49. Weththimuni, M.L.; Ben Chobba, M.; Sacchi, D.; Messaoud, M.; Licchelli, M. Durable Polymer Coatings: A Comparative Study of PDMS-Based Nanocomposites as Protective Coatings for Stone Materials. *Chemistry* **2022**, *4*, 60–76. [CrossRef]
50. Weththimuni, M.; Ben Chobba, M.; Tredici, I.; Licchelli, M. ZrO2-Doped ZnO-PDMS Nanocomposites as Protective Coatings for the Stone Materials. *ACTA IMEKO* **2022**, *11*, 5. [CrossRef]
51. Cristea, M.V.; Riedl, B.; Blanchet, P. Effect of addition of nanosized UV absorbers on the physico-mechanical and thermal properties of an exterior waterborne stain for wood. *Prog. Org. Coat.* **2011**, *72*, 755–762. [CrossRef]
52. Weththimuni, M.L.; Capsoni, D.; Malagodi, M.; Milanese, C.; Licchelli, M. Shellac/nanoparticles dispersions as protective materials for wood. *Appl. Phys. A* **2016**, *122*, 1058. [CrossRef]
53. Weththimuni, M.L.; Milanese, C.; Licchelli, M.; Malagodi, M. Improving the protective properties of shellac-based varnishes by functionalized nanoparticles. *Coatings* **2021**, *11*, 419. [CrossRef]

Article

Spectral- and Image-Based Metrics for Evaluating Cleaning Tests on Unvarnished Painted Surfaces

Jan Dariusz Cutajar [1,*], **Calin Constantin Steindal** [2], **Francesco Caruso** [3,4], **Edith Joseph** [5,6] **and Tine Frøysaker** [1]

1   Conservation Studies, University of Oslo, 0164 Oslo, Norway; tine.froysaker@iakh.uio.no
2   Cultural History Museum, University of Oslo, 0130 Oslo, Norway; c.c.steindal@khm.uio.no
3   Department of Analytical Chemistry, University of the Basque Country UPV/EHU,
    01006 Vitoria-Gasteiz, Spain; francesco.caruso@ehu.eus
4   Department of Art Technology, Swiss Institute for Art Research (SIK-ISEA), 8008 Zurich, Switzerland
5   Haute École Art Conservation-Restoration (HE-Arc CR), Haute École Spécialisée de Suisse
    Occidentale (HES-SO), 2000 Neuchâtel, Switzerland; edith.joseph@he-arc.ch
6   Laboratory of Technologies for Heritage Materials, University of Neuchâtel, 2000 Neuchâtel, Switzerland
*   Correspondence: jd_c@outlook.com or j.d.cutajar@iakh.uio.no

Abstract: Despite advances in conservation–restoration treatments, most surface cleaning tests are subjectively evaluated. Scores according to qualitative criteria are employed to assess results, but these can vary by user and context. This paper presents a range of cleaning efficacy and homogeneity evaluation metrics for appraising cleaning trials, which minimise user bias by measuring quantifiable changes in the appearance and characteristic spectral properties of surfaces. The metrics are based on various imaging techniques (optical imaging by photography using visible light (VIS); spectral imaging in the visible-to-near-infrared (VNIR) and shortwave infrared (SWIR) ranges; chemical imaging by Fourier transform infrared (FTIR) spectral mapping in the mid-infrared (MIR) range; and scanning electron microscopy coupled with energy dispersive X-ray spectroscopy (SEM-EDX) element mapping). They are complemented by appearance measurements (glossimetry and colourimetry). As a case study showcasing the low-cost to high-end metrics, agar gel spray cleaning tests on exposed ground and unvarnished oil paint mock-ups are reported. The evaluation metrics indicated that spraying agar (prepared with citric acid in ammonium hydroxide) at a surface-tailored pH was as a safe candidate for efficacious and homogenous soiling removal on water-sensitive oil paint and protein-bound ground. Further research is required to identify a gel-based cleaning system for oil-bound grounds.

Keywords: treatment evaluation; surface cleaning; soiling removal; agar gel; exposed grounds; unvarnished oil paints; Edvard Munch; spectral imaging

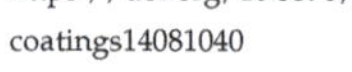

Citation: Cutajar, J.D.; Steindal, C.C.; Caruso, F.; Joseph, E.; Frøysaker, T. Spectral- and Image-Based Metrics for Evaluating Cleaning Tests on Unvarnished Painted Surfaces. *Coatings* **2024**, *14*, 1040. https://doi.org/10.3390/coatings14081040

Academic Editor: Maduka Lankani Weththimuni

Received: 8 July 2024
Revised: 9 August 2024
Accepted: 13 August 2024
Published: 15 August 2024

1. Introduction

The conservation field is a rich and dynamic profession that has continuously drawn on its interdisciplinary nature to find appropriate remedial treatments for the challenges faced by its practitioners. The surface cleaning of decorative or painted surfaces is one such challenge that occupies conservators [1]. In this paper, *surface cleaning* will be referred to as the selective removal of deposited soiling upon painted surfaces. In turn, *soiling* is used to imply atmospheric particulate matter that deposits itself on heritage surfaces [2].

Besides traditional options, the repertoire of surface cleaning treatments available to conservators features adaptions from fields such as green chemistry [3], chemical engineering [4,5], microbiology [6–11], and physics [12,13]. In recent years, notable improvements have been made with gel cleaning [14–33], which is the focus of this work.

Despite these advances, most of these cleaning treatments are still evaluated in a somewhat subjective manner. Scores according to pre-defined criteria are assigned to the results of cleaning tests, following an ever more established practice within the profession [16,34,35]. This evaluation method is relatively easy to apply, and can also be visualised as star diagrams to better comprehend outcomes [16,36–40].

On one hand, this practice reflects the inherent nature of conservation practice. Every object needs to be treated differently based on its context [41–44] and, as specialised professionals, conservators will evaluate risk and success based on their training and experience. From a stakeholder/representative perspective, the perception of what makes a successful cleaning treatment is highly subjective too, and the concept of cleanliness can be said to be values- and/or community-based, as defined by contemporary conservation theory [45,46]. From this point of view, it is possible that a fully objective evaluation is perhaps not achievable, nor wholly desirable. On the other hand, subjective scorings, albeit assigned by trained and experienced eyes, are hard to reproduce, and are reliant upon many contextual factors that might skew the judgment of the assessor in question. From another perspective, the test surfaces often used in cleaning studies are not usually homogenous (due to their complex nature) and this might introduce further bias when attempting to weigh cleaning test outcomes.

Cleaning studies in conservation often use analogous mock-ups to mimic a material surface under investigation [47–52]. In this context, mock-ups are a critical tool for conservators to explore the behaviour of a heritage object with respect to a trialled treatment, whilst respecting the integrity of the object itself. The properties of the mock-ups can be selectively tuned for the treatment (or object) under study, and the mock-ups can be consumed/reproduced according to need, or even reused for future studies on the evaluation of a treatment over time [53,54].

Surface cleaning evaluations typically select a number of salient criteria upon which to assess the treatments under consideration. Generally, these may include cleaning assessment factors such as soiling removal efficacy, homogeneity, selectivity (i.e., lack of pigment pick-up), surface disruption (e.g., by swelling, or foaming), residue deposition, and surface appearance (with respect to colour and gloss), amongst others [16,23,34,38,39]. An example of user-dependent criteria is given in Table 1.

Table 1. Example of cleaning criteria used in conservation studies [16,23,38,39].

	Scoring Criteria			
Score Rating	Cleaning Efficacy	Cleaning Homogeneity	Pigment Swelling	Selectivity (Pigment Loss)
1	No effect	Uneven removal (<30%)	Extreme, visible swelling	Unacceptable loss
2	Little effect	Inconsistent removal (<50%)	Moderate, visible swelling	Notable loss
3	Moderate effect	Consistent removal (<80%)	Sensation of swelling, invisible	Microscopic loss
4	Effective removal	Complete removal (100%)	No swelling	No loss

The above criteria are assessed visually as a function of change compared to a predetermined surface condition by a conservator's trained eye, but this introduces user bias into any evaluation, regardless of the expertise of the assessor. For this reason, the conservator's intuitive observations are often corroborated, with added empirical confidence, by measured analytical data. For example, optical and digital microscopy is usually employed to pick up differences resulting from cleaning selectivity and surface disruption; colourimetry and glossimetry are used to measure changes in surface appearance, whereas scanning electron microscopy coupled with energy dispersive X-ray spectroscopy (SEM-EDX) and Fourier transform infrared (FTIR) imaging are used to monitor chemical changes in the surface and residue deposition by treatments. Cleaning homogeneity and efficacy, however, remain based on estimates registered by the user.

Diagnostic imaging technologies (such as spectral imaging [55]) can provide opportunities to incorporate a high amount of empirical-based data that can act as a complementary aid for conservators in making more reliable, repeatable, and accurate assessments when assigning scores for novel treatment evaluations. Research efforts have already begun addressing this issue [54,56].

To address and overcome user subjectivity, the methodology presented in this article offers the choice of non-contact and/or non-invasive image, appearance, and spectral measurements, which conservators can avail of (according to access to instrumentation). By juxtaposing these techniques together, this work thus aims to contribute towards offering a robust, rigorous, and comparable evaluation framework for documenting surfaces before and after soiling removal tests.

The methodology takes advantage of the image-based outputs generated by various diagnostic imaging techniques, which can then be processed in freeware, such as FIJI (ImageJ, v. 1.54f) [57], to provide semi-quantitative percentage scores, thus offering a common platform for comparison. ImageJ is a powerful, established, open-source, freely accessible image processing package, initially developed for the biomedical sciences. Additionally, it is more user-friendly compared to other proprietary software that is typically used to process spectral imaging data (e.g., ENVI 6.0 [58], or MATLAB 2024a [59]). Furthermore, ImageJ has a vibrant online community, offering plug-ins that can be downloaded according to need. It can therefore serve as an alternative to other mainstream image data processing platforms.

The evaluation metrics developed in this study were based on optical photography and microscopy, appearance-based measurements (colourimetry and glossimetry), and imaging spectroscopy (hyperspectral imaging (HSI), micro-FTIR (μFTIR) mapping, and SEM-EDX mapping). Notably, in this paper, established post-processing techniques in imaging spectroscopy were applied to the evaluation of surface cleaning. Spectral unmixing algorithms were implemented to identify and map soiled and cleaned surfaces. These algorithms classify pure or mixed materials based on spectral similarity [60–62], either (i) using user-defined, known references or libraries (e.g., semi-supervised unmixing by *PoissonNMF* [63]), or (ii) blindly classifying them into groups (e.g., unsupervised unmixing by *LUMoS* [64]). A chemometric method for normalised difference image (NDI) mapping [65–67] of soiled and cleaned surfaces is also reported. Using principal component analysis (PCA) [68], characteristic wavelengths in the shortwave-infrared (SWIR) were selected to represent and then map soiling on the surfaces under the study.

2. Materials and Methods

Mock-ups. The case study presented in this article centres around mock-ups of Edvard Munch's (1865–1944) monumental unvarnished painting, *Kjemi* (1914–1916, Woll no. 1227), situated at the University of Oslo Aula (1911) in central Oslo. This painting forms part of a frieze of eleven monumental artworks (oil on canvas) under study within the CHANGE-ITN and Munch Aula Paintings (MAP) projects [68–70]. The mock-ups have been designed to study a new cleaning technique—agar gel spray—for unvarnished, water-sensitive painted surfaces [71] as a response to the heavy soiling history of the paintings within the Aula [53,72–77], as well as to test the applicability and suitability of the imaging techniques mentioned above for the documentation of the cleaning trials.

Figure 1 illustrates one protein-bound and one oil-bound exposed ground, and one unvarnished oil paint mock-up (5 cm × 5 cm), formulated based on material analyses of *Kjemi* [73,78]. An in-depth description of the mock-up surfaces is provided in Table 2. They were naturally and artificially aged to emulate the mechanical and water-sensitive properties of the Aula paintings. The mock-up surfaces were not intentionally contaminated with any micro-organisms, and thus were considered free of any biological contamination (however, this was not analytically confirmed). Artificial soiling was applied following a modified methodology adapted for the Aula context [51,79]. The artificial soiling recipe used is given in Table 3 (materials sourced from Rublev (Willits, CA, USA); Kremer (Aichstetten, Germany); Merck (Darmstadt, Germany); and Filippo Berio (Lucca, Italy)). The soiling comprised insoluble macroscopic particulates (iron oxide, silica, kaolin), microscopic soot particles (carbon black), water-reactive solids (Portland cement, type I), hydrophilic organics (gelatine powder, soluble starch), and hydrophobic organics (olive oil, mineral oil), all dispersed in Shellsol D40 mineral spirits. The mock-ups were soiled across three cycles of accelerated ageing (Table 2). Excess soiling was applied to facilitate soiling detection. The mock-ups before treatment (BT) were therefore slightly more heavily soiled than is currently observed on the Aula paintings.

Mock-up surface characterisation

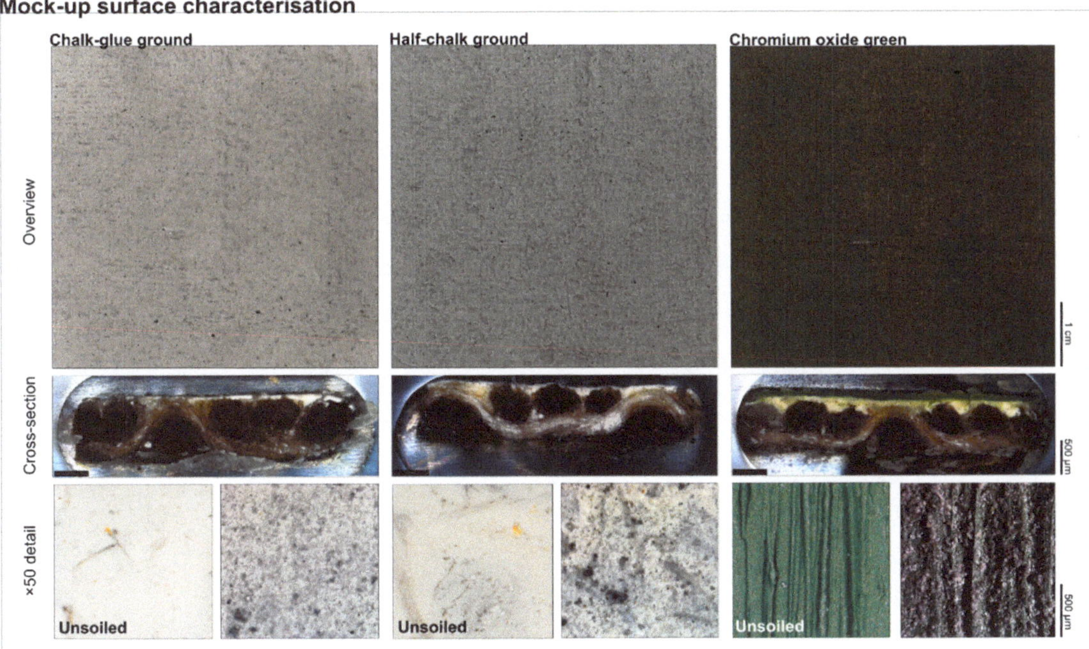

Figure 1. Overview (**top row**), cross-sections (**central row**), and magnified detail (**bottom row**) of the soiled mock-up surfaces (5 cm × 5 cm) cleaned during this study (unsoiled mock-ups are shown for comparison purposes, indicating the degree of soiling and differences in surface topography). The three mock-up classes included a chalk–glue ground, a half-chalk ground, and a chromium oxide green oil pain in linseed oil applied onto a half-chalk ground.

Agar spray and cleaning solutions. Agar spray is a novel form of agar gel cleaning whereby the agar in sol state is subjected to atomisation within a heated paint applicator, allowing it to be applied as a homogenous and ultra-fine layer onto a surface within a very short time over large areas. The agar spray technique tested was developed by Giordano [21,80], and an extensive characterisation of the new gel has been carried out by Giordano et al. [24]. Its properties make it an attractive candidate for the cleaning of soiled, textured painted surfaces, particularly those of a monumental scale, such as Munch's Aula paintings.

Four aqueous cleaning solutions at various pH and conductivity values were selected as shown in Table 4. The selection was based on free-liquid trials (evaluated by naked eye and under optical microscope—assessments are available online as an externally hosted supplementary file).

Table 2. Description and characterisation of mock-up surfaces prepared for the cleaning trials. All applications of pictorial layers were made with hog's hair brushes. All mock-up materials were sourced from Christ Engebretsen & Søn (Oslo, Norway).

Mock-Up	Stratigraphy	Ageing and Soiling [a]	Surface Properties [b]
Chalk–Glue Ground	*Canvas*: washed linen, twill weave, stretched *Size*: rabbit skin glue *Ground*: chalk, in rabbit skin glue	6 months ambient drying 3 weekly cycles of accelerated ageing: Memmert ICH110L chamber; light: 4 fluorescent lamps (6500 K (D56), 500 W); irradiance: 70 Wm^{-2}; total energy: 169,330 kJm^{-2}; 40 °C (CHT); and fluctuating RH (15%–65%) Three spraying campaigns (two for oil paint) of artificial soiling * adapted for the Aula * *Soiling layer*: 27.1 ± 2.4 μm *Min particle size*: 0.095 μm *Max particle size*: ≥10 μm	*Thickness*: 122.3 ± 39.2 μm *Water sensitivity*: 14 rolls *Chalking*: ISO 2 *pH*: 6.5 *Conductivity*: 1500 $\mu S \cdot cm^{-1}$
Half-Chalk Ground	*Canvas*: ibid. *Size*: ibid. *Ground*: chalk, zinc white, lead white in rabbit skin glue and boiled linseed oil emulsion		*Thickness*: 104.9 ± 40.2 μm *Water sensitivity*: 10 rolls *Chalking*: ISO 3 *pH*: 6.4 *Conductivity*: 500 $\mu S \cdot cm^{-1}$
Chromium Oxide Green Oil Paint	*Canvas*: ibid. *Size*: ibid. *Ground*: half-chalk ground *Pigment-binder*: undiluted chromium oxide green in linseed oil		*Thickness*: 114.9 ± 26.4 μm *Water sensitivity*: 5 rolls *Chalking*: ISO 1 *pH*: 6.4 *Conductivity*: 530 $\mu S \cdot cm^{-1}$

[a]: applies to all mock-ups; *CHT* is chamber temperature, *RH* is relative humidity, (*) indicates soiling dimensions at the bottom of the column. [b]: *Thicknesses* are reported for topmost layer i.e., the cleaning surface; *water sensitivity* and *chalking* measurements from unsoiled mock-ups, according to Mills et al. [81], and UNI EN ISO 4628-6 [82] respectively; *pH* and *conductivity* measurements from soiled mock-ups, and are averages from triplicate measurements.

Table 3. List of components within the artificial soiling used (suppliers' sources listed in-text).

Supplier	Material	Composition	Quantity	Dry Weight
			g or mL	%
Rublev	Lamp black (oil furnaces)	C	0.62	1.00
Kremer	Vine black (organic source)	C	0.62	1.00
	Burgundy ochre (fine)	$Fe_2O_3 \cdot H_2O$	1.45	2.34
	Wheat starch powder	Polysaccharide $(C_6H_{10}O_5)_n$	10.00	16.14
	Gelatin powder	Proteins and peptides	10.00	16.14
Merck	Sodium nitrate	$NaNO_3$	2.50	4.03
	Kaolin	$Al_2Si_2O_5(OH)_4$	18.00	29.06
	Portland cement (Type I)	$CaO \cdot SiO_2$ Fe, Al, MgO	17.00	27.45
	Silica, quartz	SiO_2	1.75	2.83
	Mineral oil	Hydrocarbons	5.0	-
Filippo Berio	Olive oil	Mainly triacylglycerols	2.5	-
Kremer	Shellsol D40	Hydrocarbons	1000	-

Deionised water served as a control to investigate the effect of the gel application per se. Conductivity- and pH-adjusted waters were prepared for each type of mock-up class, following methodologies in the literature [79,80,83,84], and are referred to simply as *adjusted waters* throughout the remainder of this text. Monobasic and dibasic citrate chelating solutions (pH-adjusted with (i) NaOH and (ii) NH_4OH, for comparative purposes, respectively) were selected to investigate chelating action, after being marked as promising for soiling removal in the Aula [23,85]. When using chelating solutions, a clearance solution (an appropriately adjusted water) was thereafter applied, as recommended by Stavroudis [86]. The pH and conductivity of used solutions were measured before use (deviating solutions were discarded and prepared freshly).

Table 4. The four cleaning solutions selected from the free-liquid trials for use in the cleaning tests. All chemicals were sourced from VWR International (Oslo, Norway).

Cleaning Solution	Concentration	Chalk–Glue Ground	Half-Chalk Ground	Chromium Oxide Green
Deionised water	-	pH 7.2, 20 μS·cm^{-1}	pH 7.2, 20 μS·cm^{-1}	pH 7.2, 20 μS·cm^{-1}
Adjusted water [a] (ammonium acetate)	Conductivity-related	pH 5.5, 1500 μS·cm^{-1}	pH 5.5, 500 μS·cm^{-1}	pH 5.0, 500 μS·cm^{-1}
Chelator [b] (citric acid/ sodium hydroxide)	0.5% w/v (0.026 M) CA in 10% w/v (2.5 M) NaOH	pH 5.0, 4240 μS·cm^{-1}	pH 4.5, 3240 μS·cm^{-1}	pH 4.5, 3240 μS·cm^{-1}
Chelator [b] (citric acid/ ammonium hydroxide)	0.5% w/v (0.026 M) CA in 10% w/v (5.0 M) NH$_4$OH	pH 5.0, 5320 μS·cm^{-1}	pH 4.5, 4270 μS·cm^{-1}	pH 4.5, 4270 μS·cm^{-1}
Clearance [c] (ammonium acetate)	Conductivity-related	pH 6.5, 500 μS·cm^{-1}	pH 6.5, 500 μS·cm^{-1}	pH 6.5, 500 μS·cm^{-1}

[a]: buffer prepared from 1 mL 17.4 M glacial acetic acid and adequate volume of 10% w/v (5.0 M) ammonium hydroxide. [b]: citric acid diluted from a 2.5% w/v (0.13 M) stock solution; conductivity of chelators could not be matched to surface properties as this would modify the chelator concentration. [c]: used after chelator; isotonic or hypertonic compared to mock-up surfaces at a pH where swelling is minimised.

Agar gels were prepared with the different cleaning solutions and applied in two ways: (i) as a direct spray [21,80] and (ii) as a pre-formed rigid film (prepared by spraying, as a milder modified form of cleaning). Clearance solutions were always applied as a pre-formed rigid gel (so that a double spray application did not interfere with the interpretation of results). The gel was used at a concentration of 3% w/v and sprayed evenly from a distance of 40 cm. The agar sol temperature prior to spraying was dependent on the room's and mock-up surfaces' temperature, and relative humidity. The experimental parameters measured during agar spray testing are given in Table S1. The gel was removed after two minutes, following recommendations by Giordano & Cremonesi [21,80].

Treatment evaluation metrics. A detailed description of the methodologies, including acquisition parameters, for each analytical technique is given in Appendix A. Table 5 below summarises the cleaning homogeneity and efficacy metrics, according to the range of the electromagnetic spectrum that they targeted, the multidimensional data types they employed, the concept through which they evaluated soiling removal, the equipment required to capture the data, and the post-processing steps needed to achieve results.

Table 5. Summary of (i) cleaning homogeneity and (ii) cleaning efficacy metrics in this work.

Metric [a]	Range [b]	Data Type [c]	Concept	Equipment [d]	Post-Processing [e]
(i) Cleaning homogeneity	VNIR /SWIR	2D spectral maps	Image homogeneity from grey-level co-occurrence matrix (GLCM)	DLSR camera HSI camera	Change image type to 8-bit depth for GLCM *Texture* plug-in in ImageJ (v. 1.54f)
(ii) Cleaning efficacy					

Table 5. *Cont.*

	Metric [a]	Range [b]	Data Type [c]	Concept	Equipment [d]	Post-Processing [e]
Image-based	$L*a*b*$ images	VIS	2D RGB images	Thresholded pixels representing soiling	DLSR camera (Mobile phone)	Conversion to CIELAB space; image thresholding
	Histogram skewness	VIS	2D RGB images	Histogram distribution asymmetry as function of darker soiling on lighter substrate		Spreadsheet/ statistical calculations
Appearance	Glossimetry	VIS	1D point measurements	Perceived surface texture under direct light source	Glossmeter	Spreadsheet/ statistical calculations
	Colourimetry (from HSI)	VNIR	2D $L*a*b*$ images (from 3D datacube)	Colour difference, ΔE_{2000}, before and after soiling removal	HSI camera	Conversion to CIELAB space; colourimetric and statistical calculations
Spectral-based	HSI: spectral unmixing	VNIR /SWIR	3D datacube	Spectral reflectance similarity (compa-red to unsoiled areas, or soiling)	HSI camera	Spectral calibration; algorithm pre- and post-processing
	HSI: NDI mapping	SWIR	2D normalized difference images	SWIR marker bands for soiling and surface	HSI camera	Spectral calibration; PCA; image processing
	FTIR mapping	MIR	2D chemical maps	MIR spectra (or marker bands) for soiling	FTIR spectrome-ter	Atmospheric correction; correlation map profiles
	SEM-EDX mapping	(XR)	2D chemical maps	Element signal for soiling	SEM-EDX	TruMap processing; element selection

[a]: $L*a*b*$ (CIELAB (Commission internationale de l'éclairage) colourspace coordinates representing perceptual lightness and four colours of human vision, defined in 1976); HSI (hyperspectral imaging); FTIR (Fourier transform infrared); SEM-EDX (scanning electron microscopy with energy dispersive X-ray spectroscopy). [b]: VIS (visible light); VNIR (visible to near infrared); SWIR (shortwave infrared); MIR (mid-wave infrared); XR (X-rays; XRs are detected by spectrometers). [c]: 1D (one-dimensional); 2D (two-dimensional); 3D (three-dimensional); RGB (red–green–blue colourspace). [d]: DLSR (digital single-lens reflex); [e]: PCA (principal component analysis).

For each cleaning efficacy metric, a normalised difference was calculated according to Equation (1), where x is a measured property before or after treatment (BT or AT, respectively), so as to show the percentage return to an unsoiled control's surface property as a relative marker of cleaning success; the specific values used for each technique's metric are described in Table 6, further below.

$$\text{cleaning efficacy} = \frac{x_{\text{BT}} - x_{\text{AT}}}{x_{\text{BT}}} \times 100 \tag{1}$$

All cleaning efficacy metrics were calculated in ImageJ (v. 1.54f) and then statistically treated and plotted using the Microsoft Excel 2019 spreadsheet package. The results from the metrics were then fed into the scoring criteria listed below.

Table 6. List of measured properties, x_{BT} and x_{AT}, for each metric inputted into the normalised difference equation used to calculate cleaning efficacy.

	Cleaning Efficacy Metric	Value for x_{BT}	Value for x_{AT}
Image-based	$L*a*b*$ images	Number of black pixels before treatment (black pixels represent soiling)	Number of black pixels after treatment (black pixels represent soiling)
	Histogram skewness	Difference in skewness between unsoiled (CT) and soiled (BT) mock-up ($\Delta skewness_{CT,BT}$)	Difference in skewness between unsoiled (CT) and cleaned (AT) mock-up ($\Delta skewness_{CT,AT}$)
Appearance	Glossimetry	Difference in gloss between the unsoiled (CT) and soiled (BT) mock-up ($\Delta gloss_{CT,BT}$)	Difference in gloss between the unsoiled (CT) and cleaned (AT) mock-up ($\Delta gloss_{CT,AT}$)
	Colourimetry (from HSI)	CIE2000 colour difference between the unsoiled (CT) and soiled (BT) mock-up ($\Delta E_{2000(CT,BT)}$)	CIE2000 colour difference between the unsoiled (CT) and cleaned (AT) mock-up ($\Delta E_{2000(CT,AT)}$)
Spectral-based	HSI: spectral unmixing	Mean pixel value from 100 pixel × 100 pixel area taken from unsoiled (CT) mock-up, or soiling control (sCT), in unmixing map ($\bar{x}_{CT}$, or $\bar{x}_{sCT}$)	Mean pixel value from 100 pixel × 100 pixel area taken from cleaned (AT) mock-up, in unmixing map ($\bar{x}_{AT}$)
	HSI: NDI mapping	Mean pixel value from 100 pixel × 100 pixel area taken from unsoiled (CT) mock-up, or soiling control (sCT), in NDI map ($\bar{x}_{CT}$, or $\bar{x}_{sCT}$)	Mean pixel value from 100 pixel × 100 pixel area taken from cleaned (AT) mock-up, in NDI map ($\bar{x}_{AT}$)
	FTIR mapping	Number of white pixels before treatment (white pixels represent soiling)	Number of white pixels after treatment (white pixels represent soiling)
	SEM-EDX mapping	Number of element-rich areas as counted by *Analyse particles* function in ImageJ before cleaning	Number of element-rich areas as counted by *Analyse particles* function in ImageJ after cleaning

Because each painted surface has different cleaning requirements based on its inherent properties and condition [23], the evaluation process was adapted by weighting the scoring criteria to then recalculate a weighted average score out of 1.00. Based on discussions with the conservators who have treated and monitored the paintings over the past two decades, higher importance was given accordingly to cleaning efficacy and homogeneity (both 30% weighting), followed by selectivity (20% weighting), and then colour (15% weighting) and gloss integrity (5% weighting—gloss differences on the Aula paintings are only truly perceptible when standing directly underneath the paintings, which is not how the public typically views them). Calculation tables for the weighted scores are available online as an externally hosted supplementary file. Following what has become standard practice in cleaning studies in conservation [16,23,39,85], star diagrams were also used to illustrate the ideal cleaning candidate. The combination of weighted average scores with star diagrams permitted a thorough appraisal of the cleaning results.

3. Results and Discussion

3.1. Metrics in Practice

The metrics aimed to semi-quantitatively compare treatment results to aid in decision-making in a reproducible and shareable manner, and to complement professional observations made by eye. The image- and spectral-based metrics enhanced this process through their visual inputs and outputs, examples of which are illustrated in Figure 2 below. The appearance-based metrics were calculated from point measurements (Table 5) and therefore did not have any visual inputs or outputs in terms of images or maps. It must be kept in mind that each metric measured a different property or component of the surfaces or soiling, and thus efficacy scores needed to be interpreted accordingly. Data plots for cleaning efficacy per metric are presented in Figures 3–10. Processed images from photography, microscopy, FTIR, and SEM, and data tables for all the efficacy plots discussed are available online as an externally hosted supplementary file.

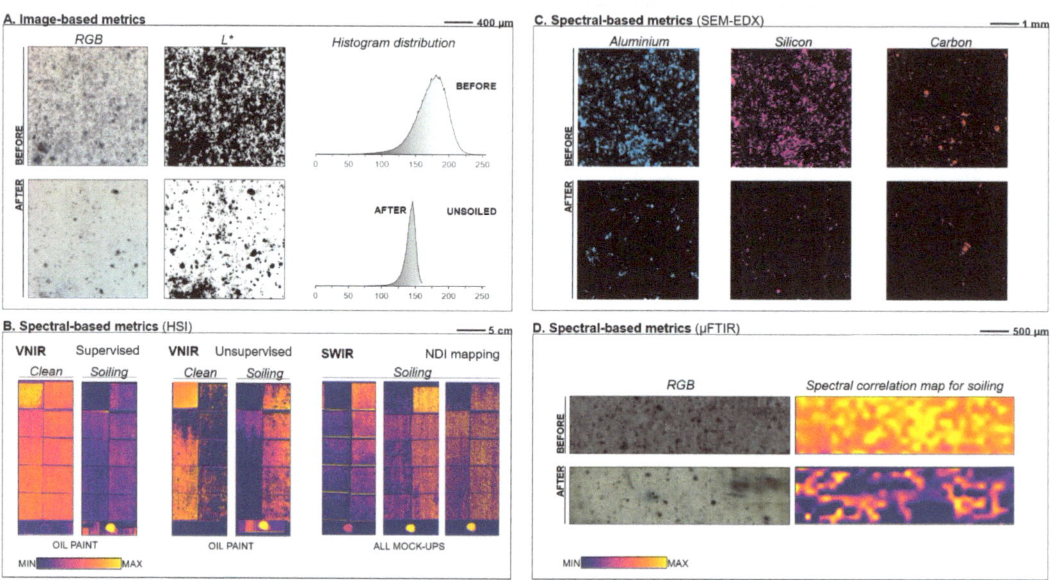

Figure 2. Examples of the visual outputs given by the image- and spectral-based metrics described. For the *image-based metrics*, figures are of the chalk–glue ground mock-ups cleaned with deionised water. The histogram distribution represents pixel values from 0 to 255 for the respective RGB images shown. For the *HSI metrics*, only oil paint mock-ups are shown for spectral unmixing maps; all mock-up classes are shown for NDI maps (chalk–glue ground, half-chalk ground, and oil paint from left to right). Each spectral map consists of a column of spray-cleaned mock-ups on the left, and a column of mock-ups cleaned by pre-formed rigid gel on the right (cleaning solutions on going from bottom to top: citric acid/NH$_4$OH, citric acid/NaOH, adjusted water, deionised water); unsoiled and soiled controls are found top left and right respectively. For the *SEM-EDX* and *FTIR metric*, the example shown is for cleaning with citric acid/NH$_4$OH.

3.1.1. Image-Based Metrics: $L^*a^*b^*$

The images serving as input data for the image-based metrics were captured using the visible (VIS) range of the electromagnetic spectrum. The image-based metrics may be considered non-invasive, non-contact, as well as more accessible and low-cost, in that the only required equipment is a digital single-lens reflex (DSLR) camera and/or a microscope. Having been converted from RGB to CIELAB colour space for this metric, the $L^*a^*b^*$ images offer insights into the perceptual lightness (L^*), and colours (as perceived by the human eye, a^* and b^*) of the mock-up surfaces. The $L^*a^*b^*$ images thus serve as an extension to the conservator's eye. The metric in turn represents soiling distribution based on the pixel intensity values of the $L^*a^*b^*$ images. It should be kept in mind that the accentuation of soiling through pixel intensity depends upon the colours of the soiling and surface being investigated.

Cleaning efficacy results for images taken at the microscale coincided better with visual observations than those taken at the macroscale, which—although comparable in trend—tended to overestimate cleaning. This could have been possible due to the resolution and scale at which the images were captured, whereby macro-imaging might have not given a relatable representation (via the image pixels) of the distribution of soiling, based on the image processing method used. Furthermore, results for macro-photography were based on one mock-up (and not a set of triplicates, due to time restrictions), and thus did not cover the heterogeneity of soiling removal results. The remaining discussion therefore focuses on the microscale results.

Out of the possible L^*, a^*, and b^* images generated for the mock-up surfaces imaged in this study, it was noticed that the b^* images gave better indications of the presence of soiling. This is because this channel did not pick up surface texture, which was prominent in L^* images (this could be taken advantage of, nonetheless, to reveal crack propagation, for example). Resultantly, the L^* channel was found to be better suited for smoother surfaces, and the b^* channel for rougher, textured surfaces.

Cleaning efficacy plots using $L^*a^*b^*$ images are given in Figure 3. The spray application superseded the pre-formed gel in terms of cleaning efficacy. However, this was only marginally so for the oil paint mock-ups, indicating that the application method of the gel did not influence soiling removal for this mock-up class, perhaps as a result of the physicochemical means of soiling-to-surface binding.

For the pre-formed gel application, there was a notable increase in the performance of soiling removal for the citrate/NH_4OH solution compared to other cleaning solutions. This trend was less apparent with respect to the spray application, but the metric indicated that the same chelating solution removed the most soiling, nonetheless. Overall, the chalk–glue ground was cleaned almost twice as much as the other mock-up types, potentially reflecting differences in soiling–surface interaction.

Based on these observations, this metric offered reasonable representations of soiling removal efficacy, although it must be noted that the thresholding of the $L^*a^*b^*$ images can be time-consuming and user-dependent.

3.1.2. Image-Based Metrics: Skewness

The use of statistical moments (describing the shape of a histogram distribution) as indicators of cleaning efficacy is based on the concept that, as darker contaminants are removed from lighter substrates, the pixel values of the respective BT and AT images reflect this change accordingly, and thus an image's histogram distribution can reflect the change taking place during cleaning. When considering a distribution's skewness in a cleaning context, a larger fraction of darker pixels (i.e., darker contaminants from soiling) will skew an image's histogram leftwards (negative skewness), whereas a larger fraction of lighter pixels (i.e., cleaner, lighter surfaces) skews the histogram rightwards (positive skewness).

The skewness metric has been successfully applied to stone cleaning [87] and paper cleaning [54] and hence was investigated for its promising application to the exposed grounds and unvarnished oil paint under study, as a fast and reliable measure. This metric was applied to the RGB micrographs, but could equally be applied to RGB macroscale photographs, as in the work by Charola et al. [87].

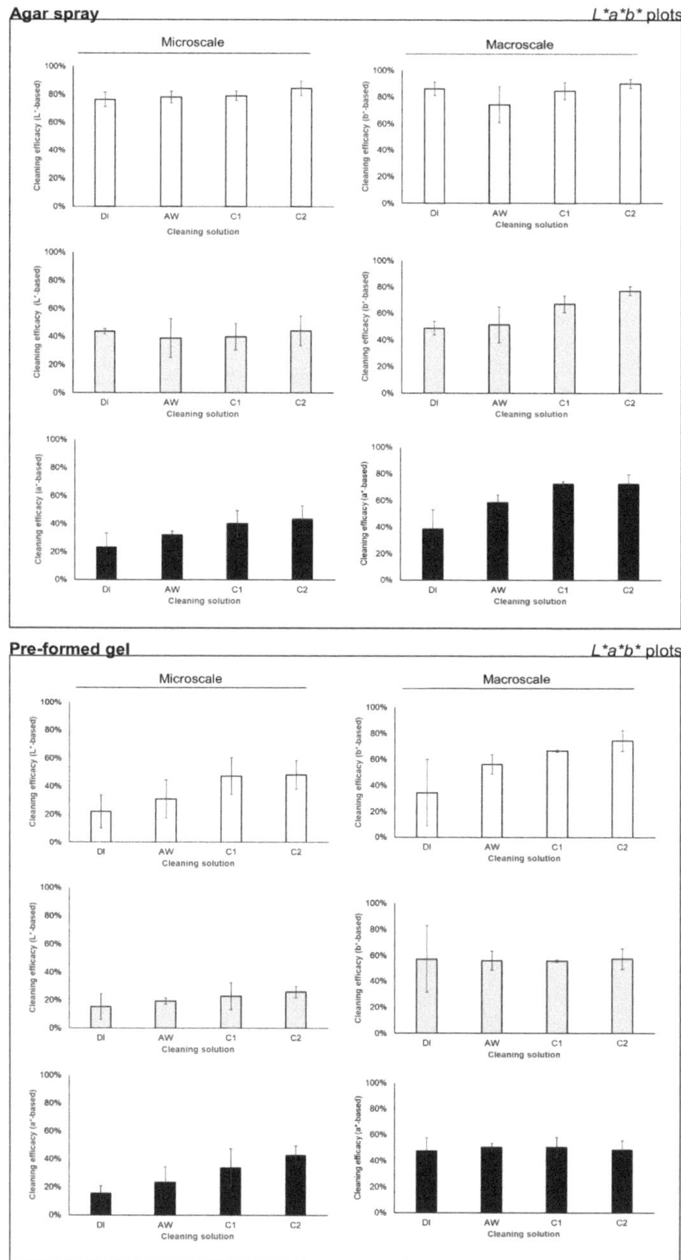

Figure 3. Cleaning efficacy plots using $L^*a^*b^*$ images at micro- (**left**) and macroscale (**right**) for the spray-cleaned mock-ups (**top**) and the mock-ups cleaned by pre-formed gel (**bottom**) (keys: white—chalk–glue; grey—half-chalk; black—chromium oxide; (DI) deionised water; (AW) adjusted water; (C1) citric acid in NaOH solution; (C2) citric acid in NH_4OH solution.

All three methods for calculating skewness were investigated during trials: they differed in how and which statistical descriptors were inputted for calculation [88]. Consequently, calculations indicated that Pearson's moment coefficient best described the skewness of histogram distributions in this case, in contrast to the cited studies that used Pearson's first and second coefficients. The selection was based on the coefficient that described the datasets best. Whereas only skewness was measured in this study, other moments such as the mean and kurtosis (flatness) of the distributions could have been additionally used to monitor changes after cleaning [87] and require further investigation for application onto unvarnished painted surfaces.

Histogram distributions of the micrographs are available online as an externally hosted supplementary file. The unsoiled mock-ups had a negative skewness (probably due to stray, darker pixels relating to canvas details, surface texture and irregularities), with a narrow distribution spread across lighter pixel values (Figure 2). The soiled ground mock-ups were characterised by less skewed and broader distributions across the centre of the plot, which narrowed and shifted towards lighter pixel values with an increased negative skewness (closer to the control's) after cleaning (Figure 2). The soiled paint mock-ups contrarily featured positively skewed distributions centred over darker pixel values (attributed to the soiling, and chromatic changes during ageing when compared to the control), which became more positively skewed as a higher percentage of lighter pixel values were present after the removal of darker soiling (the distributions remained, however, centred over darker pixels due to the inherently darker substrate).

Cleaning efficacy plots based on skewness are given in Figure 4. The efficacy plots generated through the skewness metric indicated that the agar spray application resulted in better soiling removal than the pre-formed gel, which matched visual observations. The poorer performance of both gel applications when applied to the oil paint (as opposed to the grounds) was notable and might be explained by the fact that considerable chemically imbibed soiling was observed microscopically after cleaning.

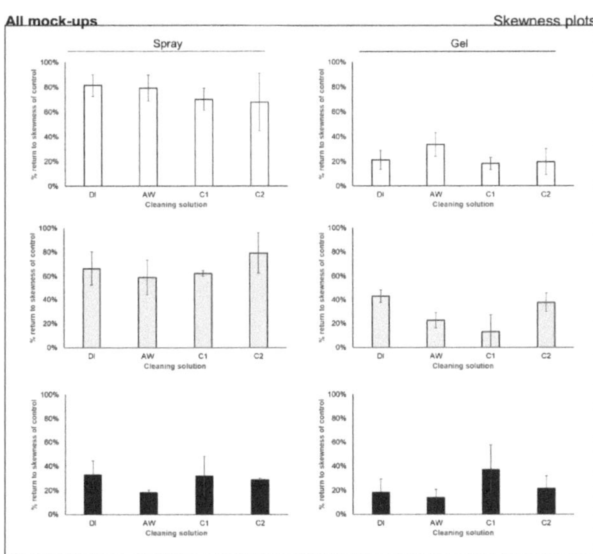

Figure 4. Cleaning efficacy plots based on skewness for the sprayed mock-ups (**left**) and the mock-ups treated by pre-formed gel (**right**) (keys: white—chalk–glue; grey—half-chalk; black—chromium oxide; (DI) deionised water; (AW) adjusted water; (C1) citric acid in NaOH solution; (C2) citric acid in NH_4OH solution.

When comparing the cleaning of both grounds, the metric seems to overestimate the efficacy for the half-chalk ground, which looked effectively more soiled than its chalk–glue counterpart on the basis of visual observations. Whilst the chelating solutions removed most soiling on the half-chalk ground and the oil paint, the adjusted water gave better results for the chalk–glue ground. In the latter case, the uncovering (or formation) of micro-cracks and other surface details by the chelating solutions may have resulted in an increase in darker pixels, which, according to this metric, will translate into less efficacious 'cleaning.

3.1.3. Appearance-Based Metrics: Colour and Gloss

The management of changes to the colour and gloss of an object's surface is typically one of the fundamental priorities of the conservator during treatment [89–92], so as not to misrepresent the aesthetic value of the piece and detract from its creator's intent [93]. These metrics are not novel in themselves but rather complement the scoring criteria selected to evaluate the agar spray treatment.

Hyperspectral data was used for the colourimetry metric. Thus, one single data acquisition simultaneously allowed for both the measurement of colour change as a result of soiling removal, and the application of the spectral-based metrics (introduced after this section). This saved time and rendered measuring colourimetric data a completely non-contact process (colourimetric measurements are usually taken by placing a colourimeter on the surface of an object).

The cleaning efficacy plots based on the measured changes in colour (ΔE_{2000}) are given in Figure 5. The plots indicated that the cleaning solutions applied by spray guaranteed a better return to the colour of the unsoiled control than the pre-sprayed counterpart. The chelating solutions once again afforded the most efficacious soiling removal with respect to colour change. However, the half-chalk ground fared most poorly of all mock-up classes, irrespective of the cleaning solution. This was likely due to soiling–surface interactions whereby remaining imbibed soiling detracted from the full appreciation of the cream-coloured ground. The pre-formed gel, however, did offer an effective option for promoting the return to the surfaces' original colours, especially with the citrate/NH_4OH solution.

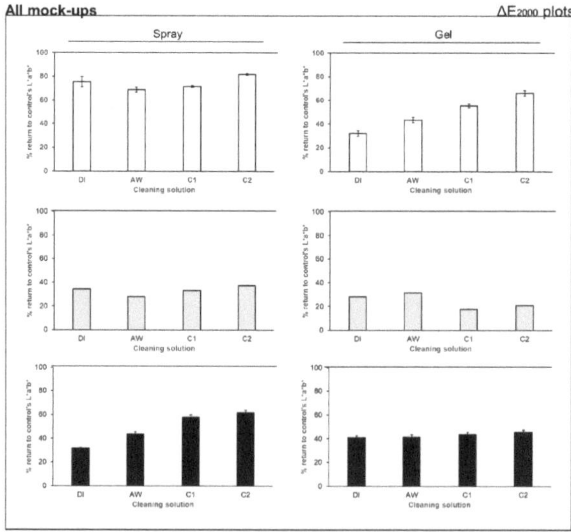

Figure 5. Cleaning efficacy plots based on ΔE_{2000} for the sprayed mock-ups (**left**) and the mock-ups treated by pre-formed gel (**right**) (keys: white—chalk–glue; grey—half-chalk; black—chromium oxide; (DI) deionised water; (AW) adjusted water; (C1) citric acid in NaOH solution; (C2) citric acid in NH_4OH solution.

For glossimetry, the metric returns the difference with respect to the control's gloss, or, if the change registered goes beyond that gloss measurement, the percentage of excess deviation beyond the control's gloss. Cleaning efficacy plots based on glossimetry are given in Figure 6.

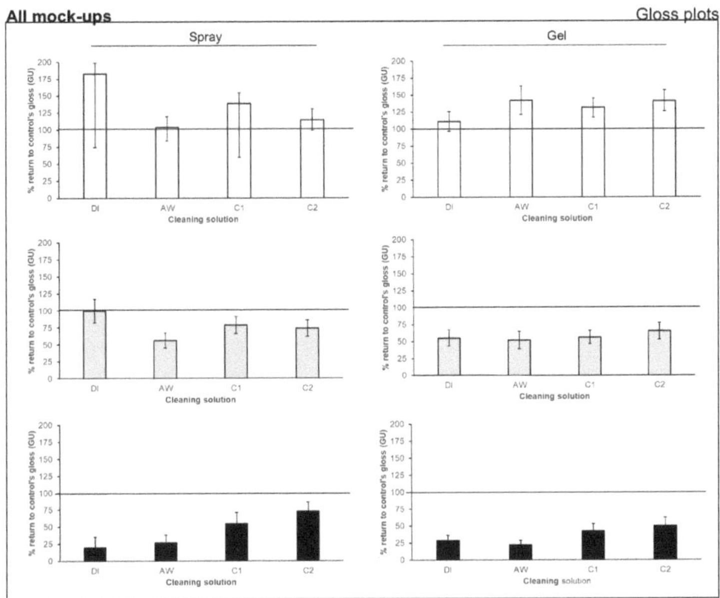

Figure 6. Cleaning efficacy plots based on glossimetry for the sprayed mock-ups (**left**) and the mock-ups treated by pre-formed gel (**right**) (keys: white—chalk–glue; grey—half-chalk; black—chromium oxide; (DI) deionised water; (AW) adjusted water; (C1) citric acid in NaOH solution; (C2) citric acid in NH$_4$OH solution.

For the chalk–glue ground, both the spray and pre-formed gel resulted in glossier surfaces after treatment. Given the large extent of matte areas of exposed chalk–glue ground in the Aula, this is a consideration worth addressing. Keeping in mind the variability of the gloss measurements, the most viable options for the cleaning of this ground were the adjusted water, followed by the citrate/NH$_4$OH solution (both applied via spray).

With respect to the half-chalk ground and the chromium oxide oil paint, spraying the agar appeared to promote an improved return to the original gloss, with the chelating solutions delivering the best results.

3.1.4. Spectral-Based Metrics: HSI

The HSI metrics illustrated the degree of soiling removal across the visible-to-near-infrared (VNIR) and SWIR ranges, thus assessing soiling removal efficacy based on the physical properties and molecular composition of the cleaned surfaces, or soiling. An example of soiling mapping is shown in Figure 2 (bottom left). All spectral maps generated from HSI post-processing are provided in Figures S1–S16. One metric used spectral unmixing algorithms to map soiling removal, whereas the other one was based on the chemometric technique of normalised difference image (NDI) mapping. As a result of HSI setup and mock-up dimensions, the major advantage of these metrics was that they permitted the simultaneous mapping of soiling across the entire surface of many mock-ups using a single acquisition. The HSI method can therefore be considered as a non-contact and relatively fast means of visual assessment.

Between the two spectral unmixing algorithms tested, the semi-supervised one gave the most reliable results, whereas the unsupervised algorithm suffered from some issues. These issues are presented and discussed in Appendix B. The remainder of the discussion here focuses on the semi-supervised algorithm.

The VNIR and SWIR spectral unmixing maps (made using the spectrum of the unsoiled reference as an endmember) demonstrate that (i) all the chalk–glue ground mock-ups were cleaned to a high degree with the spray, (ii) all half-chalk ground mock-ups were cleaned to a lesser degree, and (iii) the oil paint mock-ups gradually became cleaner on going from deionised water to adjusted water to chelating solutions. The milder cleaning effect of the pre-formed gel was also mapped.

The converse relationship between the spectral maps for cleaned areas (based on the unsoiled control's spectrum as an endmember) and for soiled areas (based on the soiled control's spectrum as an endmember) illustrated that soiling removal can be mapped using either reference spectrum. In the VNIR range, maps using the soiling's spectrum as an endmember reflected the contrary of the cleaned maps, i.e., they showed, in a complementary fashion, the presence of soiled areas instead of cleaned ones.

In the SWIR range, indirect molecular markers of the soiling were detected [68]. These markers lend great use since soiling may not always be visible, e.g., where surface contrast is poor (dark soiling on dark surfaces), or where soiling/degradation products are not necessarily visible to the naked eye. However, in the case of the SWIR maps, false positives were detected in all cases, with soiling apparently present on unsoiled controls. In the case of the chalk–glue ground, the cleaner mock-ups were shown as the most soiled. This may have been due to the unmixing algorithm confounding spectral bands common to the soiling and the mock-up surfaces (related to the binder, for example).

For this reason, the semi-supervised HSI metric was best assessed using VNIR map pixel values (when using maps based on the unsoiled reference spectrum, however, SWIR map pixel values gave results that closely matched their VNIR counterpart). Cleaning efficacy plots based on the semi-supervised algorithm are given in Figure 7. The VNIR plots showed that, when using the agar spray with the chelating solutions, (i) over 90% of soiling was removed from the chalk–glue ground mock-ups, (ii) roughly 70% of soiling was removed from the half-chalk ground mock-ups, and (iii) about 80% of soiling removal was achieved for the oil paint mock-ups. The figures for the pre-formed gel were comparably lower, with the chelating solutions performing marginally better for all mock-ups.

In contrast to the spectral unmixing metric, the soiling removal maps generated through the NDI method provided more reliability with respect to detecting the presence or absence of soiling upon surfaces. This is due to the use of specific marker bands, which were significantly indicative of the materials being mapped (cf. Figure A1). However, this method required more expertise and time to achieve reliable results, as it implied carrying out multivariate statistical analysis on the datacubes collected. The semi-supervised unmixing algorithm is thus preferable where advanced statistical expertise (or time) is lacking.

Cleaning efficacy plots based on SWIR-NDI maps are given in Figure 8. The NDI soiling removal efficacy plots matched with the observations made by eye, for both spray and pre-formed gel applications. The latter was marked as being less effective compared to its spray counterpart, irrespective of the cleaning solution (except in the case of the chalk–glue ground). When the agar was sprayed, the maps and plots indicated the chelating solutions removed most soiling, particularly for the oil paint (in the case of the half-chalk ground, performance was also matched by the adjusted water).

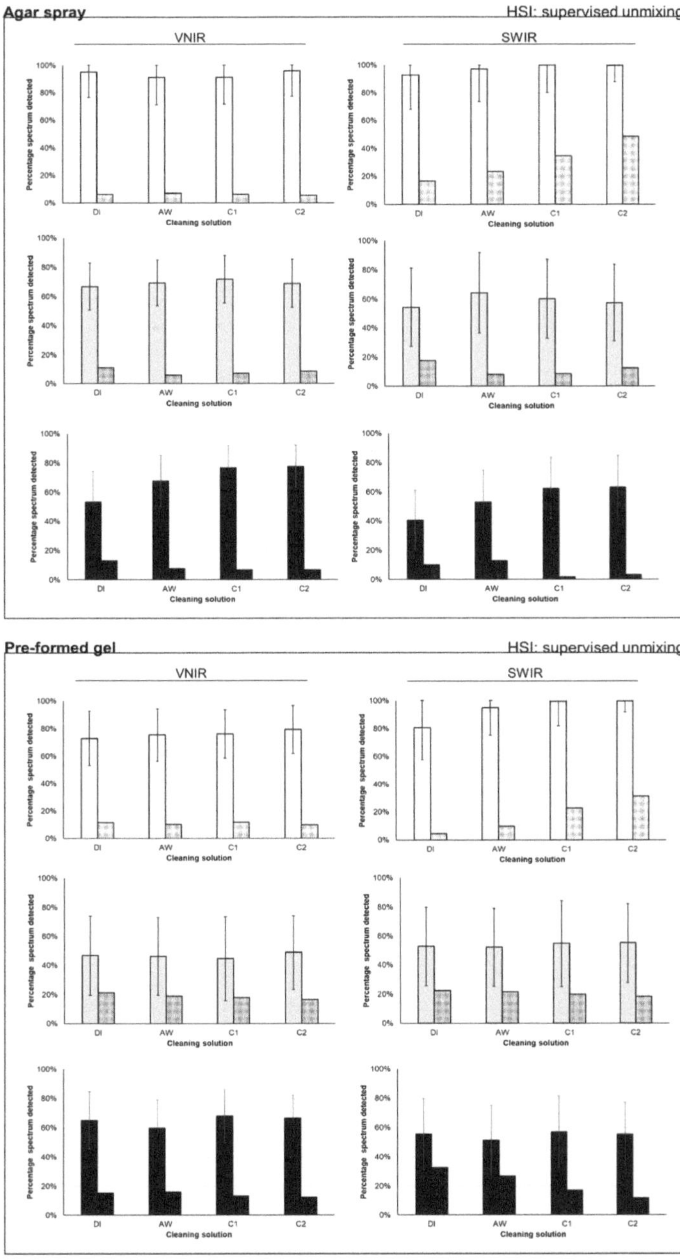

Figure 7. VNIR and SWIR metric plots as determined by the semi-supervised unmixing algorithm, based on clean (plain fill) and soiled (textured fill) reference spectra for spray-cleaned surfaces (**top**) and the pre-formed gel cleaning (**bottom**). Error plotted only for clean areas (RSD for soiled areas was above 1.00 in all cases) (keys: white—chalk–glue; grey—half-chalk; black—chromium oxide; (DI) deionised water; (AW) adjusted water; (C1) citric acid in NaOH solution; (C2) citric acid in NH$_4$OH solution).

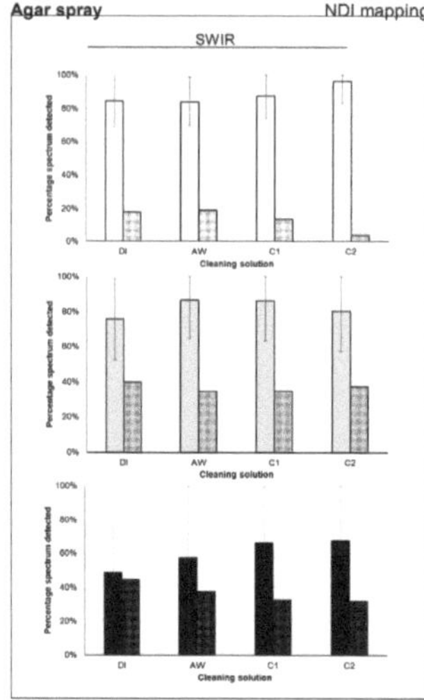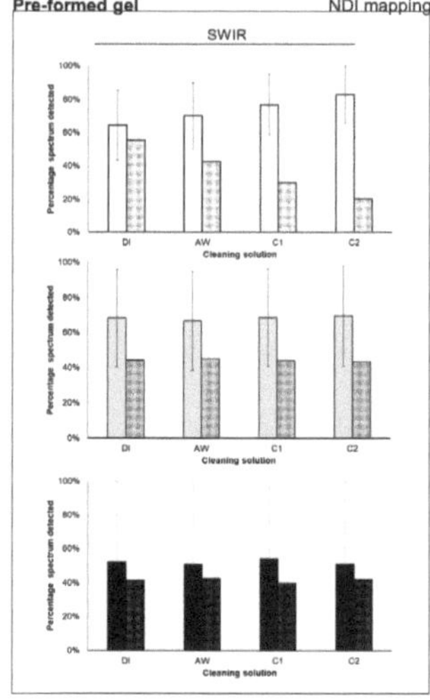

Figure 8. VNIR and SWIR metric plots as determined by the SWIR-NDI mapping, based on clean (plain fill) and soiled (textured fill) reference spectra for spray-cleaned surfaces (**left**) and the pre-formed gel cleaning (**right**). Error plotted only for clean areas (RSD for soiled areas was above 1.00 in all cases (keys: white—chalk–glue; grey—half-chalk; black—chromium oxide; (DI) deionised water; (AW) adjusted water; (C1) citric acid in NaOH solution; (C2) citric acid in NH_4OH solution).

The processed SWIR data from the NDI maps were further used to calculate a cleaning homogeneity score. As a complement to the cleaning efficacy metrics, this score was developed with the intention of empirically evaluating the degree of similar cleaning, i.e., homogeneity, achieved across the mock-ups' surfaces. In digital image analysis, a parameter for measuring image homogeneity was described by Haralick et al. [94,95], for which a plug-in in ImageJ (v. 1.54f) was developed by Cabrera [96]. This parameter is calculated from what is known as the *grey level co-occurrence matrix* (GLCM) of an image [94,97]. Since the pixels in the images of the soiling maps can represent the soiling spread, the homogeneity parameter (also known as the *inverse difference moment*) from the GLCM was deemed an appropriate measure of calculation.

3.1.5. Spectral-Based Metrics: FTIR

The FTIR metric enabled an in-depth look at the cleaned surfaces at the microscale, providing imaging and spectral information from the mid-infrared (MIR) range. Cleaning efficacy plots based on μFTIR maps are given in Figure 9. The FTIR soiling maps illustrate the distribution of points where the soiling spectrum was detected. As a result, only those points where no soiling bands were detected were considered clean, and this binary separation between soiled and cleaned could explain the lower percentages reported across the board when compared to HSI.

For the spray application, soiling removal was more effectively carried out on the exposed grounds than on the oil paint. This might reflect the nature of the soiling, which may have been more physically imbedded than chemically imbibed for the grounds. In terms of cleaning solution, the citrate/NH_4OH solution outperformed the other options, both for spray and gel. Whereas this improved efficacy gradually increased across different solutions with the pre-formed gel, less certain trends were reported for the spray.

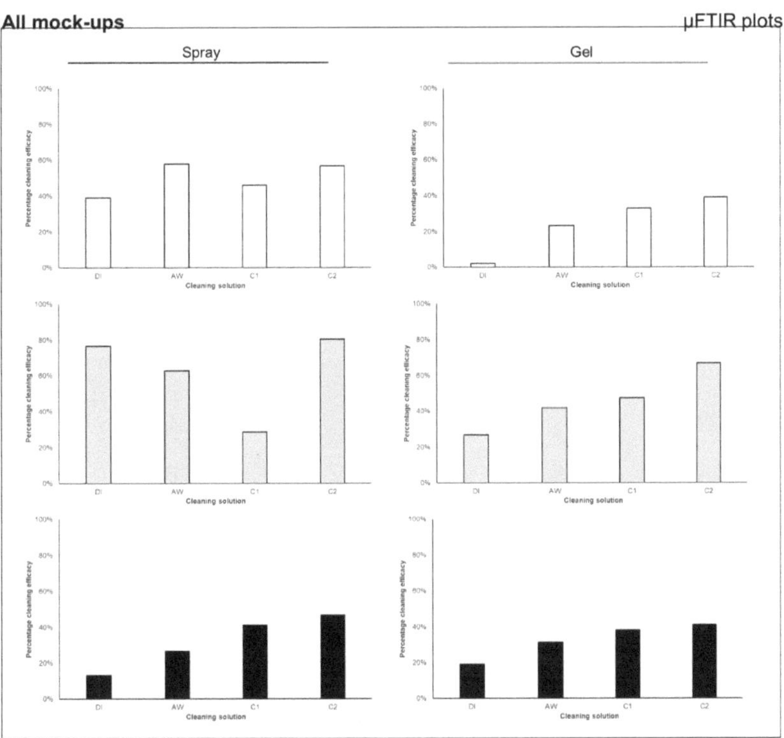

Figure 9. Cleaning efficacy plots for the sprayed gel (**left**) and plots for the pre-formed gel (**right**) based on the FTIR metric (keys: white—chalk–glue; grey—half-chalk; black—chromium oxide; (DI) deionised water; (AW) adjusted water; (C1) citric acid in NaOH solution; (C2) citric acid in NH$_4$OH solution).

3.1.6. Spectral-Based Metrics: SEM-EDX

A cleaning efficacy metric was developed using element mapping (SEM-EDX). One indirect, empirical observation that indicated the cleanliness of surfaces was the increased surface charging on AT surfaces during measurements, as there was less (carbon-containing) particulate matter to dissipate the electrons. Although the acquisition of data was slower than for HSI, the post-processing was relatively straightforward. This means that, if chemical markers for soiling can be found, as with Al and Si in this case (due to their relative abundance within the artificial soiling), then the method could be a viable technique for monitoring cleaning. However, compared to all the other techniques discussed prior, the higher cost and significant operational expertise required to run such analyses, need to be borne in mind. For this very reason, only mock-up surfaces cleaned with citric acid in NH$_4$OH were mapped.

The element maps collected indicated that soiling removal was most efficacious from the chalk–glue ground surfaces, whereas about half as much soiling was removed from the half-chalk and oil paint. Cleaning efficacy plots based on SEM-EDX element maps for Al and Si are given in Figure 10. As a result of the nature of data collected for this metric, it was also possible to calculate that the agar spray removed on average three times more soiling than its pre-formed gel counterpart.

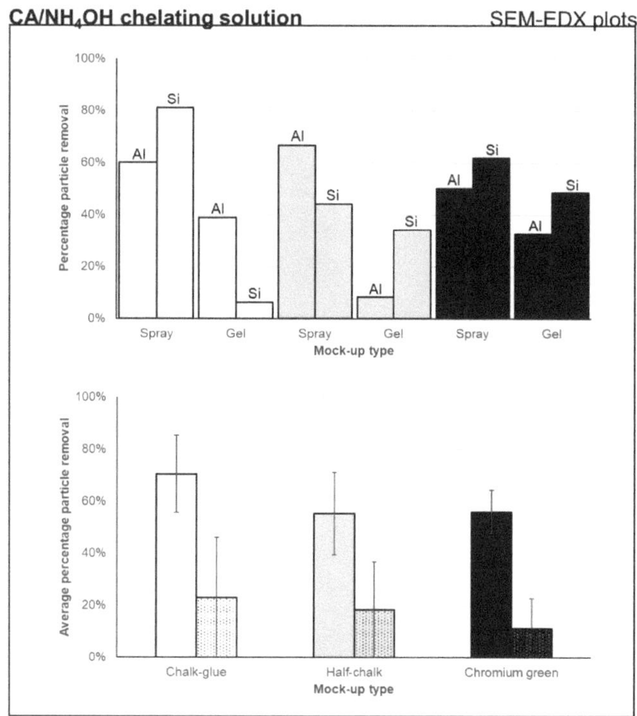

Figure 10. Cleaning efficacy plots for the SEM-EDX metric for particle removal by element (**top**) and average particle removal (**bottom**), using citric acid (CA) in an NH_4OH solution (key: white—chalk–glue; grey—half-chalk; black—chromium oxide; plain fill—agar spray; textured fill—pre-formed gel).

3.2. Evaluating Soiling Removal Scores Using the Metrics

The constellation of cleaning efficacy metrics, together with the cleaning homogeneity score, permitted a well-rounded assessment of the agar spray cleaning tests on the Aula mock-ups. Table 7 below shows an example of how the metrics compared to each other and contributed towards the scoring criteria for the spray tests. The collection of scoring tables for all mock-ups is available online as an externally hosted supplementary file. The resulting star diagrams are presented in Figure 11 (overleaf).

For the chalk–glue ground and the chromium oxide green oil paint, the star diagrams indicate that the citrate/NH_4OH solution (having the largest pentagon) satisfied most cleaning criteria. The solution's improved cleaning action was more evident when applied via spray than as a pre-formed gel, an observation that was already noticeable when evaluating the metrics and weighing in the results by eye. When applied as a pre-formed gel, the difference in base for the chelating agent was shown to be minimal, in fact for the oil paint, the citrate/NaOH solution was just about better.

The citrate/NaOH solution also seems to perform better for spray application on the half-chalk ground—however, it is interesting to note that at first glance the largest pentagons are those associated with deionised water, and, therefore, the action of the gel itself. Two implications follow from this: the first is methodological, in that combining the star diagrams with weighted scores is beneficial for dispelling vagueness that might result from having many polygons on one star diagram. For example, in the case of the star diagram for the pre-formed gel on the chalk–glue ground, the pentagons for deionised water and for the citrate/NH_4OH solution may almost seem similar in size—the weighted score however indicates the chelating agent's improved performance, reflected also in the more central spread of its pentagram towards the prioritised criteria. The second consequence is more application-based. The cleaning solutions tested for the half-chalk ground were not as effective as expected, and the fact that the control solution of deionised water performed very similarly, or even better, than other tailored solutions implies that more experimentation is required to find a better candidate for cleaning.

Table 7. Breakdown of results from the cleaning efficacy metrics (top), and overall scoring criteria (bottom) for the chalk–glue ground mock-up triplicates sprayed by agar gel. The coloured values indicate the measured values used to calculate mean values (in bold): the red values refer to image-based cleaning efficacy scores, whereas the blue values refer to spectral-based cleaning efficacy scores. The cleaning efficacy score in bold violet in the bottom table is the mean of the bold red and blue mean values.

	Cleaning Efficacy Metrics					*Mean Values*	
	Image-Based		**Spectral-Based**				
Cleaning Solution	*L*a*b**	**Skewness**	**Supervised (VNIR)**	**NDI (SWIR)**	**FTIR (MIR)**	**Image-Based**	**Spectral-Based**
Deionised water	0.76	0.81	0.95	0.85	0.39	0.79	0.73
Adjusted water	0.78	0.79	0.91	0.84	0.58	0.79	0.78
Chelator (NaOH)	0.79	0.70	0.91	0.88	0.46	0.75	0.75
Chelator (NH$_4$OH)	0.84	0.68	0.96	0.96	0.57	0.76	0.80
	Scoring Criteria						
Cleaning Solution	Cleaning Efficacy [a]	Cleaning Homogeneity [b]	Colour Integrity [c]	Gloss Integrity [c]	Selectivity [d]	Residue Absence [d]	
Deionised water	0.76	0.23	0.75	0.17	1.00	1.00	
Adjusted water	0.78	0.22	0.69	0.64	1.00	1.00	
Chelator (NaOH)	0.75	0.33	0.72	0.62	0.60	1.00	
Chelator (NH$_4$OH)	0.78	0.57	0.82	0.86	0.80	1.00	

[a]: from metrics in upper table; [b]: from GLCM; [c]: from appearance-based metrics; [d]: user-defined (Appendix A).

The star diagrams also illustrate that cleaning homogeneity was best achieved when spraying the agar directly onto the surface, and the highest homogeneity was delivered by the chelator solutions.

In the end, the suggestions for more empirical-based evaluations in this paper, combined with the tools that have developed within the profession, can further improve decision-making for conservators when planning treatment, including at a monumental scale, such as in the case of the Aula. The presentation of results in this format makes it easier to document cleaning trials and can aid in discussion with other heritage or museum colleagues, such as curators, private owners, art historians, archaeologists, and so forth.

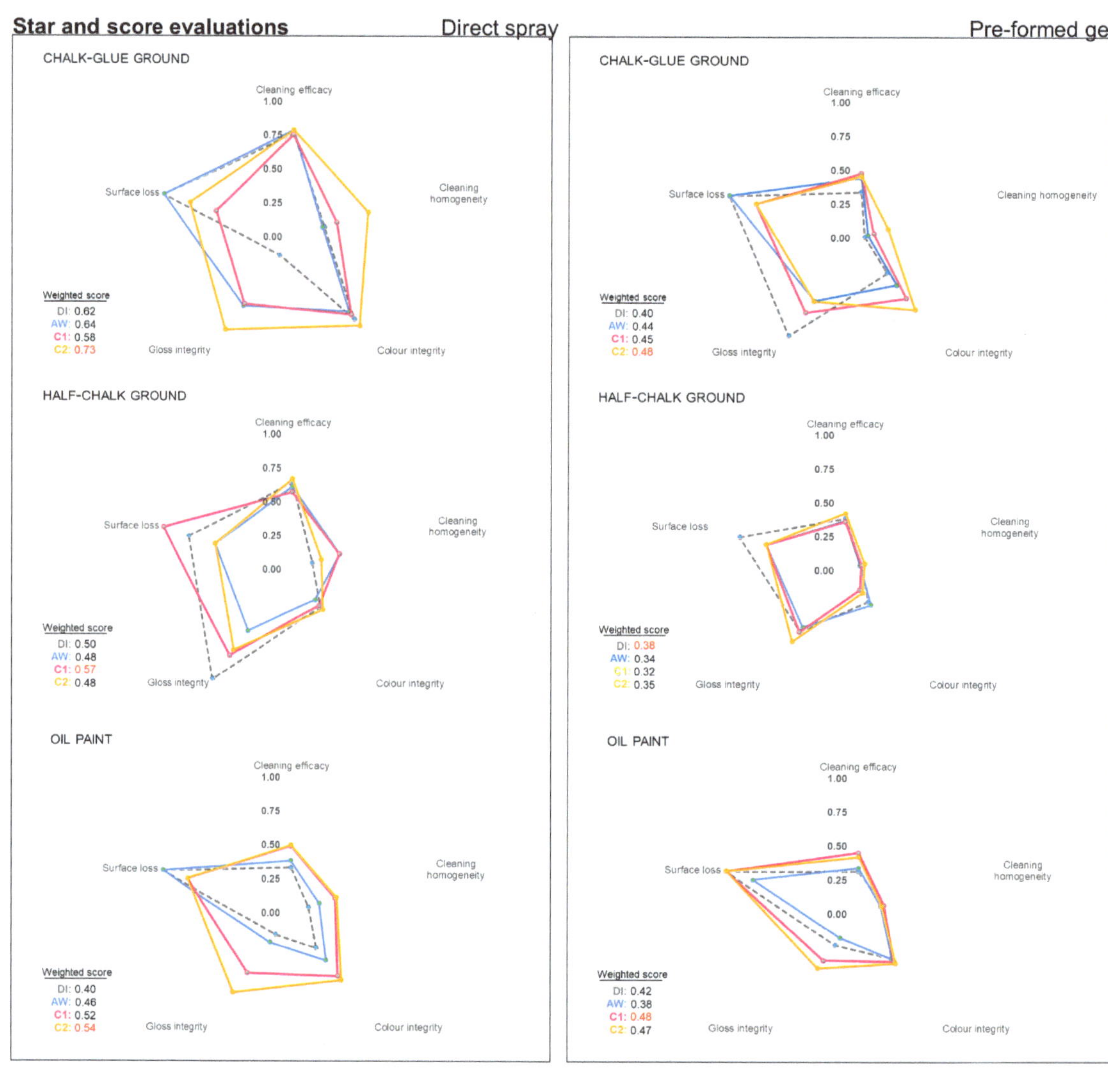

Figure 11. The star diagrams and weighted score evaluations of the agar cleaning trials (grey, dotted—deionised water (DI); blue—adjusted water (AW); crimson—citric acid in NaOH solution (C1); amber—citric acid in NH_4OH solution (C2)). The weighted score highlighted in red indicates the highest score.

4. Conclusions

The multidisciplinary approaches presented expand on the means currently available to conservators for evaluating cleaning tests within a treatment context. By measuring changes in the appearance and spectrally characteristic properties of the surfaces being cleaned, the metrics presented aimed to minimise user bias. They offer a toolkit of methods, which are relatively straightforward to execute and can promote the quality of documentation, as well as the cross-dissemination of results in the literature. Where the occasion calls for it (e.g., new soiling removal methods, preliminary studies for the cleaning of large-scale artworks, etc.), the metrics can thus be used to scientifically assess surface cleaning results, providing a more holistic evaluation than subjective, visual estimations.

Two major forms of cleaning efficacy have been defined, namely, spectral- and image-based efficacies. The spectral metrics are rooted in data collectable from a broad section of the electromagnetic spectrum, providing insights into different elemental, colourimetric, and molecular features. Where access to specialised equipment for measuring spectral-based efficacy is limited (due to budget, time restraints, etc.), the image-based efficacy metrics offer a low-cost, reliable, and effective alternative to surface evaluation. These scores, calculated using image processing methods, tend to be lower than spectral-based scores, and thus mitigate against over-estimating cleaning.

An empirical means of calculating cleaning homogeneity based on digital image analysis was also used in evaluating the cleaning tests, together with a simplified, non-contact means of measuring colour change using hyperspectral imaging. Moreover, the representativeness of each metric's outputs was assessed by comparison to visual observations made by conservators. The newly proposed metrics were integrated as weighted scores with the star diagram evaluation system already established in conservation practice. Such an integration permitted a nuanced interpretation of cleaning tests by agar spray cleaning, comparing it to its pre-formed gel counterpart.

For exposed chalk–glue grounds and chromium oxide oil paint—two water-sensitive materials used by Edvard Munch in the Aula—spraying agar prepared with citric acid in ammonium hydroxide at a surface-tailored pH was evaluated as potentially the best candidate to date for efficacious and homogenous soiling removal. Notably, the pre-formed agar can be used to deliver milder, albeit slightly less effective, cleaning of fragile areas. It remains to be seen which cleaning solution is more appropriate for the half-chalk ground, and progress can be made once the nature of soiling-surface interactions on this type of ground is further elucidated through future research efforts.

It must also be considered that the Aula paintings themselves might be more fragile, and more water-sensitive, than the mock-ups used in this study, and therefore further testing on the artworks themselves is certainly required. This applies especially to areas that have poor surface adhesion or have been previously consolidated [23]. For all the water-sensitive surfaces considered in this study, the effect of agar gel cleaning on other chemical forms of degradation, such as metal soap formation, has not been studied and should be considered in future works. This is especially true since developments in reflectance imaging spectroscopy for the detection of degradation products are picking up momentum [98–102].

Finally, the evaluation metrics themselves will benefit from further testing in other case studies to ensure their usefulness and applicability in the conservation profession. Further developments in the spectral imaging technologies and post-processing methods reported are certain to ameliorate the representativity of the metrics, which can be further tailored according to need in collaboration with conservators, conservation scientists, and imaging scientists.

Supplementary Materials: The following supporting information can be downloaded at: https://www.mdpi.com/article/10.3390/coatings14081040/s1, Figure S1: Legend to interpreting the spectral unmixing maps presented in this work; Figure S2: Spectral maps by supervised unmixing for the chalk–glue ground in the VNIR range; Figure S3: Spectral maps by supervised unmixing for the chalk–glue ground in the SWIR range; Figure S4: Spectral maps by supervised unmixing for the half-chalk ground in the VNIR range; Figure S5: Spectral maps by supervised unmixing for the half-chalk ground in the SWIR range; Figure S6: Spectral maps by supervised unmixing for the chromium oxide oil paint in the VNIR range; Figure S7: Spectral maps by supervised unmixing for the chromium oxide oil paint in the SWIR range; Figure S8: Spectral maps by unsupervised unmixing for the chalk–glue ground in the VNIR range; Figure S9: Spectral maps by unsupervised unmixing for the chalk–glue ground in the SWIR range; Figure S10: Spectral maps by unsupervised unmixing for the half-chalk ground in the VNIR range; Figure S11: Spectral maps by unsupervised unmixing for the half-chalk ground in the SWIR range; Figure S12: Spectral maps by unsupervised unmixing for the chromium oxide oil paint in the VNIR range; Figure S13: Spectral maps by unsupervised unmixing for the chromium oxide oil paint in the SWIR range; Figure S14: Normalised difference image map for the chalk–glue ground in the SWIR range; Figure S15: Normalised difference image map for the half-chalk ground in the SWIR range; Figure S16: Normalised difference image map for the chromium oxide oil paint in the SWIR range; Table S1: Environmental parameters measured during agar spray tests.

Author Contributions: Conceptualization, J.D.C. and T.F.; methodology, J.D.C., C.C.S. and E.J.; software, C.C.S.; validation, J.D.C., C.C.S., E.J., F.C. and T.F.; formal analysis, J.D.C. and F.C.; investigation, J.D.C. and C.C.S.; resources, E.J. and T.F.; data curation, J.D.C.; writing—original draft preparation, J.D.C.; writing—review and editing, J.D.C., C.C.S., F.C., E.J. and T.F.; visualization, J.D.C. and F.C.; supervision, E.J. and T.F.; project administration, J.D.C. and T.F.; funding acquisition, E.J., F.C. and T.F. All authors have read and agreed to the published version of the manuscript.

Funding: This work was funded by the Horizon 2020 research and innovation programme under the H2020 Marie Skłodowska Curie Actions grant agreement no. 813789. F.C. acknowledges his Maria Zambrano fellowship from UPV/EHU, funded by the Spanish Ministry of Universities and the European Union NextGenerationEU/PRTR, and support by grant TED2021-129299A-I00, funded by MCIU/AEI/10.13039/501100011033 and by the European Union NextGenerationEU/PRTR.

Institutional Review Board Statement: Not applicable.

Informed Consent Statement: Not applicable.

Data Availability Statement: The data supporting the reported findings as externally hosted Supplement are openly available in the CHANGE-ITN Zenodo repository at http://doi.org/10.5281/zenodo.12 656836 (accessed on 13 August 2024).

Acknowledgments: Heartfelt thanks for discussions and contributions to Teresa Duncan (cleaning efficacy), Silvia Russo (FTIR mapping), Agnese Babini (HSI algorithms), and David Lewis (skewness). Gratitude due to Conservation Studies at UiO (lab and workshops), Guro Hjulstad at KHM (glossmeter), SciCult labs at KHM (SEM), HE-Arc CR (labs, µFTIR), LATHEMA (labs, climatic chamber), and the NTNU Colourlab (HSI cameras). Further thanks to Ambra Giordano (private conservator) for her agar spray workshop that inspired this study and the encouragement to embark on it; Lena Porsmo Stoveland (conservator/researcher) for guiding mock-up production and cleaning evaluation. Final appreciation to all CHANGE-ITN colleagues for their constant support.

Conflicts of Interest: The authors declare no conflicts of interest. The funders had no role in the design of this study; in the collection, analyses, or interpretation of data; in the writing of the manuscript; or in the decision to publish the results.

Appendix A. Additional Methodological Details

Photography. The mock-ups were photographed using a Canon 6D DSLR camera with a Canon RF 100 mm F2.8L macro IS USM lens (Tokyo, Japan). They were mounted upon black velvet cloth and illuminated with Kaiser 300 W tungsten–halogen lamps (Fort Morgan, CO, USA) under incident and raking light configurations (Table A1). Colour management was carried out using ISA RezChecker targets (Image Science Associates, Williamson, NY, USA), which were used for white balance correction using a macro developed for ImageJ (v. 1.54f), once the RAW files had been converted to RGB images within the freeware [103,104].

This image-based metric was developed based on current trends in the literature [56]. It was found that the L^*, a^*, and b^* images of the mock-ups as extracted from a CIELAB1976 colour space gave faithful representations of the soiling upon the surfaces. This is likely since the soiling was captured as darker pixels on a lighter background (for L^*), bluer pixels on the yellowish background of the ground (for b^*), and redder pixels on the green background of the oil paint (for a^*). The RGB images were converted into their respective L^*, a^* and b^* images using the *Colour Space Converter* plug-in in ImageJ based on a D65 illuminant [105] (the developer warns against imprecisions of the plug-in—these were recognised and accepted during interpretation). The images were converted to 8-bit depth and thresholded to match the soiling pattern recorded on the mock-up surfaces. In the thresholded images, black pixels represented soiled pixels, and white ones, cleaned pixels. For BT images, thresholding was demarcated at the image histograms' maximum, and at the best soiling representation between 60% to 80% for AT images. Cleaning efficacy was then calculated (as shown in Tables 5 and 6 in the main text).

Table A1. Photography metadata recorded during documentation of the mock-up surfaces.

Incident Light		*Raking Light*	
ImageCapture		**ImageCapture**	
ISO	100 (ground); 200 (paint)	ISO	160 (ground); 200 (paint)
Exposure	1/125 s	Exposure	1/125 s
Aperture	f/2.8	Aperture	f/2.8
SetupGeometries		**SetupGeometries**	
Camera height	73 cm	Camera height	73 cm
Light quantity	2 (left and right)	Light quantity	1 (right)
Lights height from plane	39 cm	Lights height from plane	16 cm
Lights angle from plane	45°	Lights angle from plane	60°
Lights to lens distance	45 cm	Lights to lens distance	75 cm

Optical microscopy. The mock-ups were documented in triplicate using a benchtop Leica DM2700 M Microsystem light microscope (Wetzlar, Germany). Photomicrographs were taken using a Leica MC190 HD camera run from the Leica Application Suite v.4.13 image acquisition software. Surfaces were imaged in bright field and raking light (a Leica CLS100 lamp was used in this case) using N PLAN EPI objectives ($\times 5/0.12$ POL, and $\times 20/0.40$ POL) at regions of interest (ROIs).

The photomicrographs were used in the assessment of two separate image-based metrics. The first was similar to the $L^*a^*b^*$ metric described for photography—L^* images were used for the exposed grounds (soiling was represented as dark pixels on a bright background), and a^* images were used for the unvarnished oil paint (soiling was represented as non-green pixels on a green background).

The second metric was based on statistical moments describing image histogram distributions, following the work of Charola et al. [87] and Duncan et al. [54]. These studies used histogram skewness as a marker to monitor the removal of darker soiling on a lighter surface. In this work, Pearson's moment coefficient of skewness was used [88], as calculated by the corresponding ImageJ function. The determination was carried out for all triplicates of all mock-up categories. These values were checked for consistency in the Excel spreadsheet package by importing the image pixel values from the respective text image. The statistical descriptors (mean, mode, median, standard deviation, and skewness) were in accord, with skewness moments varying between ± 0.2.

The cleaning efficacy was then measured as the percentage return to the unsoiled control's skewness (as shown in Table 6 in the main text). The spreadsheet template used to calculate skewness is available online as an externally hosted supplementary file. The metric sometimes generated deviant numbers that were not expected, possibly because the change in skewness compared to the unsoiled reference was not always the best representative measure of cleaning efficacy for the surfaces being studied. A value of over 100% cleaning efficacy was not logical in this context, and so these values were discarded. Squaring and rooting were used to convert occasional negative values into usable positive ones, due to deviancy in the skewness calculations based on the selected controls. For these reasons, this metric should be applied with some caution.

Glossimetry. Gloss measurements on the mock-up surfaces were taken using an Elcometer 480 triple angle glossmeter. Based on recommendations for matte surfaces, an 85° measurement angle was used over an area of 4 mm × 55 mm. Ten readings were taken per mock-up triplicate using the *Batch* function of the glossmeter. Data were exported using the ElcoMaster 2.0 application and processed in the Excel spreadsheet package. A gloss metric was calculated as the percentage return to the unsoiled control's gloss (as explained in Table 6 in the main text).

Hyperspectral imaging. Spectral datacubes for each mock-up class were acquired (BT and AT) using HySpex VNIR1800 and SWIR384 pushbroom hyperspectral scanners (Norsk Elektro Optikk AS, Oslo, Norway). Acquisitions were carried out following the literature [68,106,107] (Table A2). Pre-processing of the spectral cubes was carried out using the manufacturer's acquisition software, and the associated HySpexRadV2.0 software package. Post-processing workflows were followed using the ImageJ image processing freeware package (v. 1.54f) to give datacubes with reflectance values [68,108–110].

Table A2. HSI acquisition conditions.

HSI Cameras	VNIR1800	SWIR384	Lights	Tungsten-Halogen
Spectral range, nm	407–998	951–2505	Spectral coverage, nm	c. 320–2600
Spectral bands	186	288	Quantity	2
Spectral interval [a], nm	3.26	5.45	Room lighting	Darkness
Pixels acquired	1800	384	Geometry [b]	45°, h: 130 cm, d: various
Focal length, m	0.30	0.30	Spectral reflectance	Spectralon white (99%) and grey (50%) diffuse
Field-of-view [c], cm	8.60	8.60	standard	
Spatial resolution, μm	50	220		
Mount	Fixed, perpendicular to surface		Mount	Fixed to stage
Acquisition Parameters				
HSNR [d]	0	0		
Integration time, μs				
for exposed ground	25,000	6900		
for oil paint	39,000	10,800		

[a]: known as the full width at half maximum (FWHM) of the sensor's sensitivity distribution. [b]: h represents height, d represents distance from camera (centre SWIR camera to first light: 22.5 cm left; first light to centre VNIR camera: 15.0 cm left; centre VNIR camera to second light: 30.0 cm left). [c]: known as the ground sampling distance (GSD). [d]: HSNR is the high signal-to-noise ratio (i.e., the number of times each image frame is recorded and averaged).

Two spectral unmixing algorithms (one semi-supervised, *PoissonNMF* [63], and the other unsupervised, *LUMoS* [64]) were used to spectrally identify and map soiled and cleaned areas. The unmixing algorithms' outputs featured maps, where increasing pixel value was indicative of the degree of spectral similarity (the maps were given as 32-bit, and 16-bit images for the semi-supervised and unsupervised algorithms, respectively). The mean pixel values from 100 pixel × 100 pixel areas were used (as shown in Table 6 in the main text) to indicate percentage cleaning efficacy or percentage remainder soiling, depending on which reference area the spectral mapping was calibrated upon.

Normalised difference image (NDI) mapping was carried out as reported by Malegori et al. [65], Piarulli et al. [66], and Lugli et al. [67]. Using PCA [68], characteristic wavelengths in the SWIR range were selected for each mock-up's ground or oil paint surface, and for the soiling (Figure A1).

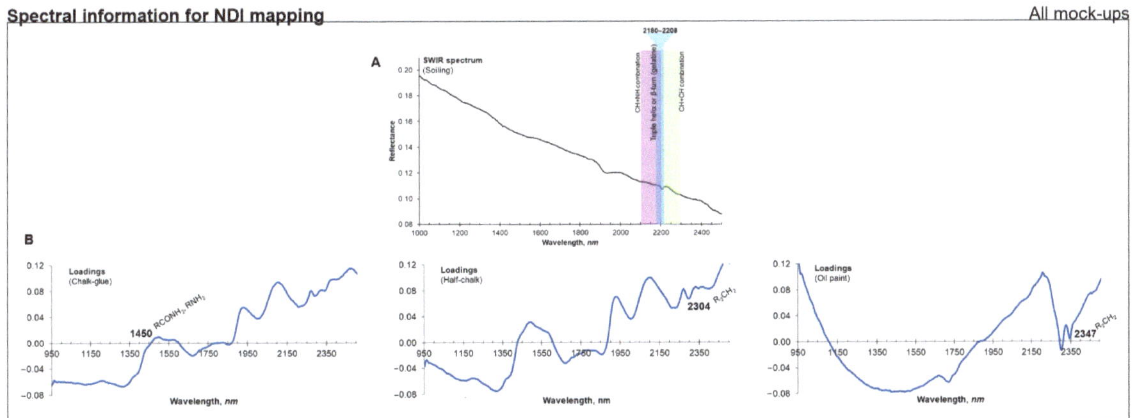

Figure A1. (**A**). SWIR spectrum for artificial soiling and characteristic peak identified for NDI mapping (highlighted sections: CH+CN combination range in violet, CH+CH combination range in green, collagen triple helix or β-turn range in blue (2180–2208 nm)). (**B**). PCA loading plots for the SWIR spectra of each mock-up surface, and the relevant peaks selected for NDI mapping (1450 nm for chalk–glue ground; 2304 nm for half-chalk ground; 2347 nm for oil paint) [111–115].

An image difference operation was carried out with image slices corresponding to the selected wavelengths, using the *Image Calculator* function within ImageJ to execute Equation (1) in the main text. The resulting 32-bit image provided a map of the soiling upon the mock-up surface. For visualisation purposes, the images were individually adjusted for brightness and contrast to make differences more visible, and an appropriate look-up table was applied. In this context, a *look-up table* is a data matrix containing information for the pixel values and can be used to change the colour scheme of the image. The same metric as described for spectral unmixing (in Table 6 in the main text) was then implemented to illustrate cleaning efficacy (based on the unsoiled mock-up), or remainder of soiling (based on a soiling control). The NDI values obtained were negative, and thus Equation (1) in the main text used to calculate the percentage efficacy based on the unsoiled control was modified as shown in Equation (A1) below.

$$\text{cleaning efficacy} = \left(\frac{\text{NDI}_{\text{cleaned}}}{\text{NDI}_{\text{unsoiled}}} \right) \times 100 \tag{A1}$$

This equation was derived as follows (Equation (A2)):

$$\begin{aligned}\text{cleaning efficacy} \; &= \left(1 - \left(\frac{\text{NDI}_{\text{unsoiled}} - \text{NDI}_{\text{cleaned}}}{\text{NDI}_{\text{unsoiled}}} \right) \right) \times 100 \\ &= \left(1 - \left(1 - \frac{\text{NDI}_{\text{cleaned}}}{\text{NDI}_{\text{unsoiled}}} \right) \right) \times 100 \end{aligned} \tag{A2}$$

$$\text{cleaning efficacy} = \left(\frac{\text{NDI}_{\text{cleaned}}}{\text{NDI}_{\text{unsoiled}}} \right) \times 100$$

Furthermore, to indicate the percentage soiling remaining on surfaces, the following equation (Equation (A3)) was used:

$$\text{cleaning efficacy} = \text{NDI}_{\text{soiling}} / \text{NDI}_{\text{cleaned}} \tag{A3}$$

Images of the NDI maps of the cleaned mock-ups were opened in ImageJ, converted to 8-bit depth (required by the plug-in), and analysed with the *GLCM Texture* plug-in, using a pixel step of 1 in the 0° direction. The value given for the inverse difference moment was taken as a homogeneity score, where a value of 1 represented an entirely homogenous image (all pixels have the same intensity), and therefore fully homogeneous cleaning (the equations used to calculate image homogeneity can be found in Albregtsen [97]).

The initial presumed overlap between mock-up and soiling identifiers was recognised, especially since considerable band overlap is common in the SWIR. For example, although collagen and gelatine, two highly similar macromolecules, each contain identical functional groups, their chemically diverse environs will have resulted in different band contributions—this has been confirmed by PCA in this case [68], and conceptually [116,117]. Furthermore, Duconseille et al. [113] confirm that aged gelatines (such as in the chalk–glue ground) have higher NH overtone vibrations than non-aged gelatines (such as in the soiling's gelatine), which feature more prominent CH+NH combination modes. For the chalk–glue ground, an attempt to use the SWIR bands for calcite (which were marked by PCA as significant) was unsuccessful, most likely due to interference from bands from other materials: two clusters resulted on the score plots representing soiled and unsoiled surfaces distinctly. The loading plots indicated that the calcite bands should be contributing to the observed variance. These bands were at 1997 nm and 2340 nm. They conflicted with proximal bands at 2009 nm and 2342 nm (associated with the gelatine, and linseed oil in the mock-ups respectively), and did not give sufficiently coherent NDI maps.

Colourimetry from HSI. A relatively quick, simplified method for calculating colour change was implemented based on selecting three bands (corresponding to R, G, and B channels) from the HSI datacube. The latter is a quick way to visualise a colour image from spectral data. As such it can be useful, but it is acknowledged that it is not a fully colourimetric method. Further colourimetric analyses in CIELAB could be carried out through the convolution of image data with XYZ colour-matching functions [118] (a comparison of these two methods was beyond the scope of this paper).

In ImageJ (v. 1.54f), $L^*a^*b^*$ values for the mock-up surfaces were generated as follows: a sub-stack of three slices from the red, green, and blue spectral regions of the datacube was produced at 16-bit depth. The three slices used were taken from the bands at 647.9, 552.2, and 466.0 nm for R, G, and B channels, respectively. The stack was then converted into an RGB image, and subsequently converted into separate L^*, a^*, and b^* images using the *Colour Space Converter* plug-in, specifying a D65 illuminant. The three images were then reconverted into a stack, and respective $L^*a^*b^*$ values were extracted from its z-axis profile.

Ten ROIs measuring 100 pixels × 100 pixels were used to extract the $L^*a^*b^*$ values for each mock-up, and processed in the Excel spreadsheet package to calculate mean and coefficient of variation values. Deviating measurements were discarded using the ISO-recommended Grubbs' test at $p = 0.05$ [119]. The formula used to measure the colour difference, ΔE, was the CIEDE2000 equation [120,121], using the calculator developed by Boronkay [122]. Error propagation was calculated according to Miller & Miller [119] and Williams [123]. The spreadsheet template used to calculate error propagation with $L^*a^*b^*$ values is available online as an externally hosted supplementary file. Propagation calculations were confirmed using the Error Propagation Calculator application [124]. Using these values, a colourimetry metric was calculated as the percentage return to the unsoiled control's colour (as shown in Table 6 in the main text).

No cleaned mock-up matched the colour nor tone of the unsoiled control; this was recognised as a highly idealistic comparison given that the surface would have continued changing during ageing. The colourimetry metric thus helped appreciate this change numerically but was based on a return to the unsoiled control, and this should be borne in mind whilst evaluating the results.

μFTIR mapping. Fourier-transform infrared (FTIR) 2D chemical maps of ROIs on the mock-ups were collected using a ThermoScientific Nicolet iNTM10 MX FTIR microscope equipped with a motorised xyz-stage. Each ROI consisted of 350 16-scan spectra collected over 5 s in attenuated total reflectance (ATR) mode (Germanium-crystal, liquid nitrogen-cooled detector, 15 arbitrary pressure units) over the range 4000–650 cm^{-1}, using an aperture of 150 μm × 150 μm, a step size of 100 μm, and 4 cm^{-1} resolution, over a total surface area of 3500 μm × 1000 μm.

ATR was used as the mode of acquisition, since spectra acquired in diffuse reflectance suffered badly from different interactions of the IR radiation with the surface, probably as a result of the non-uniform and non-homogeneous topography of the soiling [125]. As the mock-ups were expendable, the increased contact with the ATR crystal was not an issue; however, the presence of soot (carbon black) particles in the soiling mixture meant that, at certain points in the maps, background noise was registered, influencing the quality of the maps. This is possibly due to the higher refractive index of carbon black (compared to other organic materials), which, despite the higher refractive index of the germanium crystal, might have still led to light being transmitted into the sample, creating band distortions via anomalous dispersion that present as asymmetric spectral artefacts [126]. Raman point analysis carried out separately from this study confirmed that these points did contain carbon.

Visualisation of single spectra was carried out in Spectragryph 1.2.15 [127]. Pre-processing of the maps was limited to atmospheric corrections for carbon dioxide and water as carried out in Omnic 9.2.86. Post-processing began with background subtraction of the mock-up surface to intensify the soiling signal. Soiling maps were then generated using the *Correlation map* profile feature in Omnic Atlµs 9.2.91 and were based on spectral similarity to the artificial soiling's spectrum. The intensity of pixels in the soiling maps corresponded to increasing spectral matches to the molecular fingerprint of the soiling.

The latter maps were exported as greyscale .tiff files into ImageJ (v. 1.54f), assigned the desired look-up table, and adjusted for brightness and contrast to improve visualisation. The images were then automatically thresholded and converted to 8-bit binary images, so that black pixels represented cleaned areas (no soiling spectrum detected), whereas white pixels represented soiled areas (any degree of soiling spectrum detected). The percentage cleaning efficacy as measured by FTIR was then calculated (following Table 6 in the main text).

SEM imaging and SEM-EDX mapping. The mock-up surfaces were investigated by scanning electron microscopy (SEM) using a FEI Quanta 450 electron microscope equipped with an X-Max Oxford 50 mm^2 SSD detector running on Aztec 3.1 SP1 software. Each mock-up (c. 5 cm $\times$ 5 cm) was mounted on the stage of the vacuum chamber with carbon tape and no surface preparation. The same ROIs as for microscopy and FTIR were used. A slightly different area was imaged however for SEM, as the ATR diamond disrupted the surface (Figure A2), preventing monitoring.

The methodology employed implied that the acquisition process was quasi-non-contact (except for the attaching of conductive tape to the back of the mock-up). It should be borne in mind that alteration of the surface is technically possible by charge accumulation, which can shift particles around, or by burning through the surface via high electron flux (e.g., when using high acceleration voltages [128–130]).

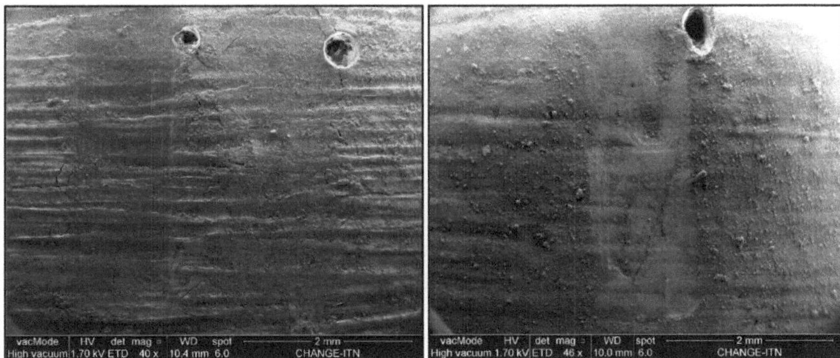

Figure A2. Disruption of the mock-up surfaces and soiling by the ATR diamond (ground (**right**); oil paint (**left**)). AT measurements were taken from the area immediately below the BT measurement, as shown on the left.

The following conditions were used for imaging based on procedures in the literature and recommendations for unvarnished painted surfaces [128,129,131]: secondary electron (SE) images were acquired at ×100, ×500 and ×1000 in high vacuum mode with an Everhart–Thornley (ETD) detector at low acceleration voltage (1.7 kV), a working distance of c. 10 mm, with a beam spot size of 6.0 mm². The magnification and contrast for each mock-up were slightly adjusted in order to obtain the sharpest image. SEM-EDX was carried on back-scattered electron (BSE) images acquired at ×500 in low-vacuum mode with a circular backscatter (CBS) detector at high acceleration voltage (20.0 kV), with the same working distance, and electron beam spot size. Element maps were processed using the TruMap feature within the Aztec® software (only the mock-ups cleaned by the citrate/NH₄OH solution were mapped due to project restrictions).

After being identified as indicative markers of the artificial soiling used, Al and Si element maps were selected for use in the SEM-EDX-based metric for cleaning efficacy (Figure A3). The maps were imported into ImageJ (v. 1.54f) as .tiff files and converted into 8-bit binary images that were automatically thresholded (applying the *Fill Holes* option to create distinct particulate units). Al- and Si-containing areas were represented as white pixels in this manner. The *Analyse Particles* function was then used to count individual, defined areas, and used to calculate cleaning efficacy (as listed in Table 6 in the main text).

Treatment evaluation scores. An image-based and spectral-based efficacy score was generated as an average of the cleaning efficacy values for each class of metric. For the image-based efficacy, the $L^*a^*b^*$ metric at the microscale and the skewness metric were considered. The total cleaning efficacy used for the scoring criteria was then calculated as a further average of these two values. For the spectral-based efficacy, the values from the semi-supervised unmixing in the VNIR, the NDI mapping in the SWIR, and the FTIR mapping in the MIR were selected, following the recommendations given in the main text, and permitting the score to cover a large swathe of the electromagnetic spectrum. Cleaning homogeneity was taken from the result of the GLCM analysis. The colourimetry and gloss metric values were fed directly into the colour and gloss integrity score fields, as they formed independent criteria for evaluation. Finally, the scores for cleaning selectivity (lack of pigment loss) and residue absence were evaluated by eye during real-time observations and were broken down into five incremental divisions (0.2 being worst, and 1.0 being best). The absence of residues was not included as a scoring criterion, as all cleaning tests had scored the best possible score, i.e., no residues were noticed when the agar technique was applied correctly.

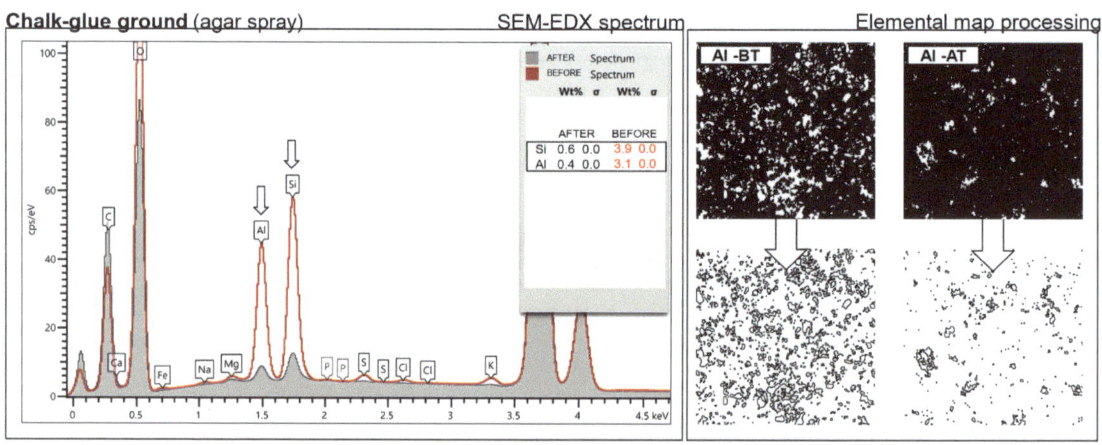

Figure A3. A comparison of SEM-EDX spectra showing the marked change in counts for Al and Si after cleaning (**left**). Examples of thresholded maps and their relevant counting output (**right**).

Appendix B. Discussion of Unsupervised Unmixing for Soiling Removal Mapping

For comparison purposes, the unsupervised unmixing algorithm was tested for mapping soiling removal. Compared to the semi-supervised algorithm (which searched for what the user specified [63]), it did not require the input of guiding regions of interest (ROIs) associated with clean or soiled areas. Although an expected number of different spectral regions was inputted as guidance, it is the algorithm that grouped all spectra into different classes and mapped similar areas accordingly [64].

The unsupervised unmixing algorithm applied to the exposed grounds resulted in somewhat inconsistent mapping for both the VNIR and SWIR ranges, particularly when attempting to map soiled areas (the algorithm did not seem to identify these surfaces as spectrally distinguishable). Mapping of cleaned areas was successful, although false positives were noted for soiling references. In this case, the general trends for soiling removal by the chelating agent observed in the VNIR range with the semi-supervised algorithm also applied here: the chalk–glue ground was almost fully cleaned, the half-chalk ground was notably less cleaned, and the oil paint was cleaned most.

Cleaning efficacy plots based on the unsupervised algorithm are given in Figure A4.

Notably, this algorithm reported the surfaces as more soiled in the SWIR range. According to calculations based on SWIR data, the spray and pre-formed gel had similar performances, with the citrate/NH_4OH solution performing notably better for the chalk–glue ground, and only marginally better for the half-chalk ground. The cleaning efficacy by spray was more marked for the oil paint than by pre-formed gel, with chelation giving markedly better results. All these results matched visual observations better than the VNIR counterparts.

However, the limited reliability of the unsupervised method did not make it ideal for assessing soiling removal. The relative standard deviation (RSD) for the cleaning efficacy values in half of the plots was large enough to cast doubt on the results (RSD for chalk–glue ground: 0.07%–13.86%; RSD for half-chalk ground: 0.06%–0.08%; RSD for oil paint: 0.55%–5.83%); whereas soiling was at times not even detected. Should it be used, the VNIR range is recommended, keeping in mind that this algorithm was designed for pigment identification [75] rather than mapping areas of cleaned and soiled surfaces.

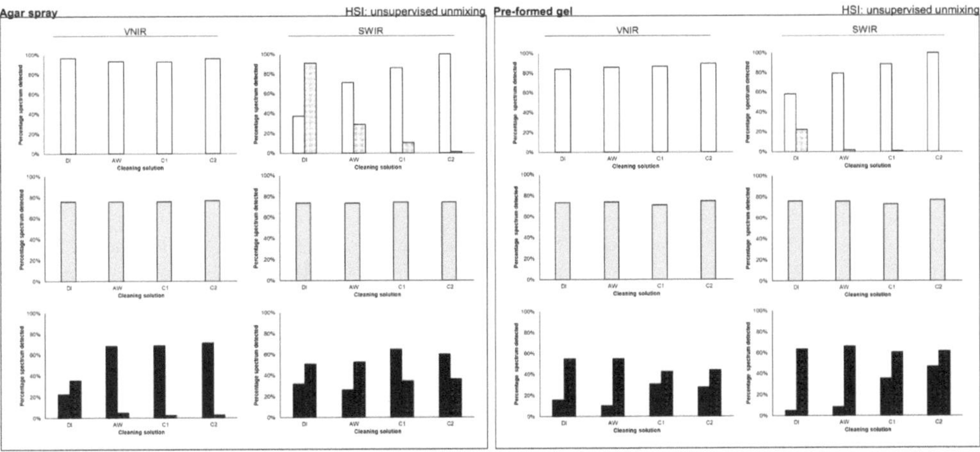

Figure A4. VNIR and SWIR metric plots as determined by the unsupervised unmixing algorithm, based on clean (plain fill) and soiled (textured fill) reference spectra for spray-cleaned surfaces (**left**) and the pre-formed gel cleaning (**right**). Error bars not plotted since RSD ranged between 1.00 and 13.86 in many cases; soiling was not detected on the half-chalk ground, and sprayed chalk–glue ground (keys: white—chalk–glue; grey—half-chalk; black—chromium oxide; (DI) deionised water; (AW) adjusted water; (C1) citric acid in NaOH solution; (C2) citric acid in NH_4OH solution).

References

1. Ormsby, B.; Bartoletti, A.; van den Berg, K.J.; Stavroudis, C. Cleaning and Conservation: Recent Successes and Challenges. *Herit. Sci.* **2024**, *12*, 10. [CrossRef]
2. Van den Berg, K.J.; Gorter, L. (Eds.) *Dirt and Dirt Removal | Paintings Conservation*; Paintings Conservation; Cultural Heritage Agency Netherlands: Amersfoort, The Netherlands, 2022.
3. Passaretti, A.; Cuvillier, L.; Sciutto, G.; Guilminot, E.; Joseph, E. Biologically Derived Gels for the Cleaning of Historical and Artistic Metal Heritage. *Appl. Sci.* **2021**, *11*, 3405. [CrossRef]
4. Angelova, L.V.; Ormsby, B.; Townsend, J.; Wolbers, R. *Gels in the Conservation of Art*; Archetype Publications: London, UK, 2017; ISBN 978-1-909492-50-9.
5. Mastrangelo, R.; Montis, C.; Bonelli, N.; Tempesti, P.; Baglioni, P. Surface Cleaning of Artworks: Structure and Dynamics of Nanostructured Fluids Confined in Polymeric Hydrogel Networks. *Phys. Chem. Chem. Phys.* **2017**, *19*, 23762–23772. [CrossRef]
6. Ranalli, G.; Zanardini, E.; Rampazzi, L.; Corti, C.; Andreotti, A.; Colombini, M.P.; Bosch-Roig, P.; Lustrato, G.; Giantomassi, C.; Zari, D.; et al. Onsite Advanced Biocleaning System for Historical Wall Paintings Using New Agar-Gauze Bacteria Gel. *J. Appl. Microbiol.* **2019**, *126*, 1785–1796. [CrossRef] [PubMed]
7. Sanmartín, P.; Bosch-Roig, P. Biocleaning to Remove Graffiti: A Real Possibility? Advances towards a Complete Protocol of Action. *Coatings* **2019**, *9*, 104. [CrossRef]
8. Bosch-Roig, P.; Regidor Ros, J.L.; Soriano Sancho, M.P.; Montes Estellés, R.; Roig Picazo, P.; Angelova, L.; Ormsby, B.; Townsend, J. Biocleaning of Wall Paintings on Uneven Surfaces with Warm Agar Gels. In *Biocleaning of Wall Paintings on Uneven Surfaces with Warm Agar Gels/Pilar Bosch-Roig; Jose Luis Regidor Ros; Maria Pilar Soriano Sancho; Rosa Montes Estellés; Pilar Roig Picazo, Gels in the Conservation of Art*; Archetype Publications Ltd.: London, UK, 2017.
9. Morlotti, M.; Forlani, F.; Saccani, I.; Sansonetti, A. Evaluation of Enzyme Agarose Gels for Cleaning Complex Substrates in Cultural Heritage. *Gels* **2024**, *10*, 14. [CrossRef]
10. Lamuraglia, R.; Campostrini, A.; Ghedini, E.; De Lorenzi Pezzolo, A.; Di Michele, A.; Franceschin, G.; Menegazzo, F.; Signoretto, M.; Traviglia, A. A New Green Coating for the Protection of Frescoes: From the Synthesis to the Performances Evaluation. *Coatings* **2023**, *13*, 277. [CrossRef]
11. Joseph, E. (Ed.) *Microorganisms in the Deterioration and Preservation of Cultural Heritage*; Springer International Publishing: Cham, Switzerland, 2021.
12. Elnaggar, A.; Nevin, A.; Castillejo, M.; Strlič, M. (Eds.) *Issue S1: Proceedings of the LACONA 10 Conference, Sharjah 2014, August 2015*; Studies in Conservation; Routledge: London, UK, 2015.
13. Alabone, G.; Carvajal, M.S. The Removal of Bronze Paint Repairs from Overgilded Picture Frames Using an Erbi-um:YAG Laser. *J. Inst. Conserv.* **2020**, *43*, 107–121. [CrossRef]
14. Duncan, T.T.; Chan, E.P.; Beers, K.L. Maximizing Contact of Supersoft Bottlebrush Networks with Rough Surfaces to Promote Particulate Removal. *ACS Appl. Mater. Interfaces* **2019**, *11*, 45310–45318. [CrossRef]
15. Freese, S.; Diraoui, S.; Mateescu, A.; Frank, P.; Theodorakopoulos, C.; Jonas, U. Polyolefin-Supported Hydrogels for Selective Cleaning Treatments of Paintings. *Gels* **2019**, *6*, 1. [CrossRef] [PubMed]
16. Bartoletti, A.; Barker, R.; Chelazzi, D.; Bonelli, N.; Baglioni, P.; Lee, J.; Angelova, L.V.; Ormsby, B. Reviving WHAAM! a Comparative Evaluation of Cleaning Systems for the Conservation Treatment of Roy Lichtenstein's iconic Painting. *Herit. Sci.* **2020**, *8*, 9. [CrossRef]
17. Mastrangelo, R.; Chelazzi, D.; Poggi, G.; Fratini, E.; Pensabene Buemi, L.; Petruzzellis, M.L.; Baglioni, P. Twin-Chain Polymer Hydrogels Based on Poly(Vinyl Alcohol) as New Advanced Tool for the Cleaning of Modern and Contemporary Art. *Proc. Natl. Acad. Sci. USA.* **2020**, *117*, 7011–7020. [CrossRef]
18. Sansonetti, A.; Bertasa, M.; Canevali, C.; Rabbolini, A.; Anzani, M.; Scalarone, D. A Review in Using Agar Gels for Cleaning Art Surfaces. *J. Cult. Herit.* **2020**, *44*, 285–296. [CrossRef]
19. Al-Emam, E.; Motawea, A.G.; Caen, J.; Janssens, K. Soot Removal from Ancient Egyptian Complex Painted Surfaces Using a Double Network Gel: Empirical Tests on the Ceiling of the Sanctuary of OSIRIS in the Temple of Seti I—Abydos. *Herit. Sci.* **2021**, *9*, 1–10. [CrossRef]
20. Bertasa, M.; Canevali, C.; Sansonetti, A.; Lazzari, M.; Malandrino, M.; Simonutti, R.; Scalarone, D. An In-Depth Study on the Agar Gel Effectiveness for Built Heritage Cleaning. *J. Cult. Herit.* **2021**, *47*, 12–20. [CrossRef]
21. Giordano, A.; Cremonesi, P. New Methods of Applying Rigid Agar Gels: From Tiny to Large-Scale Surface Areas. *Stud. Conserv.* **2021**, *66*, 437–448. [CrossRef]
22. Jia, Y.; Sciutto, G.; Botteon, A.; Conti, C.; Focarete, M.L.; Gualandi, C.; Samorì, C.; Prati, S.; Mazzeo, R. Deep Eutectic Solvent and Agar: A New Green gel to Remove Proteinaceous-Based Varnishes from Paintings. *J. Cult. Herit.* **2021**, *51*, 138–144. [CrossRef]
23. Stoveland, L.P.; Frøysaker, T.; Stols-Witlox, M.; Grøntoft, T.; Steindal, C.C.; Madden, O.; Ormsby, B. Evaluation of Novel Cleaning Systems on Mock-Ups of Unvarnished Oil Paint and Chalk-Glue Ground within the Munch Aula Paintings Project. *Herit. Sci.* **2021**, *9*, 144. [CrossRef]
24. Giordano, A.; Caruso, M.R.; Lazzara, G. New Tool for Sustainable Treatments: Agar Spray—Research and Practice. *Herit. Sci.* **2022**, *10*, 123. [CrossRef]

25. Delattre, C.; Bearman, G.; Choi, Y.L.; McPherson, L.; Stiglitz, M. The use of Enzymatic Gels in the Conservation Treatment of Mendelssohn's "Green Books". In Proceedings of the ICOM-CC 20th Triennial Conference Preprints, Valencia, Spain, 18–22 September 2023; ICOM: Paris, France, 2023; pp. 1–8.
26. Husby, L.M.; Andersen, C.K.; Pedersen, N.B.; Ormsby, B. Selecting, Modifying, and Evaluating of WATER-Based Methods for the Removal of Dammar Varnish from Oil Paint. *Meddelelser Konserv.* **2023**, *2023*, 51–65.
27. Husby, L.M.; Andersen, C.K.; Pedersen, N.B.; Ormsby, B. Evaluating Three Water-Based Systems and One Organic Solvent for the Removal of Dammar Varnish from Artificially Aged Oil Paint Samples. *Herit. Sci.* **2023**, *11*, 244. [CrossRef]
28. Ortiz Miranda, A.S.; Lehmann Banke, P.; Ludvigsen, L. Non-Invasive Imaging Systems as Tools for Evaluating Treatments: The Case "Bathers" by Henri Matisse. In Proceedings of the ICOM-CC 20th Triennial Conference Preprints, Valencia, Spain, 18–22 September 2023; Bridgland, J., Ed.; ICOM: Paris, France, 2023; pp. 1–11.
29. Tobin, G.; Sawicki, M. Developing Conservation Practices for Cleaning Gilded Surfaces: Application for xPVOH-Borax Organogels to Clean Two Gilded Frames. In Proceedings of the ICOM-CC 20th Triennial Conference Preprints Valencia, Valencia, Spain, 18–22 September 2023; Bridgland, J., Ed.; ICOM: Paris, France, 2023; pp. 1–16.
30. Cuvillier, L.; Passaretti, A.; Guilminot, E.; Joseph, E. Agar and Chitosan Hydrogels' Design for Metal-Uptaking Treatments. *Gels* **2024**, *10*, 55. [CrossRef]
31. Giraud, T.; Gomez, A.; Lemoine, S.; Pelé-Meziani, C.; Raimon, A.; Guilminot, E. Use of Gels for the Cleaning of Archaeological Metals. Case study of silver-plated copper alloy coins. *J. Cult. Herit.* **2021**, *52*, 73–83. [CrossRef]
32. Guilminot, E. The Use of Hydrogels in the Treatment of Metal Cultural Heritage Objects. *Gels* **2023**, *9*, 191. [CrossRef] [PubMed]
33. Khaksar-Baghan, N.; Koochakzaei, A.; Hamzavi, Y. An Overview of Gel-Based Cleaning Approaches for Art Conservation. *Herit. Sci.* **2024**, *12*, 248. [CrossRef]
34. Macchia, A.; Biribicchi, C.; Carnazza, P.; Montorsi, S.; Sangiorgi, N.; Demasi, G.; Prestileo, F.; Cerafogli, E.; Colasanti, I.A.; Aureli, H.; et al. Multi-Analytical Investigation of the Oil Painting "Il Venditore di Cerini" by Antonio Mancini and Definition of the Best Green Cleaning Treatment. *Sustainability* **2022**, *14*, 3972. [CrossRef]
35. Frøysaker, T. Unintended Contamination? A Selection of Munch's Paintings with Non-Original Zinc White. In *Public Paintings by Edvard Munch and His Contemporaries. Change and Conservation Challanges*; Frøysaker, T., Streeton, N.L.W., Kutzke, H., Hanssen-Bauer, F., Topalova-Casadiego, B., Eds.; Archetype Publications: London, UK, 2015; pp. 132–140.
36. Daudin-Schotte, M.; Bisschoff, M.; Joosten, I.; van Keulen, H.; van den Berg, K.J. Dry Cleaning Approaches for Unvarnished Paint Surfaces. In *New Insights into the Cleaning of Paintings: Proceedings from the Cleaning 2010 International Conference Universidad Politécnica de Valencia and Museum Conservation Institute*; Mecklenburg, M., Charola, A.E., Koestler, R.J., Eds.; Smithsonian Institution Scholary Press: Washington, DC, USA, 2013; pp. 209–219.
37. Gillman, M.; Lee, J.; Ormsby, B.; Burnstock, A. Water-Sensitivity in Modern Oil Paintings: Trends in Phenomena and Treatment Options. In *Conservation of Modern Oil Paintings*; van den Berg, K.J., Bonaduce, I., Burnstock, A., Ormsby, B., Scharff, M., Carlyle, L., Heydenreich, G., Keune, K., Eds.; Springer International Publishing: Cham, Switzerland, 2019; pp. 477–494; ISBN 978-3-030-19253-2.
38. Ormsby, B.; Lee, J.; Bonaduce, I.; Lluveras-Tenorio, A. Evaluating Cleaning Systems for Use on Water Sensitive Modern Oil Paints: A Comparative Study. In *Conservation of Modern Oil Paintings*; van den Berg, K.J., Bonaduce, I., Burnstock, A., Ormsby, B., Scharff, M., Carlyle, L., Heydenreich, G., Keune, K., Eds.; Springer International Publishing: Cham, Switzerland, 2019; pp. 11–35; ISBN 978-3-030-19254-9.
39. Bartoletti, A.; Maor, T.; Chelazzi, D.; Bonelli, N.; Baglioni, P.; Angelova, L.V.; Ormsby, B.A. Facilitating the Conservation Treatment of Eva Hesse's Addendum through Practice-Based Research, including a Comparative Evaluation of Novel Cleaning Systems. *Herit. Sci.* **2020**, *8*, 35. [CrossRef]
40. Chung, J.; Ormsby, B.; Burnstock, A.; Van den Berg, K.; Lee, J. An Investigation of Methods for Surface Cleaning Unvarnished Water-Sensitive Oil Paints based on Recent Developments for Acrylic Paints. In Proceedings of the ICOM-CC 18th Triennial Conference, Copenhagen, Denmark, 4–7 September 2017; ICOM: Copenhagen, Denmark, 2017; pp. 4–8.
41. Sully, D. Conservation Theory and Practice: Materials, Values, and People in Heritage Conservation. In *The International Handbooks of Museum Studies*; John Wiley & Sons, Ltd.: New York, NY, USA, 2015; pp. 293–314; ISBN 978-1-118-82905-9.
42. Cutajar, J.D.; Duckor, A.; Sully, D.; Fredheim, L.H. A Significant Statement: New Outlooks on Treatment Documentation. *J. Inst. Conserv.* **2016**, *39*, 81–97. [CrossRef]
43. Fredheim, L.H.; Khalaf, M. The Significance of Values: Heritage Value Typologies Re-Examined. *Int. J. Herit. Stud.* **2016**, *22*, 466–481. [CrossRef]
44. Avrami, E.; Mason, R. *Mapping the Issue of Values. Values in Heritage Management: Emerging Approaches and Research Directions*; Getty Publications: Los Angeles, CA, USA, 2019; pp. 9–33.
45. Lithgow, K.; Golfomitsou, S.; Dillon, C. Coming Clean about Cleaning. Professional and Public Perspectives: Are Conservators Truthful and Visitors Useful in Decision-Making? *Stud. Conserv.* **2018**, *63*, 392–396. [CrossRef]
46. Coming Clean. Available online: http://www.comingcleanucl.com (accessed on 29 October 2022).
47. Carlyle, L.; Witlox, M. Historically Accurate Reconstructions of Artists' Oil Painting Materials. *Tate Pap.* **2005**, *7*, 1–9.
48. Carlyle, L. Historically Accurate Reconstructions of Oil Painters Materials: An Overview of the HART Project 2002–2005. In *Reporting Highlights of the De Mayerne Programme: Research Programme on Molecular Studies in Conservation and Technical Studies in*

History; Ferreira, E.S.B., Jaap, J.B., Eds.; Netherlands Organisation for Scientific Research: The Hague, The Netherlands, 2006; pp. 63–76. ISBN 90-77875-14X.

49. Lawson, L.; Cane, S. Do Conservators Dream of Electric Sheep? Replicas and replication. *Stud. Conserv.* **2016**, *61*, 109–113. [CrossRef]

50. Pugliese, M.; Ferriani, B.; Ratti, I. Materiality and Immateriality in LUCIO Fontana's Environments: From Documentary Research to the Reproduction of Lost Artworks. *Stud. Conserv.* **2016**, *61*, 188–192. [CrossRef]

51. Stoveland, L.P.; Ormsby, B.; Stols-Witlox, M.; Frøysaker, T.; Caruso, F. Designing Paint Mock-Ups for a Study of Novel Surface Cleaning Techniques for Munch's Unvarnished Aula Paintings. In *Conservation of Modern Oil Paintings*; van den Berg, K.J., Bonaduce, I., Burnstock, A., Ormsby, B., Scharff, M., Carlyle, L., Heydenreich, G., Keune, K., Eds.; Springer International Publishing: Cham, Switzerland, 2019; pp. 553–563; ISBN 978-3-030-19254-9.

52. Stoveland, L.P.; Stols-Witlox, M.; Ormsby, B.; Streeton, N.L.W. Mock-Ups and Materiality in Conservation Research. In Proceedings of the Transcending Boundaries: Integrated Approaches to Conservation, ICOM-CC 19th Triennial Conference BEIJING Preprints, Beijing, China, 18 January 2024; Bridgland, J., Ed.; ICOM-CC: Beijing, China, 2021; pp. 1–14.

53. Frøysaker, T.; Miliani, C.; Grøntoft, T.; Kleiva, I. Monitoring of Surface Blackening and Zinc Reaction Products on Prepared Samples Located Adjacent to Munch's the Source in the Aula at the University of Oslo. In *Public Paintings by EDVARD Munch and His Contemporaries. Change and Conservation Challanges*; Frøysaker, T., Streeton, N.L.W., Kutzke, H., Hanssen-Bauer, F., Topalova-Casadiego, B., Eds.; Archetype Publications: London, UK, 2015; pp. 126–131.

54. Duncan, T.T.; Vicenzi, E.P.; Lam, T.; Brogdon-Grantham, S.A. A Comparison of Dry Cleaning Materials for the Removal of Soot from Rough Papers. *J. Am. Inst. Conserv.* **2023**, *39*, 152–167. [CrossRef]

55. Striova, J.; Dal Fovo, A.; Fontana, R. Reflectance Imaging Spectroscopy in Heritage Science. *Riv. Nuovo C.* **2020**, *43*, 515–566. [CrossRef]

56. Duncan, T.T.; Chan, E.P.; Beers, K.L. Quantifying the 'Press and PEEL' removal of Particulates Using Elastomers and Gels. *J. Cult. Herit.* **2021**, *48*, 236–243. [CrossRef]

57. Schindelin, J.; Arganda-Carreras, I.; Frise, E.; Kaynig, V.; Longair, M.; Pietzsch, T.; Preibisch, S.; Rueden, C.; Saalfeld, S.; Schmid, B.; et al. Fiji: An Open-Source Platform for Biological-Image Analysis. *Nat. Methods* **2012**, *9*, 676–682. [CrossRef] [PubMed]

58. Exelis Visual Information Solutions ENVI Remote Sensing Software for Image Processing & Analysis. 2023. Available online: https://www.nv5geospatialsoftware.com/Products/ENVI (accessed on 6 August 2024).

59. The MathWorks Inc. MathWorks-Makers of MATLAB and Simulink. 2022. Available online: https://se.mathworks.com/ (accessed on 6 August 2024).

60. Drumetz, L.; Chanussot, J.; Jutten, C. Chapter 2.7—Variability of the Endmembers in Spectral Unmixing. In *Data Handling in Science and Technology*; Amigo, J.M., Ed.; Elsevier: Amsterdam, The Netherlands, 2020; Volume 32, pp. 167–203; ISBN 0922-3487.

61. Nascimento, J.; Martín, G. Chapter 2.6-Nonlinear spectral unmixing. In *Data Handling in Science and Technology*; Amigo, J.M., Ed.; Elsevier: Amsterdam, The Netherlands, 2020; Volume 32, pp. 151–166; ISBN 0922-3487.

62. Quintano, C.; Fernández-Manso, A.; Shimabukuro, Y.E.; Pereira, G. Spectral Unmixing. *Int. J. Remote Sens.* **2012**, *33*, 5307–5340. [CrossRef]

63. Neher, R.A.; Mitkovski, M.; Kirchhoff, F.; Neher, E.; Theis, F.J.; Zeug, A. Blind Source Separation Techniques for the Decomposition of Multiply Labeled Fluorescence Images. *Biophys. J.* **2009**, *96*, 3791–3800. [CrossRef]

64. McRae, T.D.; Oleksyn, D.; Miller, J.; Gao, Y.-R. Robust Blind Spectral Unmixing for Fluorescence Microscopy Using Unsupervised Learning. *PLoS ONE* **2019**, *14*, e0225410. [CrossRef]

65. Malegori, C.; Alladio, E.; Oliveri, P.; Manis, C.; Vincenti, M.; Garofano, P.; Barni, F.; Berti, A. Identification of Invisible Biological Traces in Forensic Evidences by Hyperspectral NIR Imaging Combined with Chemometrics. *Talanta* **2020**, *215*, 120911. [CrossRef] [PubMed]

66. Piarulli, S.; Sciutto, G.; Oliveri, P.; Malegori, C.; Prati, S.; Mazzeo, R.; Airoldi, L. Rapid and Direct Detection of Small Microplastics in Aquatic Samples by a New Near Infrared Hyperspectral Imaging (NIR-HSI) Method. *Chemosphere* **2020**, *260*, 127655. [CrossRef] [PubMed]

67. Lugli, F.; Sciutto, G.; Oliveri, P.; Malegori, C.; Prati, S.; Gatti, L.; Silvestrini, S.; Romandini, M.; Catelli, E.; Casale, M.; et al. Near-Infrared Hyperspectral Imaging (NIR-HSI) and Normalized Difference Image (NDI) Data Processing: An Advanced Method to Map Collagen in Archaeological Bones. *Talanta* **2021**, *226*, 122126. [CrossRef] [PubMed]

68. Cutajar, J.D.; Babini, A.; Deborah, H.; Hardeberg, J.Y.; Joseph, E.; Frøysaker, T. Hyperspectral Imaging Analyses of Cleaning Tests on Edvard Munch's Monumental Aula Paintings. *Stud. Conserv.* **2022**, *67*, 59–68. [CrossRef]

69. CHANGE. Available online: https://change-itn.eu/ (accessed on 1 July 2024).

70. The Munch Aula Paintings Project (MAP)-Department of Archaeology, Conservation and History. Available online: https://www.hf.uio.no/iakh/english/research/projects/aula-project/index.html (accessed on 1 July 2024).

71. Lee, J.; Ormsby, B.; Burnstock, A.; van den Berg, K.J. Modern Oil Paintings in Tate's Collection: A Review of Analytical Findings and Reflections on Water-Sensitivity. In *Conservation of Modern Oil Paintings*; van den Berg, K.J., Bonaduce, I., Burnstock, A., Ormsby, B., Scharff, M., Carlyle, L., Heydenreich, G., Keune, K., Eds.; Springer International Publishing: Cham, Switzerland, 2019; pp. 495–522; ISBN 978-3-030-19254-9.

72. Frøysaker, T. The Paintings of Edvard Munch in the Assembly Hall of Oslo University. Their Treatment History and the Aula-Project. *Restauro Forum Für Restaur. Konserv. Denkmalpfleger* **2007**, *113*, 246–257.

73. Frøysaker, T.; Liu, M. Four (of eleven) Unvarnished Oil Paintings on Canvas by Edvard Munch in the Aula of Oslo University. Preliminary Notes on Their Materials, Techniques and Original Appearances. *Restauro Forum Für Restaur. Konserv. Denkmalpfleger* **2009**, *115*, 44–62.

74. Frøysaker, T.; Miliani, C.; Liu, M. Non-Invasive Evaluation of Cleaning Tests Performed on "Chemistry" (1909–1916). A Large Unvarnished Oil Painting on Canvas by Edvard Munch. *Restauro Forum Für Restaur. Konserv. Denkmalpfleger* **2011**, *117*, 53–63.

75. Frøysaker, T.; Liu, M.; Miliani, C. Extended Abstract—Noninvasive Assessments of Cleaning Tests on an Unvarnished Oil Painting on Canvas by Edvard Munch. In *New Insights into the Cleaning of Paintings: Proceedings from the Cleaning 2010 International Conference, Universidad Politécnica de Valencia and Museum Conservation Institute*; Mecklenburg, M., Charola, A.E., Koestler, R.J., Eds.; Smithsonian Institution Scholary Press: Washington, DC, USA, 2013; pp. 119–123.

76. Mengshoel, K.; Liu, M.; Kempton, H.M.; Frøysaker, T. Moving monumental Munch: From Listed Building to Temporary Studio and Back Again. In *Moving Collections. Processes and Consequences*; Bronken, I.A.T., Braovac, S., Olstad, T.M., Ørnhøi, A.A., Eds.; Archetype Publications: London, UK, 2012; pp. 65–72.

77. Scharffenberg, K.S. Investigations of a Tide-Line and Its Influences on the Painting Materials in The Source. In *Public Paintings My Edvard Munch and His Contemporaries. Change and Conservation Challanges*; Frøysaker, T., Streeton, N.L.W., Kutzke, H., Hanssen-Bauer, F., Topalova-Casadiego, B., Eds.; Archetype Publications: London, UK, 2015; pp. 117–125.

78. Frøysaker, T.; Schönemann, A.; Gernert, U.; Stoveland, L.P. Past and Current Examinations of Ground Layers in Edvard Munch's Canvas Paintings. *J. Art Technol. Conserv.* **2019**, *34*, 285–300.

79. Ormsby, B.; Soldano, A.; Keefe, M.; Phenix, A.; Learner, T. An Empirical Evaluation of a Range of Cleaning Agents for Removing Dirt from Artist's Acrylic Emulsion Paints. In *The AIC Painting Specialty Group Postprints*; Buckley, B., Ed.; AIC: Washington, DC, USA, 2013; Volume 23, pp. 77–87.

80. Giordano, A.; Cremonesi, P. *Gel Rigidi Polisaccaridici Per Il Trattamento Dei Manufatti Artistici*; Il Prato Edizioni: Prato, Italy, 2019.

81. Mills, L.; Burnstock, A.; Keulen, H.; Duarte, F.; Megens, L.; Van den Berg, K.J. Water Sensitivity of Modern Artists' Oil Paints. In Proceedings of the ICOM Committee for Conservation, 15th Triennial Meeting, Preprints, Valencia, 22–26 September 2008; ICOM-CC: Rome, Italy, 2008; Volume 2, pp. 651–659.

82. *ISO 4628-6: 2011*; Paints and Varnishes—Evaluation of Degradation of Coatings—Designation of Quantity and Size of Defects, and of Intensity of Uniform Changes in Appearance—Part 6: Assessment of Degree of Chalking by Tape Method. ISO: London, UK, 2011.

83. Keynan, D.; Hughes, A. Testing the Waters: New Technical Applications for the Cleaning of Acrylic Paint Films and Paper Supports. *Book Paper Group Annu.* **2013**, *32*, 43–51.

84. Stavroudis, C.; Doherty, T. The Modular Cleaning Program in Practice: Application to Acrylic Paintings. In Proceedings of the Cleaning 2010 International Conference, Valencia, Spain, 26–28 May 2010; Universidad Politecnica de Valencia and Museum Conservation Institute: Valencia, Spain, 2013; pp. 139–145.

85. Stoveland, L.P. Soiling Removal from Painted MOCK-ups. Evaluation of Novel Surface Cleaning Methods on Oil Paint and Chalk-glue Ground in the Context of the Unvarnished Aula Paintings by Edvard Munch. Ph.D. Thesis, University of Oslo, Oslo, Norway, 2021.

86. Stavroudis, C. Gels: Evolution in Practice. In *Gels in the Conservation of Art*; Angelova, L.V., Ormsby, B., Townsend, J., Eds.; Archetype Publications: London, UK, 2017; pp. 209–227.

87. Charola, A.E.; Wachowiak, M.; Webb, E.K.; Grissom, C.A.; Chong, W.; Szczepanowska, H.; DePriest, P. Developing a Methodology to Evaluate the Effectiveness of a Biocide. In Proceedings of the 12th International Congress on the Deterioration and Conservation of Stone, Paris, France, 22–26 October 2012; Columbia University: New York, NY, USA, 2012.

88. Doane, D.P.; Seward, L.E. Measuring Skewness: A Forgotten Statistic? *J. Stat. Educ.* **2011**, *19*, 1–18. [CrossRef]

89. Mills, J.S.; Smith, P. *Cleaning, Retouching and Coating: Technology and Practice for Easel Paintings and Polychrome Sculpture*; International Institute for Conservation of Historic and Artistic Works: London, UK, 1990.

90. Szczepanowska, H.M. *Conservation of Cultural Heritage: Key Principles and Approaches*; Routledge: London, UK, 2013; ISBN 978-0-415-67474-4.

91. Van den Berg, K.J.; Burnstock, A.; de Keijzer, M.; Krueger, J.; Learner, T.; de Tagle, A.; Heydenreich, G. (Eds.) *Issues in Contemporary Oil Paint*; Springer International Publishing: Cham, Switzerland, 2014; ISBN 978-3-319-10099-9.

92. Stoner, J.H.; Rushfield, R.A. (Eds.) *Conversation of Easel Paintings*, 2nd ed.; Routledge Series in Conservation and Museology; Routledge: New York, NY, USA, 2021; ISBN 978-0-367-02379-9.

93. Quabeck, N. Reframing the Notion of "The Artist's Intent:" A Study of Caring for Thomas Hirschhorn's Intensif-Station (2010). *J. Am. Inst. Conserv.* **2021**, *60*, 77–91. [CrossRef]

94. Haralick, R.M.; Shanmugam, K.; Dinstein, I. Textural Features for Image Classification. *IEEE Trans. Syst. Man Cybern.* **1973**, *SMC-3*, 610–621. [CrossRef]

95. Murata, S.; Herman, P.; Lakowicz, J.R. Texture Analysis of Fluorescence Lifetime Images of AT- and GC-Rich Regions in Nuclei. *J. Histochem. Cytochem. Off. J. Histochem. Soc.* **2001**, *49*, 1443–1451. [CrossRef] [PubMed]

96. Cabrera, J.E. GLCM Texture Analyzer. 2006. Available online: https://imagej.nih.gov/ij/plugins/texture.html (accessed on 11 November 2022).

97. Albregtsen, F. Statistical Texture Measures Computed from Gray Level Coocurrence Matrices. 2008. Available online: https://www.uio.no/studier/emner/matnat/ifi/INF4300/h08/undervisningsmateriale/glcm.pdf (accessed on 10 November 2022).

98. Rosi, F.; Harig, R.; Miliani, C.; Braun, R.; Sali, D.; Daveri, A.; Brunetti, B.G.; Sgamellotti, A. *Mid-Infrared Hyperspectral Imaging of Painting Materials*; Pezzati, L., Targowski, P., Eds.; SPIE: Munich, Germany, 2013; p. 87900Q.

99. Rosi, F.; Miliani, C.; Braun, R.; Harig, R.; Sali, D.; Brunetti, B.G.; Sgamellotti, A. Noninvasive Analysis of Paintings by Mid-Infrared Hyperspectral Imaging. *Angew. Chem. Int. Ed.* **2013**, *52*, 5258–5261. [CrossRef]

100. Sandak, J.; Sandak, A.; Legan, L.; Retko, K.; Kavčič, M.; Kosel, J.; Poohphajai, F.; Diaz, R.H.; Ponnuchamy, V.; Sajinčič, N.; et al. Nondestructive Evaluation of Heritage Object Coatings with Four Hyperspectral Imaging Systems. *Coatings* **2021**, *11*, 244. [CrossRef]

101. Russo, S.; Brambilla, L.; Thomas, J.-B.; Joseph, E. 2D Chemical Imaging for the Monitoring of the Formation of Metal Soaps on Oil-Painted Copper and Zinc Substrates. In Proceedings of the Metal 2022 Proceedings of the Interim Meeting of the ICOM-CC Metals Working Group, Helsinki, Finland, 5–9 September 2022; ICOM-CC: Helsinki, Finland, 2023.

102. Knez, D.; Toulson, B.W.; Chen, A.; Ettenberg, M.H.; Nguyen, H.; Potma, E.O.; Fishman, D.A. Spectral Imaging at High Definition and High Speed in the Mid-Infrared. *Sci. Adv.* **2022**, *8*, eade4247. [CrossRef]

103. Bindokas, V.; Mascalchi, P. ImageJ/Fiji Macro to Automatically Correct White Balance in RGB Images. 2017. Available online: https://github.com/pmascalchi/ImageJ_Auto-white-balance-correction (accessed on 1 November 2022).

104. Image.sc Auto White Balance of Stack-Image Analysis. 2019. Available online: https://forum.image.sc/t/auto-white-balance-of-stack/22439/1 (accessed on 1 November 2022).

105. Schwartzwald, D. Color Space Converter. 2012. Available online: https://imagej.nih.gov/ij/plugins/color-space-converter.html (accessed on 1 November 2022).

106. Pillay, R.; Hardeberg, J.Y.; George, S. Hyperspectral Imaging of Art: Acquisition and Calibration Workflows. *J. Am. Inst. Conserv.* **2019**, *58*, 3–15. [CrossRef]

107. Deborah, H.; George, S.; Hardeberg, J.Y. Spectral-Divergence Based Pigment Discrimination and Mapping: A Case STUDY on *The Scream* (1893) by Edvard Munch. *J. Am. Inst. Conserv.* **2019**, *58*, 90–107. [CrossRef]

108. Babini, A.; George, S.; Hardeberg, J.Y. Hyperspectral imaging Workflow for the Acquisition and Analysis of Stained-Glass Panels. In Proceedings of the Optics for Arts, Architecture, and Archaeology VIII, Online, 21–25 June 2021; Groves, R., Liang, H., Eds.; SPIE: Bellingham, WA, USA, 2021; p. 51.

109. Babini, A. Hyperspectral Imaging of Stained Glass. Ph.D. Thesis, NTNU, Gjøvik, Norway, 2023.

110. Grillini, F. Reflectance Imaging Spectroscopy: Fusion of VNIR and SWIR for Cultural Heritage Analysis. Ph.D. Thesis, NTNU, Gjøvik, Norway, 2023.

111. Vagnini, M.; Miliani, C.; Cartechini, L.; Rocchi, P.; Brunetti, B.G.; Sgamellotti, A. FT-NIR Spectroscopy for Non-Invasive IDENTIFICATION of Natural Polymers and Resins in Easel Paintings. *Anal. Bioanal. Chem.* **2009**, *395*, 2107–2118. [CrossRef]

112. Hourant, P.; Baeten, V.; Morales, M.T.; Meurens, M.; Aparicio, R. Oil and Fat Classification by Selected Bands of Near-Infrared Spectroscopy. *Appl. Spectrosc.* **2000**, *54*, 1168–1174. [CrossRef]

113. Duconseille, A.; Andueza, D.; Picard, F.; Santé-Lhoutellier, V.; Astruc, T. Molecular Changes in Gelatin Aging Observed by NIR and Fluorescence Spectroscopy. *Food Hydrocoll.* **2016**, *61*, 496–503. [CrossRef]

114. Eldin, A.B. *Near Infra Red Spectroscopy*; IntechOpen: London, UK, 2011; ISBN 978-953-307-683-6.

115. Amato, S.R.; Burnstock, A.; Michelin, A. A Preliminary Study on the Differentiation of Linseed and Poppy Oil Using Principal Component Analysis Methods Applied to Fiber Optics Reflectance Spectroscopy and Diffuse Reflectance Imaging Spectroscopy. *Sensors* **2020**, *20*, 7125. [CrossRef] [PubMed]

116. Li, X.; Sun, C.; Zhou, B.; He, Y. Determination of Hemicellulose, Cellulose and Lignin in Moso Bamboo by near Infrared Spectroscopy. *Sci. Rep.* **2015**, *5*, 17210. [CrossRef] [PubMed]

117. Catelli, E.; Sciutto, G.; Prati, S.; Chavez Lozano, M.V.; Gatti, L.; Lugli, F.; Silvestrini, S.; Benazzi, S.; Genorini, E.; Mazzeo, R. A New Miniaturised Short-Wave Infrared (SWIR) Spectrometer for on-Site Cultural Heritage Investigations. *Talanta* **2020**, *218*, 121112. [CrossRef] [PubMed]

118. Magnusson, M.; Sigurdsson, J.; Armansson, S.E.; Ulfarsson, M.O.; Deborah, H.; Sveinsson, J.R. Creating RGB Images from Hyperspectral Images Using a Color Matching Function. In Proceedings of the IGARSS 2020—2020 IEEE International Geoscience and Remote Sensing Symposium, Waikoloa, HI, USA, 26 September–2 October 2020; pp. 2045–2048.

119. Miller, J.N.; Miller, J.C. *Statistics and Chemometrics for Analytical Chemistry*, 6th ed.; Prentice Hall/Pearson: Harlow, UK, 2018; ISBN 978-0-273-73042-2.

120. Luo, M.R.; Cui, G.; Rigg, B. The development of the CIE 2000 Colour-Difference Formula: CIEDE2000. *Color Res. Appl.* **2001**, *26*, 340–350. [CrossRef]

121. Habekost, M. Which Color Differencing Equation Should Be Used. *Int. Circ. Graph. Educ. Res.* **2013**, *6*, 20–33.

122. Boronkay, G. Colour Conversion Centre. Available online: http://ccc.orgfree.com/ (accessed on 1 November 2022).

123. Williams, J.H. Guide to the Expression of Uncertainty in Measurement (the GUM). In *Quantifying Meas. Tyranny Numbers*; Morgan & Claypool Publishers: San Rafael, CA, USA, 2016. [CrossRef]

124. Wienand, J.; Wuest, L. Error Propagation Calculator (Online Tool for Any Formula). Available online: http://www.julianibus.de/ (accessed on 1 November 2022).

125. Larkin, P. Chapter 3-Instrumentation and Sampling Methods. In *Infrared and Raman Spectroscopy*; Larkin, P., Ed.; Elsevier: Oxford, UK, 2011; pp. 27–54; ISBN 978-0-12-386984-5.

126. Campbell, W. TN21-03. ATR Crystal Choice and Quest Puck Guide. 2022. Available online: https://specac.com/news-atr-spectroscopy-of-carbon-black/ (accessed on 13 June 2023).
127. Menges, F. Spectragryph-Optical Spectroscopy Software. 2021. Available online: http://www.effemm2.de/spectragryph/ (accessed on 13 June 2023).
128. Botton, G.; Prabhudev, S. Analytical Electron Microscopy. In *Springer Handbook of Microscopy*; Hawkes, P.W., Spence, J.C.H., Eds.; Springer Handbooks; Springer International Publishing: Cham, Switzerland, 2019; pp. 345–453; ISBN 978-3-030-00069-1.
129. Erdman, N.; Bell, D.C.; Reichelt, R. Scanning Electron Microscopy. In *Springer Handbook of Microscopy*; Hawkes, P.W., Spence, J.C.H., Eds.; Springer Handbooks; Springer International Publishing: Cham, Switzerland, 2019; pp. 229–318; ISBN 978-3-030-00069-1.
130. Hawkes, P.W.; Spence, J.C.H. (Eds.) *Springer Handbook of Microscopy*; Springer Handbooks; Springer International Publishing: Cham, Switzerland, 2019; ISBN 978-3-030-00068-4.
131. Morrison, R.; Bagley-Young, A.; Burnstock, A.; van den Berg, K.J.; van Keulen, H. An Investigation of Parameters for the Use of Citrate Solutions for Surface Cleaning Unvarnished Paintings. *Stud. Conserv.* **2007**, *52*, 255–270. [CrossRef]

Review

Recent Advances in the Application of Metal Oxide Nanomaterials for the Conservation of Stone Artefacts, Ecotoxicological Impact and Preventive Measures

Marwa Ben Chobba [1], **Maduka L. Weththimuni** [2,3,*], **Mouna Messaoud** [1], **Clara Urzi** [4] and **Maurizio Licchelli** [2,3]

[1] Laboratory of Advanced Materials, National School of Engineering, University of Sfax, P.O. Box 1173, Sfax 3038, Tunisia; marwa.benchobba@enis.tn (M.B.C.); mouna.messaoud@enis.tn (M.M.)
[2] Department of Chemistry, University of Pavia, Via Taramelli 12, 27100 Pavia, Italy; maurizio.licchelli@unipv.it
[3] Research Center for the Conservation of Cultural Heritage (CISRiC), University of Pavia, Via A. Ferrata 3, 27100 Pavia, Italy
[4] Department of Chemical, Biological, Pharmaceutical and Environmental Sciences, University of Messina, Viale F. Stagno d'Alcontres, 31, 98166 Messina, Italy; urzicl@unime.it
* Correspondence: madukalankani.weththimuni@unipv.it

Abstract: Due to the ongoing threat of degradation of artefacts and monuments, the conservation of cultural heritage items has been gaining prominence on the global scale. Thus, finding suitable approaches that can preserve these materials while keeping their natural aspect of is crucial. In particular, preventive conservation is an approach that aims to control deterioration before it happens in order to decrease the need for the intervention. Several techniques have been developed in this context. Notably, the application of coatings made of metal oxide nanomaterials dispersed in polymer matrix can be effectively address stone heritage deterioration issues. In particular, metal oxide nanomaterials (TiO_2, ZnO, CuO, and MgO) with self-cleaning and antimicrobial activity have been considered as possible cultural heritage conservative materials. Metal oxide nanomaterials have been used to strengthen heritage items in several studies. This review seeks to update the knowledge of different kinds of metal oxide nanomaterials, especially nanoparticles and nanocomposites, that have been employed in the preservation and consolidation of heritage items over the last 10 years. Notably, the transport of nanomaterials in diverse environments is undoubtedly not well understood. Therefore, controlling their effects on various neighbouring non-target organisms and ecological processes is crucial.

Keywords: metal oxide; nanoparticles; nanocomposite; protective coatings; heritage conservation; stone artefacts

check for **updates**

Citation: Ben Chobba, M.; Weththimuni, M.L.; Messaoud, M.; Urzi, C.; Licchelli, M. Recent Advances in the Application of Metal Oxide Nanomaterials for the Conservation of Stone Artefacts, Ecotoxicological Impact and Preventive Measures. *Coatings* **2024**, *14*, 203. https://doi.org/10.3390/coatings14020203

Academic Editor: Joaquim Carneiro

Received: 27 December 2023
Revised: 30 January 2024
Accepted: 2 February 2024
Published: 4 February 2024

1. Introduction

Building materials are subject to a complex environment in which biological, chemical, and physical variables often work in concert to induce degradation and significantly alter their appearance [1]. In general, cultural heritage items made using stone materials are composed of different pores, with pore sizes varying from one item to another [2]. This feature, associated with bad conservation status, enhances the capacity of stone to absorb water. Decay processes of stone materials are strongly related to their porosity properties due the fact that decay agents (including water) may easily penetrate under their surface [3]. Moreover, air pollution and climate changes participate in material deterioration as well [4]. Along with the presence of water and increase in surface roughness, these abiological factors enhance the chances of microorganism colonization and growth. To preserve tangible cultural heritage items worldwide, three main approaches are followed: restoration, remedial conservation, and most importantly, preventive conservation. In general terms, restoration is the repair of damage; such treatment seeks to restore the work of art as close as possible at the original status [5–8]. Remedial conservation is based on techniques to arrest current damage and reinforce the structure in order to preserve it [9]. Examples of remedial conservation include the stabilization of corroded metals, consolidation of mural paintings, dehydration of wet archaeological materials, etc. Instead, preventive conservation is generally carried out after conservation treatment in order to apply a "set of operations aiming to prolong the life of the material by avoiding/preventing future damage" [10]. In fact, preventive conservation is a methodology that intends to control the unavoidable deterioration of cultural goods by minimizing the most expensive interventions. Very often, these different approaches are consequential and are all applied together in a well-planned conservative intervention to preserve cultural heritage items.

At a high level, the preventive conservation of stone materials which have been used in cultural heritage items includes the application of protective coatings [11]. Protective materials that are frequently employed in the preservation of cultural heritage items include acrylic resins [12], silicone resins (polyalkyl alkoxysiloxanes or alkylalkoxysilanes) [13,14], acrylic resins combined with silicones [15], fluorinated polymers [16–18], and other synthetic organic polymers. Due to their water repellence features, these polymer-based coatings prevent water penetration inside pores and stop microorganisms from adhering to stone surfaces. Despite each of them exhibiting advantages, none of them can be considered the best solution for every cultural heritage item. Indeed, they also present severe limitations, such as migration inside the stone, porosity with time, consequent loss of protection effectiveness, low permeability to water vapour, poor durability, and insufficient compatibility of organic materials with substrates [3].

Alternatively, scientists have proposed several inorganic consolidants (water saturated solutions of $Ca(OH)_2$, ammonium oxalate, and diammonium hydrogenphosphate), which usually display better durability and higher physico-chemical compatibility with stone than organic ones [19]. Although they show good properties, there are other limitations (insufficient penetration depth, unsatisfactory strengthening effect, and need for many applications). In the last decade, coatings based on nanomaterials ($Ca(OH)_2$, $Sr(OH)_2$, $Mg(OH)_2$, and nano-apatite nanoparticles) have been used in the conservation of wall paintings, bas-reliefs, and ancient monuments due to their distinctive properties [19–22].

The optical, physical, and magnetic qualities provided by nanostructured materials make it possible to create transparent hydrophobic and/or antifouling protective layers on surfaces. Moreover, nanomaterials have larger surface areas compared to similar masses of larger-scale materials, thereby enhancing their chemical reactivity. They have high reactivity even when kept on the surface of restored materials. Furthermore, thanks to their tiny particle size these nanomaterials can penetrate deeply inside damaged stone materials. Recently, treatments of monuments materials have shifted into the application of nanomaterials with self-cleaning properties instead of coatings with combined biocidal and antipollution properties. In fact, the self-cleaning properties of these coatings could be more effective, having by design the aim of avoiding any attachment of inorganic and organic particles, including microorganisms. Many studies have been performed in this context [23–27]. In particular, heterogeneous photocatalysis using metal oxide nanomaterials such as TiO_2, ZnO, CuO, AgO, MgO, etc. has been considered as an interesting method and used extensively for cleaning applications [28,29]. The particular activity of these oxides, based on the generation of reactive oxygen species (ROS) when they are exposed to UV light, makes them very attractive compounds [30].

However, although on the one hand these nanomaterials possess discrete self-cleaning properties, they possess a certain toxicity as well. In fact, over the years an increasing number of studies have pointed out the risks of nanotoxicity, including mortality rates, new diseases linked to contact between nanomaterials and the human body, and commonly, impacts on other living beings [31].

The present review paper intends to first present the main factors that contribute to the deterioration of stone materials used in cultural heritage items. Next, we exhibit the main features of metal oxide nanomaterials and the recent developments in their use for heritage preventive conservation that have occurred in the last ten years. Finally, we end the review by discussing the impacts of nanomaterials on the environment, human beings, and other living beings along with the precautions that must be taken when working with them.

2. Decay Mechanisms of Stone Materials

Most of our cultural and historical heritage items are in danger due to different combined processes of physical, chemical, and biological decay. In outdoor settings, such deterioration often follows a process of loss, deterioration, corrosion, or fracturing, which is shown by a deep modification of the initial physical, chemical, and visive features. Deterioration can be caused by:

- Intrinsic factors: depending on the nature of the material, orientation, manufacturing technique, and procedures used to affect the work.
- Extrinsic factors: due to external sources such as environmental factors (temperature, air pollutants, relative humidity, and light), anthropogenic factors (handling, misuse, vandalism, tourism, etc.), catastrophic factors (earthquakes, fires, floods, etc.), and not less important, biological factors such as macro- and micro-organisms [1].

Among the main mechanisms of deterioration, three processes are the most known. The first mechanism consists of physical or mechanical processes in which material behaviour is altered because of various mechanical forces (tensile, compressive, etc.) without changing the chemical composition of the substance. The second process consists of chemical processes in which the matter is altered as a result of a chemical reaction. The third process involves biotic processes in which living organisms engage in mechanically and chemically attacking the material; this process is known as biodeterioration [10].

In addition to intrinsic factors, the following main causes of decay involved in the deterioration of cultural heritage items are explicated in the following parts: water, microorganism colonization, weather, and climate changes.

2.1. Intrinsic Properties of Stone Materials

Several kinds of stone are utilized for heritage buildings around the world, such as sandstone (e.g., Angkor monuments, Cambodia) [32], limestone (e.g., Megalithic Temples, Malta) [33], calcarenite (as in the Baroque town of Noto in Sicily, Italy) [34], Candoglia marble (e.g., Milan Cathedral, Italy) [35], granite (e.g., Évora Cathedral, Portugal) [36], and volcanic rock (e.g., the Churches of Lalibela, northern Ethiopia) [37]. This diversity may explain the variability in the intrinsic assessment of the features of stone materials, particularly their crystal structure, physical properties, and mineral constituents. Pires et al. (2022) have stated that the presence of clay minerals may affect the properties of building stones [38]. When exposed to water or moisture, common deterioration patterns can include color changes, increased fracture, and swelling those results in loss of material. On the other hand, porosity is a key factor in the lifetime of stone. In fact, when it comes to the processes of stone degradation, porosity is a crucial factor. In addition to measuring pore space volume, porosity provides information on pore connectivity [39]. Pore connectivity rises with porosity and is correlated with the pore space skeleton coordination number. Pore diameters in porous stone materials can range from millimeters to nanometers. Depending on how the pores are connected, the existence of very large pores may affect how quickly stone deteriorates [40]. Indeed, the degradation of stone is highly influenced by the existence of particularly large pores. Thus, pore size emerges as the key pore structure parameter in terms of heritage stone materials degradation. Different decay processes can take place in stone depending on pore size, including capillarity imbibition, water adsorption, capillarity condensation, and crystallization pressure due to salt and ice. Many physical and chemical degradation processes are caused by the penetration of water inside the pore network.

2.2. Water

Water is often the main factor behind the processes that alter and degrade cultural heritage items. Historical buildings are often made from highly porous materials with average porosity ranging from 30% to 40% [41]. Over the years, historical buildings constructed from porous materials experience increased porosity due to physical and chemical degradation processes, leading to additional water absorption and consequent weathering [42].

Stone in contaminated urban environments is subject to alteration processes due to significant temperature swings and the crystallization of water-soluble salts within the porous stone, which can impair its coherence and mechanical characteristics [43]. Salt crystallization is one of the most damaging abiotic mechanisms involved in the rapid disintegration of porous stone [1,19,44]. When stone is highly permeable to water, water is able to move through the layers of stone, with salt crystallizing where the water is evaporated. When water contained in stone evaporates, the salt concentration increases until precipitation occurs [45]. The types and sources of these salts vary and have changed over time, primarily as a result of human activity in urban areas [46]. Salt crystallization rate of decay is influenced by a variety of factors, including the kind of salt, the temperature and relative humidity the heritage materials' location, the pore size and pore density of the stone used for construction, and the frequency of the wetting/drying cycles [47]. Eventually, these factors may result in moisture retention, slow carbonation beneath the surface, and mechanical weakness in porous stone. The majority of earlier studies on salt weathering have only looked at a single salt composition; however, there are always a variety of salts present in structures [48].

Pollution can affect marble and limestone buildings through the dissolution of calcite due to acid rain [1]. Acid rain refers to rain with a pH equal to 4 or lower. When the calcite in marble or limestone reacts with the sulfuric, sulfurous, and nitric acids from contaminated air and rain, the calcite dissolves [49]. Consequently, the removal of material and the loss of carved details can be observed in the exposed portions of buildings and statues [50]. In the past, volcanic activity was the main cause of acidification. Nowadays, air pollution due to human activity contributes to the dissolution of monuments and structures through acid rain.

In "cold regions" characterized by air temperatures below the freezing point, freeze–thaw cycles are another mechanism that contributes to the deterioration of stone [19]. Stone naturally contains pores that allow it to hold water both on the surface and inside pores for a long time [51]. The primary mechanism of rock deterioration results from the dual action of the water–ice phase transition. Alterations in the water state (liquid, solid, gas) in the pores due to variations in the ambient temperature induce deterioration of the rock structure due to the alternation between cold contraction and thermal expansion process [51]. The freezing–thawing action of the water inside the pores and fissures plays a central role in the alteration of rocks due to freezing when the temperature oscillates across 0 °C [52]. Water freezes at temperatures below 0 °C and converts into ice, which can cause internal stresses on the stone and lead to damage. In fact, crystallization pressure is produced when ice is created inside the pores, which consequently creates pressure against the material walls of the pores [1]. Previous studies have demonstrated that the characteristics of the stone (i.e., porosity, lithology, and water content) and the environment where heritage materials are located, including the frequency of the freeze–that cycle, amount of temperature variation, and resulting stress levels all have an impact on the freeze–thaw failure phenomenon [53].

On the other hand, atmospheric and soil moisture may increase the concentration of weathering agents, allowing water to enter the nucleus in the case of porous sandstone. The damaging effects of moisture on historical buildings can be observed in different ways, starting with salt erosion, chemical corrosion of the historical artefact, and the development of microorganisms such as fungi, algae, and moss, with cracks on the stone surfaces ultimately observed due to swelling and shrinkage [54]. High water availability may promote microbial growth by increasing intracellular water potential, consequently stimulating hydration and enzyme activity.

2.3. Microorganism Colonization

While stone artefacts decay as a result of physical and chemical processes, biological processes can cause aggressive damage as well [55]. In particular, microorganisms on stones can cause what is known as "biodeterioration", which was initially established by Polynov in 1945 while researching soil formation [56]. In the following studies, "biodeterioration" has been broadly accepted as any biologically-caused unfavorable change in material appearance and properties [48,56]. Cultural heritage assets are a highly diverse environment and are inhabited by a variety of microorganisms. These have an undeniable impact on the degradation of works of art [57]. Biodeterioration is a global issue, as microorganisms can colonize cultural assets in any country regardless of the climatic conditions and the features of the item [58]. Different kinds of microorganism groups can adhere to and develop on stone materials depending on environmental factors and the chemical composition of stone artefacts. Bioreceptivity is affected by the intrinsic characteristics of stone, such as its surface roughness, open porosity, chemical composition, capillary water, and abrasion pH [59,60]. Moreover, depending on the in situ microclimate, ambient climatic conditions, and physio-chemical and chemical item features, various microorganisms (i.e., bacteria, fungi archaea, algae, cyanobacteria, and lichens) can be deposited on stone heritage materials and develop into epilithic and/or endolithic biofilms [61].

Biological organisms are able to cause aggressive aesthetic and structural damage. Aesthetically, microorganisms result in the appearance of spots, discolouration, encrustation, streaks, and coloured patinas. Structurally, artefacts can be deteriorated through water retention, the disintegration, degradation, and breakage of materials, and alkaline dissolution [1]. To ensure efficient protection of monuments and temples, identification of the specific active microorganisms and their physiological roles is required for better knowledge of the degradation mechanisms and processes that cause deterioration. The mechanisms of degradation through biological processes are different, and depend on factors such as climatic conditions, stone features, and living microbial communities. As mentioned before, Pollutants are able to decay stone artwork through chemical weathering (salt crystallization/dissolution) and may in turn encourage microbial biodeterioration. On the other hand, it has been reported that fungi are some of the most active microorganisms that may utilize organic support as nutrition [62]. Due to their heterotrophic nature, fungi perform biodeterioration by transferring this support and creating specific chemicals, including inorganic and organic acids [62]. In particular, black meristematic fungi are an example of microorganisms well adapted to live on rock surfaces in harsh environmental condition, including stone monuments exposed outdoors [63]. Their resilience is related to polyextremotolerant characteristics such as the presence of melanin in the cell wall, meristematic development, oligotrophy, and slow growth. Moreover, the capacity of eukaryotic microalgae and prokaryotic cyanobacteria to chromatically adjust to various types of light enables them to grow on stone in archaeological sites with low light intensities, i.e., caves crypts, and catacombs [64–66]. Complex microbial communities need to be investigated and identified due to their high contribution to the decay process [65,67]. More details on the mechanisms of degradation of heritage stone materials through biodeterioration can be found in the reported literature [34,56,68,69].

2.4. Weather and Climate Changes

Cultural heritage places additionally suffer from the effects of weather and climatic changes. Weather conditions (e.g., temperature, humidity, etc.) play an important role in the degradation process. The negative effects of improper temperatures (too hot or too cold) lead to gradual degradation that may only become apparent over time, and the results could consequently be underestimated. Elevated temperatures can result in the desiccation of organic materials, causing them to become less flexible and break. Lamp et al. (2016)

stated that thermal stress weathering contributes to flaking, cracking, and exfoliation of porous rocks [70]. The same authors declared that thermal stresses, hydration of clays, crystallization, subsequent hydration of salts, and freezing of pore water can all contribute to spalling in porous rocks found in temperate locations. In addition, stone may experience both macro- and micro-degradation processes under conditions of intense heat, which can result in structural instability (e.g., stone cracking, color change, textural alteration, mineralogical changes, etc.) [71]. Intense heatwaves often cause fires. Fire is a major threat to stone-built cultural heritage. It has been stated that a rise in cultural heritage exposure to fire in the Europe is highly predicted [72]. Moreover, alterations in temperature and rainfall affect the distribution and abundance of lichens and other organisms that can develop on stone substrates [71,73]. Relative humidity, or RH, is another factor that needs to be taken into consideration when studying decay factors. It is generally recommended that RH values at heritage sites be between 40% and 65%. Environment with inappropriate RH is likely to affect stone heritage materials and participate in their degradation. It has been revealed that biological deterioration of cultural materials is amplified by a rise in relative humidity in warmer climates [74]. On the other hand, for stone heritage materials, natural moisture condensation is considered to be a significant cause of degradation [75].

Climate change has a harmful impact on human beings, natural systems, and both natural and cultural world heritage items [76]. Experts from the Intergovernmental Panel on Climate Change (IPCC) claim that climate change has an impact on the frequency and severity of dangerous occurrences, including floods, landslides, and droughts [77]. Gradual variations in temperature, air moisture, wind speed, fire due to heatwaves, sea level rise, desertification, ocean alteration of properties, and other climate-related changes have a destructive impact on cultural heritage sites [71,76]. All of these threats and others have already been named by the United Nations Educational, Scientific, and Cultural Organization (UNESCO) [78]. For historical buildings, water is the primary cause of degradation. If precipitation increases due to climate change, soils may become saturated and downpipes and gutters may be overloaded, resulting in a higher danger of dampness penetrating into art or worked materials such as masonry walls [71]. Moreover, because soluble salts are present in stone materials, fluctuating variations in temperature and precipitation result in more salt crystallization cycles, and consequently more damage [71,79]. On the other hand, alteration of wind speed or direction during storms can destroy historic structures and archaeological sites [80]. Figure 1 summarizes the impact of the extrinsic degradation factors discussed in this section on cultural building items.

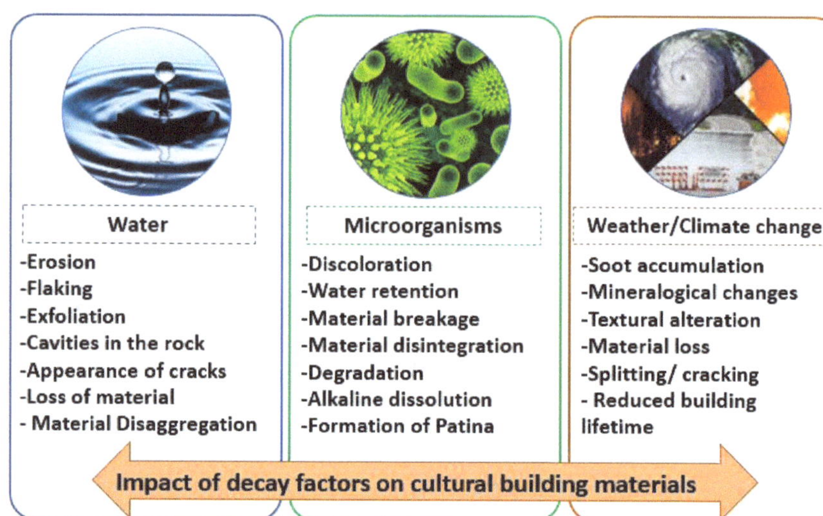

Figure 1. Impacts of the main deterioration factors (water, microorganism colonization, weather and climate change) on artefacts.

3. Metal Oxide Nanomaterials: Properties and Applications

Metal oxide nanomaterials with small crystallite sizes and high surface areas have attracted a great deal of interest due to their wide range of applications. Because of their distinctive features, synthetic metal oxide nanoparticles are among the most widely produced nanomaterials [81]. As the size of the nanoparticles decreases, more surface and interface atoms are produced. The specific size of the nanoparticle can determine its magnetic, conductive, chemical, and electrical capabilities [81]. Certain metal oxides possess photocatalytic features, making them able to absorb light, induce the charge separation process, generate electrons and holes, and oxidize organic pollutants. A photocatalyst is a substance that can absorb light and generate electron–hole pairs, which allow the participants to perform chemical transformations. The key characteristics of a photocatalytic system are an appropriate band gap, suitable shape, large surface area, stability, and reusability [82]. In this review, four kinds of different metal oxide NPs based on titanium, zinc, magnesium, and copper are discussed in detail in relation to their frequent use in treatment applications for stone heritage structures.

Titanium dioxide is a naturally occurring titanium compound. In the early 20th century, this material was used as a white pigment because of its high refractive index [83]. Titanium dioxide has been extensively used for many industrial applications, including as a white pigment included in paper, plastic, medicinal and cosmetic products, toothpastes, food coloring, and many others instances in which white coloration is desired [83]. The photocatalytic activity of TiO_2 was later discovered and published in Nature, and is called the "Honda Fujishima effect" [84]. Many studies have focused on its photocatalytic effectiveness under different conditions. These unique properties have widened the application of TiO_2 to environmental and ecological applications such as air purification, water treatment, self-cleaning coatings, and non-spotting glass, as well as bio-medicinal applications such as self-sterilizing coatings. The use of TiO_2 has received wide attention thanks to its biological and chemical stability, low cost, ease of production, and harmlessness to the environment [83]. Because of its distinctive properties, particularly its capacity for photocatalysis, TiO_2 is regarded as an interesting semiconductor [85]. When titanium dioxide absorbs enough energy, all photoinduced phenomena, including photocatalysis, superhydrophilicity, and photovoltaics, are initiated. The photocatalytic mechanism of TiO_2 is illustrated in Figure 2.

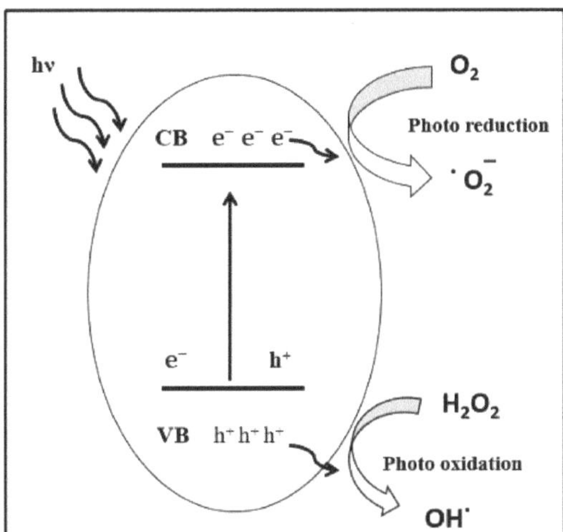

Figure 2. Schematic illustration of photocatalytic activity and photogeneration of charge carriers in a photocatalyst.

On the other hand, it has been claimed that titania has a substantially slower rate of charge carrier recombination compared to other semiconductors. This is advantageous because it has been proposed that chemical processes require photo-generated charge carriers (e^-/h^+) with a lifetime of at least 0.1 ns [30]. When a photon with an energy higher than the band gap energy is absorbed by a semiconductor, an electron passes from the valence to the conduction band, creating a valence band hole (h_{Vb}^+) and conduction band electron (e_{cb}^-) according to the following equation.

$$\mathrm{TiO_2} + h\upsilon \rightarrow e_{cb}^- + h_{vb}^+ \tag{1}$$

Hence, surface-based oxidation/reduction reactions can start as photogenerated electron–hole pairs, diffuse to the surface, and transfer to absorbed species (i.e., O_2, H_2O_2, OH^-), resulting in the production of reactive oxygen species (ROS). These created species are able to degrade contaminants, resulting in their mineralization [86]. On the other hand, they have the ability to partially degrade bacterial outer membranes, penetrate cytoplasmic membranes, and induce peroxidation of membrane lipids, which results in cell death [87]. Jin et al. (2011) reported that TiO_2 NPs in the anatase phase contribute to ultrastructural damage to cells by the generation of ROS [88]. Indeed, the anatase phase causes both cytotoxicity and genetic toxicity, as the ROS are internalized in the cytoplasm, and some are lodged within the mitochondria and nucleus. More recently, Henningham et al. (2015) announced that ROS can effectively deteriorate bacterial nucleic acids, proteins, and cell membranes [89]. A report by Joost et al. (2015) considered that ($HO^\bullet$) radicals play a significant role in seriously oxidatively altering the components of bacterial cellular membranes by starting the process that causes the peroxidation of fatty acids (radical chain reaction) [90]. Indeed, it was assumed that the formation of peroxides in oleic and linoleic acids was probably driven by ($HO^\bullet$) radicals that attack a hydrogen atom in R-H and create a carbon radical $R^\bullet$, then, molecular oxygen is added to R, creating a peroxyl radical $ROO^\bullet$. Finally, the peroxyl radical abstracts a hydrogen from the R-H bond, creating a lipid hydroperoxide ROOH.

Zinc oxide (ZnO) NPs, which are insoluble in water and have the appearance of a white powder, have excellent chemical, electrical, and thermal stability due to their energy band of 3.37 eV and bonding energy of 60 meV [91]. They are safe, inexpensive, and simple to prepare [92]. Additionally, thanks to their optical, electrical, and photocatalytic features, ZnO NPs are used in several applications, such as chemical sensors, solar cells, and notably, photocatalysis [93]. It has been stated that ZnO reveals antimicrobial activity at the nanoscale [94–98]. Indeed, Mokammel et al. (2019) stated that the antibacterial effects of ZnO NPs can be ascribed to their capacity to break down bacterial cell walls and interfere with DNA replication [99]. Reactive oxygen species are produced when NPs release metal ions inside the cells. Several studies have demonstrated that ZnO NPs have a high capacity for producing ROS, which makes them very active in damaging cell walls and making the membrane more permeable. Brayner et al. (2006) suggested that ZnO NPs can occasionally become separated into Zn^{2+} ions; these ions are able to interact with intracellular components by diffusing through damaged cell membranes [100]. Additionally, it has been hypothesized that Zn^{2+} ions released during ZnO dissolution bind to the membranes of microorganisms, extending the lag phase of the microbial growth cycle [101]. Another explanation for the strong antibacterial activity of ZnO NPs is their high surface-to-volume ratio and surface abrasiveness. The low toxicity and great UV absorption of ZnO NPs, in addition to their self-cleaning and antibacterial activity, make them an excellent choice for use in many domains, in particular in the field of heritage building conservation.

Copper oxide (CuO) NPs have stable chemical and physical characteristics, are relatively inexpensive, and are photocatalytic. CuO NPs have the potential to be used as anti-infective agents thanks to their attractive crystal morphologies and incredibly large surface area. The antibacterial effect of CuO NPs may be attributed to different proposed mechanisms. Several research works suggests that CuO NPs may stick to bacterial cell walls due to their positive charge and interact with the carboxyl and amine groups that exist on the surfaces of microbial cells. Consequently, bacteria that possess a high density of these ionic groups, such as *Bacillus subtilis*, have greater affinity and are more prone to copper oxide NPs [102]. On the other hand, CuO NPs have less impact on Gram-negative bacteria such as *Proteus* spp. and *P. aeruginosa* [103]. Copper is an essential component of several enzymes found in living microorganisms. Consequently, Cu^{2+} must be present in relatively high concentrations in order to have toxic effects on microbial pathogens. Cu^{2+} ions have a role in the production of ROS, which interact with DNA and intercalate nucleic acid strands when present in large doses [103]. The production of Cu^{2+} may prevent many microorganisms from synthesizing amino acids. Additionally, the production of ROS may cause bacteria to experience membrane damage brought on by oxidative stress [104].

Thanks to special and distinctive features that make them useful in several fields (e.g., catalysis, sensing, medicinal and biological applications), magnesium oxide-based NPs (MgO NPs) have been the focus of interest over the past 20 years. MgO NPs are typically used in the medical field because of their resistance, biocompatibility, and high stability [105]. In addition to their biocompatibility, MgO NPs have been documented to display improved photocatalytic activity. Several studies have reported their efficiency in degrading organic dyes such as methylene blue, methyl orange, and Congo red stains [106–108]. On the other hand, MgO NPs have demonstrated anticancer, antioxidant, and antibacterial capabilities against both Gram-negative and Gram-positive bacteria in vitro, including *Staphylococcus aureus* and *Escherichia coli* [109,110]. Interestingly, MgO possesses antimicrobial activity without the need for activation by a light source (photo-activation). It has been reported that MgO NPs and bacterial cells interact to generate cell membrane permeability, oxidative stress, leakage of intracellular contents, and ultimately cell death [111,112]. In addition to their photocatalytic properties, metal oxide NPs, notably oxides of titanium, zinc, magnesium, and copper, possess many interesting features which are illustrated in Figure 3.

Figure 3. Properties of metal oxide NPs: TiO_2, ZnO, MgO, and CuO NPs.

An important aspect to establish the effectiveness and the suitability of treatments based on nanoparticles is to consider the synthesis method used to obtain the nanomaterials. The experimental synthesis parameters can determine the morphology, particle sizes, agglomeration level, and crystalline structure of the resulting nanoparticles [27]. The synthesis process plays a critical role in controlling and obtaining ultra-fine nanopowders. On the other hand, the preparation procedure affects the physical and chemical features of metal oxides NPs [113]. Several methods have been developed to synthesize nano-size metal oxide particles, in particular TiO_2, ZnO, MgO, and CuO. The different processes are illustrated in Figure 4.

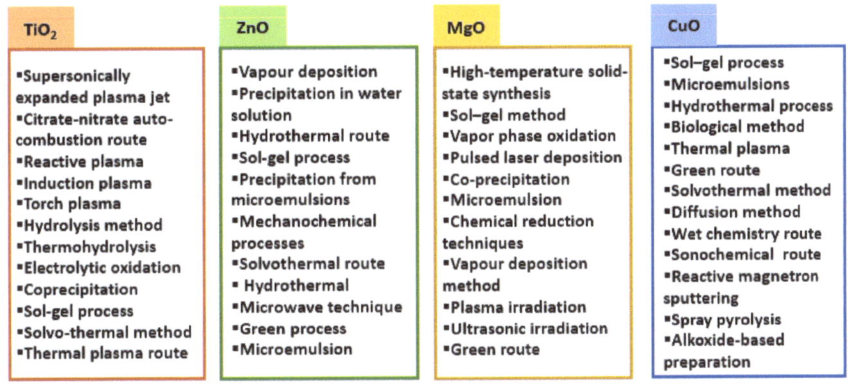

Figure 4. Different synthesis routes stated in the literature for preparing TiO_2, ZnO, MgO, and CuO NPs.

4. Application of Metal Oxide-Based Nanocomposite Coatings on Stone Building Materials

Metal oxide nanomaterials have been widely used in the field of conservation. Many studies performed in the field of conservation of heritage building materials deal with TiO_2 nanomaterials thanks to their superior photocatalytic activity. In this regard, D'Orazio and Grippo (2014) prepared a water-dispersed TiO_2/poly(carbonate urethane) nanocomposite through cold mixing of single components in a sonication process [114]. The prepared nanocomposite with the presence of 1% (w/w) of TiO_2 nanoparticles showed good self-cleaning properties by the degradation of methyl orange dye after 3 h of irradiation under UV light (Osram Ultra-Vitalux lamp, 300 W, 230 V) and was considered a promising coating for use on porous deteriorated stone such as Neapolitan yellow tuff, a common stone building material in the South of Italy since antiquity. It is worth noting that the type of photocatalyst is not the only factor that needs to be considered; a number of factors need to be taken into consideration in order to produce an acceptable protective coating, including the intrinsic characteristics of the stone (i.e., open porosity, composition, type of stone, etc.), the deterioration mechanisms, climatic factors, the compatibility between the developed product and the stone, the impact of the applied coating on the properties of the stone (i.e., breath function, water penetration, esthetic features, etc.), the coating procedure, and more. Among these factors, binders are crucial for assuring the effectiveness of the coating. A suitable binder must interact well with the substrate to prevent the polymer film from altering after prolonged exposure to sunlight, guarantee good dispersion of nanomaterials, adhere the photocatalyst properly to stone surfaces, have good water-repellent features, and be resistant against ageing. Different binders have been mixed with TiO_2 NPs and used to protect heritage materials, such as perfluoropolyether, acrylic polymer, and methyl acrylate copolymer [115]. Treatments such as Fosbuild, which is a commercial product composed of an aqueous emulsion of acrylic polymer (4 wt.%) mixed with TiO_2 (0.3 wt.% of anatase, 25 nm), show high water repellent features. However, there are no indications concerning the effects of the treatment on the breath function of the tested stone substrates. Recently, Polydimethylsiloxane (PDMS) has been employed due to its advantageous properties, such as high water-repellent property, resistance to fading from sunlight and UV rays, and resistance to deterioration from heat, water, or oxidizing chemicals, among others [116]. Kapridaki and Maravelaki (2013) worked on using PDMS as a binder to prepare nanocomposite containing TiO_2-SiO_2 NPs for stone protection [117]. They declared that PDMS offers hydrophobic coating, improves the durability and flexibility of the silica network, and prevents the gel from cracking during drying. Developed nanocomposites exhibited interesting properties (aesthetic modifications, water repellence behavior, acceptable water vapor permeability, good self-cleaning activity, and high antibiofilm performances). Tavares et al. (2014) elaborated a PDMS/TiO_2 nanocomposite by varying the amount of nanoparticles in the mixture at 0%, 0.5%, and 1% by weight [118]. In more detail, TiO_2 NPs were first synthesized by a microwave-assisted hydrothermal process; then, nanocomposite preparations were obtained by simple mixing of NPs with PDMS at different ratios and the coatings were processed by the spray method. The photocatalytic activity of different nanocomposite formulations was evaluated by the degradation of methylene blue dye at a suspension concentration of 1×10^{-3} mol/L under UVC lamps (254 nm, $\approx$4.9 eV) for three hours. Commercial TiO_2 (P25) was studied in research work for comparison purposes. The results demonstrated that TiO_2 concentration affects the photodegradation of methylene blue stain, as coating with 1% (by weight) TiO_2 showed a significantly faster degradation rate while pure PDMS films showed insignificant reduction of the dye. However, the P25 and 1% TiO_2 coatings exhibited similar behaviors. Moreover, the study lacked data on the impact of the coating on stone properties such as the breath function and on the durability of the coating. Kapridaki et al. (2018) [119] developed a hydrophobic and photoactive hybrid nanocoatings composed of three layers: (i) a tetraethoxysilane (TEOS)-nano-Calcium Oxalates consolidant; (ii) a hydrophobic layer composed of TEOS-PDMS; and (iii) TiO_2 nanoparticles as a self-cleaning layer. This method of conservation was proposed for lithotypes and mortars with various levels of porosity and petrographic features. The results showed that the treatment adhered well to stone surfaces, with only a small amount of material lost during the peeling test. All the substrates under study showed improved hydrophobic characteristics after treatment as demonstrated by the contact angle and capillary water absorption values. For the majority of the examined lithotypes, the permeability of water vapor was ensured to an acceptable level. The self-cleaning test was performed through discolouration of methylene blue dye on different lithotypes. The results demonstrated that the treatment displayed an enhanced self-cleaning activity, accomplishing the total degradation of MB on many of the tested specimens owing to the incorporation of nano-TiO_2 into the silica network, as shown by the creation of stable Si-O-Ti bonds detected by FTIR analysis. In addition to PDMS, other binders have been proposed; for example, Corcione et al. (2018) [120] elaborated a protective nanocomposite coating which was tested on Lecce stone. The coating was made of a hybrid methacrylic–siloxane resin modified with 1 wt.% oleic acid (OLEA) and 3-(trimethoxysilyl) propylmethacrylate (MEMO)-coated TiO_2 nanorods (NRs). OLEA-coated TiO_2 NRs were created in the anatase phase using a colloidal method, allowing for control of the size, shape, and crystalline quality of the NRs. TiO_2 NRs coated with two different amounts of OLEA/MEMO (1 and 3 wt.%, respectively) were dissolved in MEMO and stirred at room temperature. Brushing was used to apply the specified amount of the produced formulations, which was calculated by weighing the sample before and after the treatment. The discoloration capacity of the developed treatment was evaluated through Methyl Red (MR) as a model stain by dropping 200 μL of MR dye solution dispersed in isopropanol (3.5×10^{-3} M). The durability of the applied treatment was investigated by exposing stone substrates under outdoor conditions for one year. The results showed that treatment provided excellent surface hydrophobicity, as suggested by the considerable contact angle value (about 136°). According to the stone vapour water permeability test, the difference between treated and untreated stone was only 22%. This finding confirms that the treatment does not block the natural water vapour permeability of stone to the outside environment. In addition, the treatments induced chromatic variations of ΔE between 2.10 and 4.30. These results show that these coatings can be used without creating adverse aesthetic alterations of the stone substrate. The protective efficacy (PE) was evaluated by comparing the weight of water adsorbed by the treated stone specimens after 8 days compared to untreated specimens. The findings showed that the PE was almost 90% for the coating containing 1 wt.% of TiO_2 NRs (i.e., 10 mg cm^{-2}), while it reached a value of only 20% for the treatment composed only by resin. The self-cleaning test showed that, due to the inclusion of the TiO_2 NRs, the nanocomposite coating exhibited a higher discolouration percentage (34%) than the resin-coated sample (20%) after 24 h of irradiation. In 2020, Azadi et al. developed multifunctional inorganic–organic hybrid coatings with self-cleaning, hydrophobic, thermal stability, and weathering resistance features for application to outdoor stone building artefacts [121]. Hybrid coatings were prepared from fluorinated acrylic copolymers containing a suitable silane functional group (as the organic substance) and TiO_2 nanoparticles (as the inorganic compound). The authors stated that TiO_2 NPs serve to enhance the thermal resistance and hardness of the coating in addition to inducing self-cleaning properties. The presence of organofluorine substances improves the water-repellent characteristics of the coatings and their resistance to weathering. Tetraethyl orthosilicate (TEOS) was included in the treatment to ameliorate the thermal resistance of the protective coatings. Nanocomposite thin films revealed acceptable colour variation ($\Delta E^* < 5$), hydrophobic behavior (i.e., a contact angle around 131°), low water absorption percentage in the first three hours, better mechanical properties, and good photocatalytic activity through the degradation of methylene blue dye. The results showed that the presence of the organofluorine–titania hybrid led to better hydrophobicity by increasing the roughness of the surface. Interestingly, the coatings exhibited good resistance against aging.

As the main factor causing weathering is water, researchers have suggested using superhydrophobic materials with a static contact angle higher than 150° coupled with small contact angle hysteresis (typically, a roll-off angle below 10°). The capacity of superhydrophobic materials to inhibit water penetration allows them to significantly reduce the impact of water erosion. Moreover, the low roll-off angles of surfaces efficiently reduce the accumulation of pollutants and microorganisms, demonstrating promise for protecting stone artifacts [122]. Suitable roughness and low surface energy are required for the development of superhydrophobicity in substances. For this, researchers have suggested preparing nanocomposite coatings obtained from nanoparticles combined with organic coatings, with the nanoparticles offering a rough structure and the organic component providing small surface energy. In this context, Peng et al. (2022) [123] elaborated nanocomposite superhydrophobic coatings composed of silicon dioxide and titanium dioxide NPs, which were used to modify a waterproof coating obtained from dodecyltrimethoxysilane (DTMS). DTMS is utilized for many applications, such as fabric waterproofing, alloy anticorrosion, and notably, sandstone protection [123]; however, reports have declared that DTMS has weak resistance to light and low durability. It is expected that the introduction of NPs into the structure of organic coatings will be able to enhance features such as substrate adherence, thermomechanical characteristics, chemical stability, resilience to wear, self-cleaning, and notably, resistance against UV light degradation, which would extend the durability and the hydrophobic character of DTMS. Peng and co-workers elaborated a superhydrophobic coating named DST composed of SiO_2 and TiO_2 with concentrations of 0.5% (w/w) and 0.01 (w/w), respectively. Developed nanocomposites were then tested on red sandstone collected from the Daming Place Building Material Market, Xi'an, China. The coatings showed overall chromatic variation of less than 2, which is acceptable for use in conservation. The results revealed that the NPs were well-dispersed in DTMS, with about 70% of the particle sizes ranging between 60 and 90 nm (Figure 5a). The nanoscale surface roughness (see SEM image in Figure 5b) induced superhydrophobic behavior, as suggested by very large static contact angle values (up to about 152°). The hydrophobic character of the coatings was further demonstrated by measuring their water repellence efficiency, which was more than 92% after 72 h of absorption compared to untreated samples (Figure 5c). Interestingly, their findings revealed that the simultaneous effects of TiO_2 and SiO_2 NPs results in amelioration of the thermal and chemical stability of DTMS in addition to improving its UVA resistance, consequently guaranteeing better durability (Figure 5d).

Further studies concerning the use of TiO_2 nanomaterials in coatings applied on different stone substrates have been developed over the last ten years. Applications in the field of heritage conservation are summarized in Table 1.

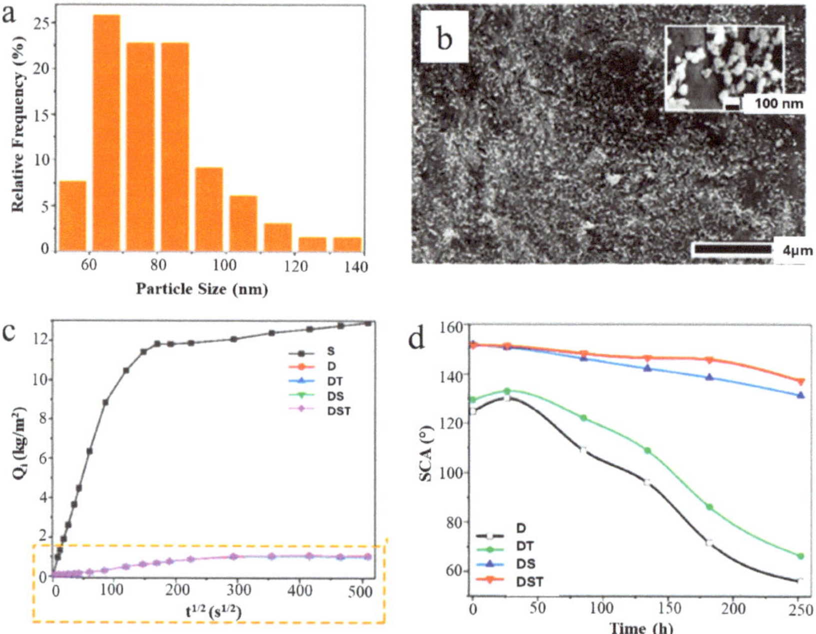

Figure 5. (**a**) Particle size distribution, (**b**) SEM images, (**c**) capillary water absorption curves, and (**d**) variation in static contact angles of the prepared nanocomposite during the thermal ageing test. D refers to dodecyltrimethoxysilane (DTMS); DT and DS are composed of DTMS-modified TiO_2 and SiO_2, respectively; DST corresponds to TiO_2 and SiO_2 co-modified DTMS, where isopropanol was the solvent, OP-10 was the emulsifier, and the concentrations of TiO_2 and SiO_2 were 0.01% (w/w) and 0.5% (w/w), respectively [123].

Table 1. Application of metal oxide NPs as coatings for the treatment of heritage stone materials (references are presented in a chronological order).

Nanomaterials Composition	Material Size	Substrate	Experiment Conditions	Obtained Results	Ref.
TiO_2	–	Travertine, a natural limestone	• Aqueous TiO_2 colloidal suspension (NPs amount: 1 wt.%) • NPs prepared by sol–gel process • Application procedure: spray method • Applied amount: 0.20 g/m (SL: single layer) and 0.60 g/m (ML: Three layers) • Self-cleaning test: Rhodamine B (RhB)	• Acceptable colour variation ($\Delta E^* = 2.15$–2.5) • Contact angle values (α) for SL ~70° and for ML ~85° • Water absorption of treated surfaces decreased about 50% • TiO_2-based coatings-decolorized RhB up to 75%	[124]
SiO_2-TiO_2	40–100 nm	Greek marbles from Naxo	• SiO_2–TiO_2 composites were prepared by sol–gel route • Application procedure: brushing method • Photocatalytic activity test: Methyl orange (MO)	• Acceptable chromatic variation ($\Delta E^* = 0.6$–1.9) • Contact angle (α): 93°–106° (Hydrophobic coatings) • [1] WCA decreased by 88% • Water vapor coefficients decreased by around 23% • High self-cleaning activity: faster degradation of MO-$\Delta E^*/\Delta E^*_0$ ~0.05 after exposition for 120 min	[125]

Table 1. *Cont.*

Nanomaterials Composition	Material Size	Substrate	Experiment Conditions	Obtained Results	Ref.
TiO_2	Anatase: 3–6 nm Brookit: 5–10 nm	Pietra di Lecce	• TiO_2 NPs obtained by sol–gel • Application procedure: brushing method • Photocatalytic activity test: methyl orange (MO)	• Acceptable colour variation ($\Delta E^* = 2.7$–4.9) • Slight reduction in capillarity absorption • Good MO stain degradation	[126]
TiO_2	10–15 nm	-Apuan marble (AM) -Ajarte limestone (AL)	• Two kinds of nanocomposite were prepared: ✓ WNC: alkyl alkoxy silane oligomers (15% w/w) with TiO_2 NPs (0.96% w/w) in water ✓ ANC: alkyl alkoxy silane monomers (40% w/w) with TiO_2 NPs (0.12% w/w) in 2-propanol • Application procedure: capillary absorption method	• Acceptable chromatic variation ($\Delta E^* = 1$–3.5). • Contact angles (α): WNC = 114° $\pm$ 1° (AM); WNC = 122° $\pm$ 7° (AL); ANC = 130° $\pm$ 10° (AM), ANC = 141° $\pm$ 2° (AL) • ANC provided the largest reduction in water uptake • Higher photocatalytic activity: discoloration levels ~90% with both treatments against Rhodamine B.	[127]
SiO_2-TiO_2-PDMS	25 nm	Modica stone	• The amount of polymer on stone surface ~50 g/m^2 • Optimization was performed based on TiO_2 NPs amounts (2.5, 10, 20 and 40 g/m^2) • Application procedure: brushing • Self-cleaning efficiency: (MB, 10% w/w in water solution)	• Acceptable chromatic variation: amount of $TiO_2 < 7$ g/m^2 • The stone surface saturation was obtained as 23.7 g/m^2 and amount of TiO_2 as >20 g/m^2 • Maximum discoloration rate: TiO_2 ~20 g/m^2; degradation of MB ~80%	[128]
TiO_2	–	Trani stone: Low porosity (2%)	• TiO_2 NPs/fluoropolymer coatings • Application procedure: brushing • Photocatalytic activity test: stain (Rhodamine B)	• Acceptable colour variation ($\Delta E^* = 0.5$–3.5) • Contact angle measurements (α): 100°–130° before exposition to outdoor condition; $\alpha = 20°$–60° after exposition • Rhodamine B degradation ~95% (7.5 h)	[129]
SiO_2-TiO_2	20 nm	Portland cement (WC) paste	• SiO_2/TiO_2 prepared by an improved sol–gel method • Photocatalytic activity test: stain (Rhodamine B)	• The photodegradation capacity of nanocomposite coatings was not considerably altered, while it dropped to around 9% in the case of P25	[130]
TiO_2@SiO_2 Core-Shell	Shell thicknesses: 1.95– 6.13 nm	White Portland cement past	• Commercial TiO_2 (P25) was employed as the core structure, and the typical Stöber process was used to create the SiO_2 shell • TiO_2@SiO_2 suspensions were sprayed on specimen surfaces • Self-cleaning test: Rhodamine B solution (20 mL, 10 mg/L)	• The degradation rates of TiO_2@SiO_2 coatings were higher than that of pure P25 • The maximum Rhodamine B degradation rates (68.1%) Shells thickness $\approx$ 3.92 nm; Surface areas $\approx$ 96.63 m^2/g	[131]
MgO/TiO_2	24 to 56 nm	-Red bricks -Gypsum mortars	• Coatings were obtained through immersion for 15 min • Hydrophobic composite coatings: oxide powders (0.5 wt.%) in modified sodium polyacrylate (NaPAC$_{16}$) aqueous solution (0.1 wt.%) • Antimicrobial activity test was performed against *Staphylococcus aureus*, *Candida albicans*, and *Aspergillus niger* • Photo-catalytic test through MO photodegradation	• Acceptable colour modifications ($\Delta E^* < 5$) • MO degraded: over 80% by NaPAC$_{16}$-TiO_2 and 93.7% by NaPAC$_{16}$-MgO • NaPAC16-oxide NPs showed antimicrobial activity against all tested strains	[111]

Table 1. *Cont.*

Nanomaterials Composition	Material Size	Substrate	Experiment Conditions	Obtained Results	Ref.
TiO_2 nanosheets-SiO_2	Thickness $\approx$6.5 nm	Capri limestone (open porosity: 9%–12%)	• Self-Cleaning test: methylene blue (MB) • Soot photo-elimination test: soot water dispersion (80 mg/L)	• Acceptable colour variations ($\Delta E^* < 3$) • Better photocatalytic activity of Titania nanosheets • Coating containing nanocomposite removed more than 50% of the soot stains after 600 h of irradiation	[132]
TiO_2-SiO_2	100 nm	Concrete	• Soot was applied at a rate of 20 µL/cm^2 • Methylene blue (MB solution: 0.5 mM) • Durability test: indoor conditions by MB dye degradation test	• MB degradation (1 h): 95% with nanocomposite coating. • The coated surfaces showed an almost unaltered structure after the outdoor durability tests; long-lasting effectiveness	[133]
TiO_2-SiO_2-PDMS	-	Cement mortar	• Application procedure: spray method • Self-cleaning test: MO stains	• Hydrophobic coatings (static contact angle = 152.6°) • The surface of the uncoated wall was shown to be easily polluted by MO and was difficult to clean with water, in contrast to treated specimens	[134]

[1] Water capillary absorption coefficients (WCA); Methylene blue (MB); Methyl orange (MO); Rhodamine B (RhB).

Biofouling plays a main role in the deterioration of submerged heritage stone. Indeed, underwater archaeological sites are deteriorating as a result of marine fouling [135]. After prolonged exposure times, a diverse community composed of plants, animals, and bacteria known as marine fouling takes place as a result of accumulation processes. For this reason, Roveri et al. (2018) prepared a nanocomposite coating composed principally from hydrophobized silica with AgO and ZnO nanoparticles (0.2% w/w) [136]. The tested stone substrates belonged to three different lithotypes: Apuan marble, Balegem limestone, and Schlaitdorf sandstone, with open porosities of 0.7 ± 0.1, 9.9 ± 0.8, and 16 ± 1 %vol, respectively. The nanocomposite coating showed acceptable chromatic variation ($\Delta E^* < 5$), good effectiveness in decreasing water uptake, and an important decrease in surface wettability with a contact angle of about 140° but had high permeability reduction (about 70%–80%). On the treated Schlaitdorf and Balegem stone samples, it showed good ability to inhibit microorganism growth, while on the treated Apuan marble samples exhibited insignificant reduction of the CFU number for both *Bacillus cereus* and *Pseudomonas putida*. These results may be explained by the small amount of protective coating that the marble was able to absorb and the small quantity of bacteria that able to adhere to marble surfaces. To examine the durability of nanocomposites, three alternative artificial ageing protocols that account for the distinct impacts of heat, UV light irradiation, and meteoric run-off were carried out in laboratory settings. The findings showed good durability of the nanocomposite coating under different ageing conditions.

Factors such as weathering and mechanical damage can lead to partial or total collapse of heritage stone materials. Furthermore, microbial colonization, particularly by different strains of fungi, can affect the physical and mechanical properties of stone. In this context, Van der Werf et al. (2015) developed nanocomposites based on ZnO NPs embedded in water-repellent siloxanes-based ESTEL1100 and SILO111 matrices as antimicrobial coatings [137]. The ZnO NPs were elaborated through a reproducible electrochemical process, and different mixtures were applied on a real monument, specifically, the outside of the 12th century Church of San Leonardo di Siponto in Manfredonia, Italy. The results showed that the consolidant products ESTEL and SILO were able to successfully suppress the growth of *Aspergillus niger* (*A. niger*) fungus strains after treatment with 0.4% w/w ZnO NPs while barely altering the color of the stones. Marble is one of the most significant and fundamental architectural materials in heritage buildings. To completely protect exposed ancient marble architectural components (e.g., columns) from fungi, experts are looking to new technologies to achieve the ideal approach. In particular, ZnO photocatalytic inorganic nanoparticles have been used to produce protective surface coatings and prevent microbial and fungal growth on exposed marble columns. Aldosari et al. (2019) [138] developed ZnO nanoparticles that were mixed with a synthetic acrylic polymer prepared by emulsion polymerization to design a coating combining biocidal and consolidating properties for use on old marble columns. The prepared nanocomposite coating was tested on marble samples collected from several Egyptian archeological sites. The results showed that ZnO NPs were obtained with a spherical morphology and a diameter ranging from 15 to 50 nm. SEM micrographs showed that samples treated with polymer coatings containing the ZnO NPs exhibited homogeneous distribution of the particles that were uniformly dispersed on the surface without producing cracks or segregation, contrary to specimens that were coated with only polymer. Combined with good distribution of the NPs, the polymer provided a hydrophobic coating with a contact angle of 140° and acceptable chromatic variation. The findings showed that the fungal strains, *A. niger* and *Penicillium* sp. grew quickly and diffusely on the untreated marble specimens. The chemical composition, rougher surface, high initial porosity, and mineralogical features of marble made untreated surfaces more susceptible to microbial attack. Samples treated with only the polymer showed low capacity to inhibit fungal growth, which was attributed to the features of the polymer. Because the polymers are organic compounds, they can be decomposed by microbial action, and can theoretically even encourage the growth of certain microorganism species. Interestingly, the nanocoating exhibited high antifungal activity against the tested strains due to the high photo-killing performance of the ZnO particles. In fact, as discussed in the previous section, the ROS generated during the photocatalytic process can damage the membranes of microbial cells. It is worth noting that the effectiveness of the applied treatment was greatly influenced by the uniform distribution of the NPs across the entire treated surface. Moreover, the results revealed that incorporating ZnO NPs into polymer can improve the resistance of stone surfaces to changes in relative humidity and temperature in addition to improving their durability against UV aging.

The ability of microorganisms to produce biofilms is the main cause of monument biodeterioration. Organic and inorganic materials can be colonized by bacterial biofilms, which use them as chemical and energy sources, causing substantial and irreparable damage to artifacts regardless of their composition. Moreover, bacteria have the ability to adhere to surfaces and even to each other, forming colonies of cells that produce an extracellular matrix made of DNA, proteins, and polysaccharides [139]. In this regard, Schifano et al. (2020) [21] evaluated the efficiency ZnO nanorod-decorated graphene nanoplatelets (ZNGs) to inhibit biofilm development by bacterial species (*Arthrobacter aurescens* and *Achromobacter spanius*) isolated from deteriorated historical monuments, precisely, the Temple of Concordia in the Valley of the Temples in Agrigento, Sicily, Italy. Zinc oxide nanorods (ZnO NRs) were prepared by thermally decomposing zinc acetate dihydrate followed by sonication. Commercially available graphite intercalation compound was thermally expanded for 30 s at 1050 °C to create graphene nanoplatelets. The prepared nanocomposite showed high antibacterial activity against *Streptococcus mutans* in a previous work [140]. Next, ZnO NRs with rod-shaped particles, a diameter of around 36 nm, and a length ranging from 300 to 400 nm were prepared directly on the planar face of pristine graphene nanoplatelets to produce ZNGs. Specimens of Noto stone, Carrara marble, and yellow brick were spray-coated with 2 mL of ZNGs aqueous suspension using an airbrush. The substrate stones differed in terms of their qualities, including hardness, porosity, and alkalinity, which influence the resulting distribution. A reduction in bacterial viability of about 60, 70, and 90% was obtained for the treated Noto stone (porosity $\approx$ 38%), Carrara marble (porosity $\approx$ 0.4%), and yellow brick (porosity $\approx$ 28.5%), respectively. These results show a relationship between the antibiofilm performance of the ZNGs and the distribution of the nanostructure in relation to specimen porosity. For example, the antibiofilm impact was found to be slightly decreased by the high porosity of Noto stone. Additionally, the results showed that the structure of the produced nanocomposite favored the inhibition of biofilm growth as can be observed through high-resolution field emission scanning electron microscopy (FE-SEM) (Figure 6). The ZnO NRs acted as nanoneedles, breaching the bacterial wall, while the nanosheets of graphene nanoplatelets provided a wide surface area for the oriented development of the ZnO NRs over the graphene surface. More precisely, the shape of the ZnO NRs increased their ability to pass through cell membranes by enhancing the adherence of nanostructures to the cell wall and increasing their ability to damage bacterial surfaces.

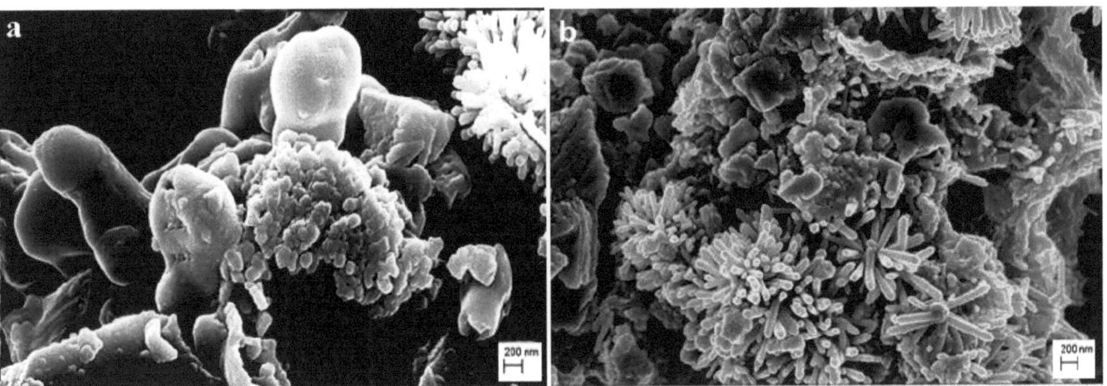

Figure 6. FE-SEM micrograph of *A. aurescens* TC4 cells after exposure to Noto stone covered with ZNGs. Panel (**a**) shows a bacterial cell after 24 h exposure, while panel (**b**) shows a brick covered by ZNGs. Bar, 200 nm [21].

Copper oxide NPs have been used in the field of heritage stone protection as well. CuO NPs are selected for this application due to their encouraging benefits, in particular their low cost compared to conventional biocidal agents. Furthermore, they are more stable than Ag^0 and Cu^0 NPs, which are susceptible to oxygen and sunlight, and most importantly, they do not require a source of light in order to be active. In this context, Zarzuela et al. (2016) [141] elaborated multifunctional nanocomposites composed of CuO and SiO_2 NPs through the sol–gel route as a protective coating for building stones. In their study, n-octylamine (n-8, 99%) was used as a surfactant to minimize surface tension and regulate pore size in order to produce xerogels free of cracks. The tested stone had a porosity of about 19% and was principally composed of calcite (45%), dolomite (>37%), and quartz (>7%). CuO nanoparticles with spherical shapes and sizes varying from 50 to 60 nm were integrated into silica matrix. The findings showed that nanocomposites containing an intermediate amount of CuO (i.e., 0.15% w/v) exhibited acceptable chromatic variation ($\Delta E \approx 2.5$), good biocidal activity against *Escherichia coli* and the yeast *Saccharomyces cerevisiae*, and good water repellence features. In fact, the static contact angle values were higher than 90° for the treated species, and the total water uptake (%TWU) dropped from 5.69 ± 0.05 for untreated stone to 0.14 ± 0.01 for treated stone. The mechanical properties of the coatings were tested through drilling resistance, peeling, and the Vickers hardness test. The results showed that the mechanical features of the stone were enhanced, proving the efficiency of the protective coating as a consolidant. This study showed promising results; however, the authors did not study the durability of the coating or its resistance in severe conditions. The same nanocomposite was used to create a novel kind of manufactured stone with thermal, chemical, and antibacterial resistance for possible use in cladding and tiles [142]. The authors proposed replacing the resin matrix with an amorphous silica matrix made via a sol–gel process utilizing quartz sand as the aggregate and adding copper oxide nanoparticles as a biocide agent. The authors suggested that this kind of stone could be used in the restoration of heritage buildings. CuO NPs were first introduced in concentrations ranging from 0.00% to 1.00% *w/v* relative to the silica oligomer. The prepared mixture was mixed with n-octylamine and de-ionized water, then sonicated using an ultrasonic probe (2 W/mL for 10 min). The quartz/SiO$_2$ sol paste was made using quartz particles of three distinct sizes (Figure 7). The paste was cast and cured into silicone molds under laboratory conditions (20 °C, 45% RH), then different tests were performed to investigate its performance. The results showed that the prepared materials had excellent surface hardness, heat resistance, and antifungal capabilities against yeast and *Aspergillus carbonarius* spores. Nevertheless, they showed reduced mechanical strength compared to a manufactured stone made of only resin matrix. The sol–gel kinetics and quartz sedimentation were affected by the concentration of CuO and the amount of water in the matrix, which influenced the structure and mechanical properties. In the same context, Kahrizsangi et al. (2016) [143] studied the impact of Cr$_2$O$_3$ nanoparticles on the mechanical and physical characteristics of an MgO-CaO refractory composition, with a focus on the enhancement of hydration resistance. Their results showed that the addition of 1.5 wt.% Cr$_2$O$_3$ NPs produced the best results, achieving the maximum improvement in mechanical and physical performance.

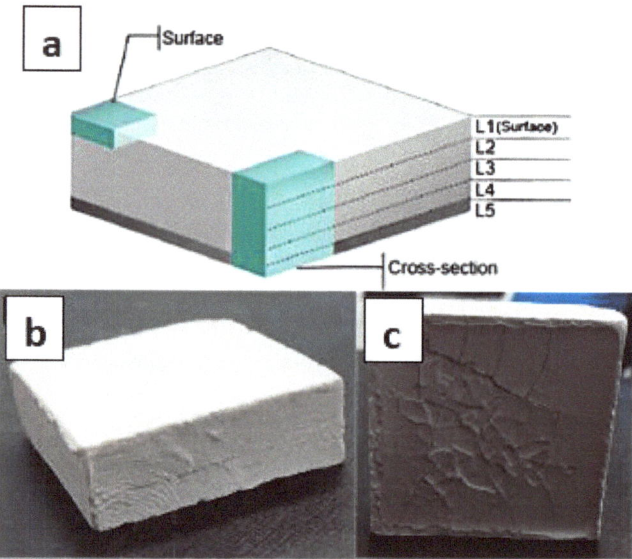

Figure 7. (a) Schematic representation of samples (L1 corresponds to the surface in contact with the mold); (b) isometric view of the engineered stone specimen; (c) front view of the detached crust formed by bleeding of the excess sol. It is worth to note that the materials cross sections were divided into several portions, designated L1 through L5, based on their depth using a diamond cutter blade in order to examine the component distribution. Figure reproduced with permission from [142].

5. Metal Oxide NPs with Enhanced Photo-Response Activity

Despite the interesting properties of metal oxide NPs, because of their large bandgaps they can generally absorb only UV light, which makes up about 4% of the entire solar spectrum. Thus, developing NPs that can effectively remove contaminants and absorb visible light is highly desirable. Metal doping is one of several suggested methods for improving the ability of metal oxide NPs to absorb visible light [1].

Because nano-sized TiO$_2$ has been largely studied and tested for the conservation of cultural heritage materials, considerable efforts have been developed in order to shift the absorption region of titanium dioxide to the visible region by modifying its band gap energy. Many research groups have focused on developing synthesis routes to prepare visible light-responsive TiO_2 as well as to enhance its antibacterial activity by doping with noble metals. Localized surface plasmon resonance (LSPR) has attracted a great deal of attention in recent years in the field of photocatalysis. The excitation of plasmonic metals by LSPR permits the production of energetic electrons at the metal surface. Under visible light, they gain enough energy to facilitate the transition to a semiconductor's conduction band and take part in the chemical reaction. In particular, doping TiO_2 NPs with silver has been shown to be a good method for improving the photo-response activity of TiO_2. In fact, silver ions have attracted considerable attention due to their amazing photocatalytic and antibacterial activity [144]. Silver ions represent a non-specific bactericide, contrary to antibiotics, as they can act against a large spectrum of bacterial and fungal species [145]. On the other hand, a novel approach has emerged as a feasible solution to produce effective materials by doping with rare earth ions (i.e., lanthanide ions). In this context, our research group developed TiO_2 NPs with enhanced photo-response properties elaborated through the sol–gel technique to produce a thin layer capable of protecting monumental stone [146,147]. Doping with lanthanide ions, more precisely Gadolinium ions (Gd^{3+}), was chosen as a method to enhance the photoactivity of TiO_2 NPs by reducing the recombination of photo-generated charge carriers and employing the visible region of the solar spectrum. Doping with the more well-known silver ions was performed as well. Silver is a widespread choice due to its high reactivity and capacity to produce surface plasmons at desired wavelengths. Lecce stone, a fossiliferous biocalcarenite quarried near Lecce in the South of Italy, was used as a model of highly porous stone. Pure and Gd^{3+}/Ag^+ doped TiO_2 nanoparticles (NPs) were prepared by the hydrothermal assisted sol–gel route with different doping percentages (0, 0.1, 1, 3 and 5 mol%). Morphological observations showed that all prepared NPs had a spherical shape, with particles sizes varying from 10 to 30 nm. First, we started by performing preliminary analyses in order to optimize the powder/binder ratios. Nanocomposites containing pure TiO_2 NPs with different powder/binder ratios (0.1%, 0.2%, 0.5% and 1% w/v TiO_2 in polydimethylsiloxane, PDMS) were applied on Lecce stone (LS) specimens. Preliminary analyses were based on measuring overall chromatic variations and static contact angles after application of nanocomposite on the stone surface. Based on the obtained results, it was decided to perform the subsequent tests with a nanopowder/binder ratio equal to 1% (w/v), as this was considered the optimal ratio. However, during the tests it was noticed that the PDMS binder altered the original color of the treated stone surfaces. To resolve this problem, t-Butanol (TBA) was added to PDMS (1:1 PDMS/TBA equivalent ratio was used) to reduce the darkness caused by the polymer. Acceptable chromatic variations were observed after application of NPs/PDMS diluted with TBA. The preliminary tests indicated that the nanocomposite containing TiO_2 NPs doped with 3 and 5 mol% Ag induced excessive chromatic variations; thus, investigation of these concentrations was discontinued [146]. Moreover, it was decided to stop investigation of materials containing 3 and 5 mol% Gd-TiO_2 based on the results of the preliminary study concerning their photocatalytic and antibacterial activity. The next step was to apply different coatings composed of the synthesized nanoparticles and binder on samples of Lecce stone in order to verify their efficiency in protecting stone substrates from biodeterioration. Coatings with multifunctional properties composed of the synthesized nanoparticles (pure TiO_2, 0.1–1 mol% Gd-TiO_2, 0.1–1 mol% Ag) and binder were applied to Lecce stone (LS) specimens. In a comparative study, a series of samples was treated only with PDMS:TBA (1:1) and another series was kept untreated. The coatings were evaluated by performing contact angle, chromatic variations, SEM-EDS, optical observation, capillary absorption, water vapour permeability, self-cleaning, antimicrobial, and ageing tests (Figure 8). The results revealed that all the applied coatings showed acceptable chromatic variations with $\Delta E^* < 5$ and exhibited water repellent properties. Optical microscope analyses suggested that the nanoparticles and binder were homogeneously distributed on the surface of the LS. SEM-EDS analysis further confirmed the homogeneous distribution of the nanoparticles and polymer on the Lecce stone surface, with localization of some NPs in the pores, confirming that the protective coating was successfully deposited (Figure 8b). The kinetics of capillary suction were only affected by the polymer and NPs treatments during the first 30 min, as shown by the CA and Qf values, which indicate the amount of water absorbed in 96 h and the coefficient of average absorption in 30 min, respectively. In addition, it was found that the treatment decreased the vapour permeability of the LS; however, the breath function of the original material was not dramatically affected (permeability reduction lower than 40%). Additionally, the surface hardness was increased by adding NPs to the surface of the LS samples. An interesting finding was that the PDMS coating with 1 mol% Gd/Ag-doped TiO_2 NPs was much tougher than the coating containing undoped NPs. The self-cleaning test was carried out by applying methylene blue dye (0.1% w/v in ethanol solution) to LS specimens. Colour variation was measured before and after applying MB and again after 48 and 96 h of UV irradiation. The results exhibited that the samples treated with NPs displayed faster degradation of the organic dye pollutant, with higher activity of doped materials, except for 1 mol% Gd-TiO_2. Moreover, the coating composed of binder and 1% Ag-doped TiO_2 nanoparticles showed the best self-cleaning activity, with total discoloration after only 6 h (Figure 8i,l). The antimicrobial analysis showed that treatment inhibited the overall spreadability and colonization of different microorganisms on the LS surface. In particular, the nanocomposite coatings composed of PDMS and 1 mol% Ag-doped TiO_2 NPs displayed the highest activity (Figure 8r). Additionally, the durability of the nanocomposite coatings was evaluated following exposure to various ageing cycles (high humidity, solar light, and long-term microbial incubation) in order to evaluate their stability. The effect of solar irradiation on chromatic variations caused by the protective films was carried out by performing accelerated ageing tests. The results showed that the samples treated with only polymer and the specimens coated with undoped and 1% Ag-doped TiO_2 nanoparticles showed unacceptable chromatic variations after 1000 h of irradiation under solar light. Solar irradiation did not affect the water repellent properties of applied coatings, despite a slight decrease in contact angle measurements after 1000 h of irradiation under a visible lamp. Additional findings revealed the stability of the coatings, notably the 0.1 mol% Gd-TiO_2 NPs and 1 mol% Ag-TiO_2 NPs, in terms of self-cleaning and antimicrobial activity. According to Normal 20/85 (1996) [148], one of the requirements for accepting the application of such treatments is that a protective coating should not result in noticeable alteration of the heritage material and must be stable over time. In conclusion, the 0.1 mol% Gd-TiO_2 and 0.1 mol% Ag-TiO_2 materials can be considered good candidates to protect monument surfaces made from this kind of stone, while coatings containing nanoparticles doped with a concentration of 1 mol% Ag could be useful for other kinds of stone.

Next, our research group worked on preparing a nanocomposite coating for the preservation of Serena stone (SS) historic artifacts [149]. The study was carried out in order to determine the ideal application conditions of synthesized silver-doped TiO_2 NPs distributed in a PDMS binder as a protective coating for Serena sandstone materials. Ag-TiO_2 NPs were mixed with PDMS at a powder/solvent ratio equal to 1% w/v, while the binder contained PDMS (M.W. 4200) and Tert-Butyl alcohol (TBA) at a ratio equal to 1:10 (PDMS:TBA). The nanocomposite coating was then applied at an amount of 2 g/m^2 on SS specimens (Figure 9). After application of the nanocomposite protective material on historical stone surfaces by brushing, a coating with good water repellent features, self-cleaning activity, and antimicrobial performance was obtained.

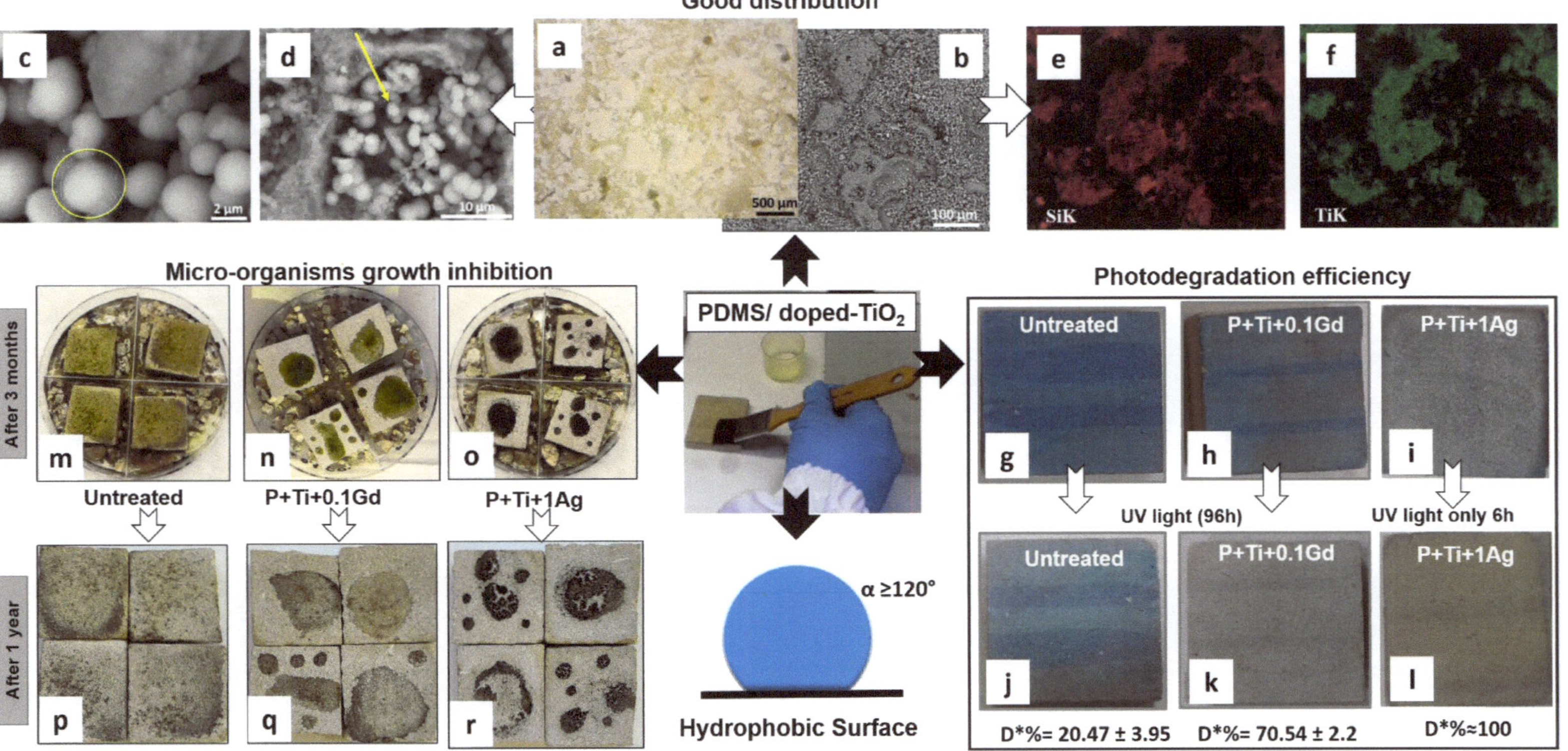

Figure 8. (**a**) Optical microscopic images of coated sample taken by normal light and (**b–d**) SEM micrographs of surfaces coated with nanocomposite coating at different magnification. Corresponding EDS mapping of silicon (**e**) and titanium (**f**). Photographs of the surfaces of untreated and treated specimens before (**g–i**) and after (**j–l**) the self-cleaning test. Overview of untreated and treated samples of the considered stone specimens after incubation for 90 days (**m–o**) and for one year (**p–r**) with pre-set MB discoloration percentage (D* (%)). P + Ti + 0.1 Gd and P + Ti + 1 Ag correspond to nanocomposites composed of PDMS (P) containing 0.1 mol% Gd-TiO$_2$ and 1 mol% Ag-TiO$_2$, respectively [146,147].

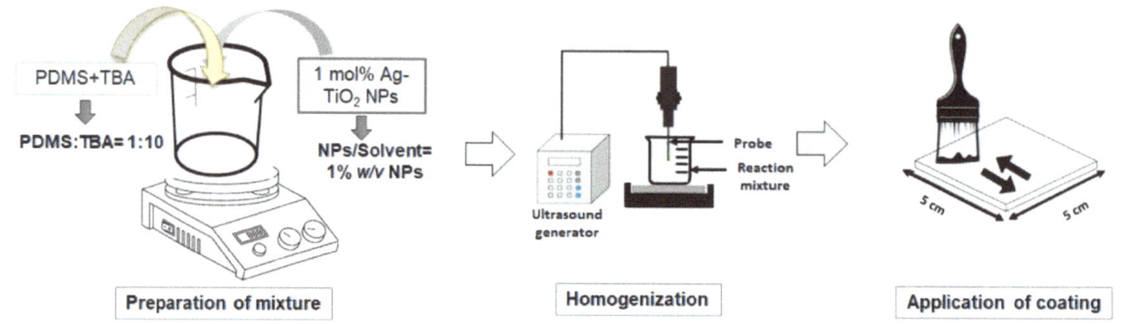

Figure 9. Schematic illustration of the experimental protocol for protective coating application.

Despite the enhancement of the photo-reactivity of TiO_2 NPs induced by doping, Ag can be easily oxidized upon coming into touch with TiO_2, because of its chemical reactivity [150]. To get around this issue, Ag NPs can be encapsulated in a material such as SiO_2 to balance the distance between TiO_2 and Silver. Indeed, incorporating the photocatalyst inside a mesoporous silica coating is considered an intriguing way to maintain strong surface adhesion and long-term wear resistance. In this context, Pinho et al. (2014) [151] elaborated mesoporous Ag-TiO_2-SiO_2 photocatalytic coatings for outdoor applications by integrating TiO_2 and Ag NPs into SiO_2 matrix in the presence of a surfactant (n-octylamine). The team of Cádiz investigated different sol formulations prepared at different TiO_2 and Ag concentrations. Prepared sols were applied onto limestone surfaces entirely made of calcite through a spraying approach until apparent refusal. All applied coatings showed acceptable colour variation except for a formulation composed by 4% (w/v) TiO_2 and 5% (w/w) Ag, which displayed undesired colour variation as a result of the high amount of Ag. A peeling test was performed to test the degree of adhesion of the applied nanocomposite coatings. The findings revealed that the specimens treated with sols and 1% (w/w) Ag in addition to samples containing 1% (w/v) TiO_2 exhibited only practically insignificant weight loss in contrast to untreated samples. These results demonstrate that TiO_2 and Ag were bonded to the SiO_2 matrix, which in turn was strongly attached to the tested stone. Total water uptake (TWU) values measured after 48 h proved that the applied coatings successfully inhibited water penetration inside the stone pores, as it was near zero for all of the treated samples and much lower than the corresponding untreated samples. The outcomes of the self-cleaning test revealed that the coating composed of silica, 1% (w/v) TiO_2, and 10% (w/w) Ag (named S1T10Ag) showed the highest activity, degrading most of the methylene blue dye, in contrast to coatings containing only SiO_2 and Ag NPs incorporated into silica matrix without the presence of TiO_2 NPs (Figure 10a). This confirms that the presence of titanium dioxide nanoparticles is indispensable for the self-cleaning activity. The authors deduced that the reduced average size and high dispersion of Ag NPs present in the S1T10Ag nanocomposite, along with the high absorption of visible light, are the key factors making these coatings highly effective in degrading the tested stain. TEM analysis revealed that the TiO_2 and Ag NPs included in the prepared nanocomposites were separated by a thin SiO_2 interlayer; thus, the thickness of the SiO_2 interlayer between the TiO_2 and Ag could probably be lowered by improving their dispersion, as the Ag nanoparticles would be located closer to TiO_2 NPs (Figure 10b). The authors stated that the surfactants help to reduce the amount of Ag species in the sols and increase the stability of the TiO_2 and Ag NP dispersion during the sol–gel transition. Although this study investigated the water repellence and self-cleaning performance of the coatings, more analysis is required to examine the biocidal activity and durability of the coatings in real conditions.

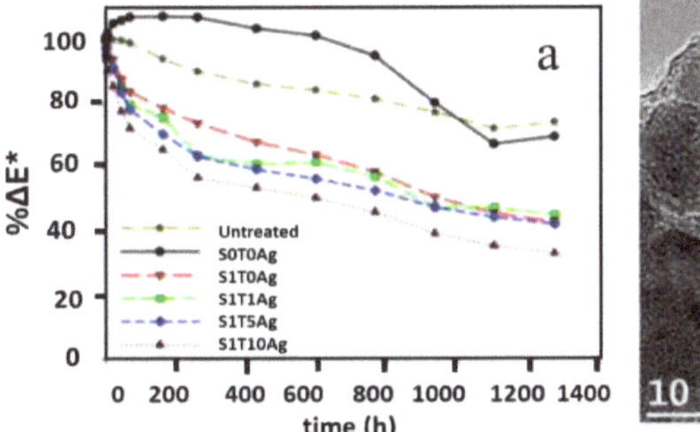
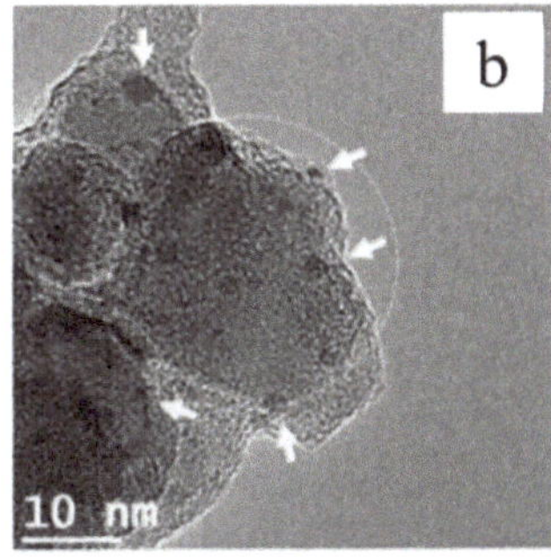

Figure 10. (**a**) The photocatalytic effect of methylene blue on stained limestone specimens based on measuring the %ΔE^* evolution. Treated and untreated samples were irradiated by UV light (λmin = 365 nm) for 1000 h. (**b**) Representative transmission electron microscopy image of the prepared S1T10Ag nanocomposite. Arrows and dashed circles are used to indicate the presence of Ag nanoparticles and TiO_2 domains, respectively. Figure reproduced with permission from [151].

Cádiz' team also used a sol–gel approach to develop a nanocomposite with self-cleaning and de-polluting activity based on a Gold (Au)-Titanium dioxide NP photocatalyst embedded in a silica matrix to protect heritage building materials from air contamination, [152]. Gold NPs were selected because of their excellent efficacy in improving the photo-response activity of TiO_2 NPs. In fact, the localized surface plasmon resonance (LSPR) of gold NPs varies from 500 to 600 nm, compared to silver NPs which have LSPR maximum absorption in the range of 400 nm (i.e., near to UV light); thus, they can absorb more solar light [153]. Despite studies concerning the use of gold nanoparticles for conservation of building materials being scarce because of its high price, the authors incorporated low amounts of gold (0.5% w/w Au/TiO_2) to overcome this issue. Gold NPs with two different average sizes (13 and 38 nm) were considered in the study. Three formulations with different Au/TiO_2 compositions (0.25%, 0.5% and 1% w/w) were dispersed in silica matrix in the presence of n-octylamine at a TiO_2/SiO_2 ratio of around 2.5% w/w. The prepared sols were first applied on Capri limestone, an oolitic limestone quarried from Cabra, Spain with open porosity of 12%. The findings revealed that coatings with different preparations preserved the appearance of the original material. The self-cleaning test showed that better MB stain removal efficacy was obtained after the addition of AuNPs, proving the improvement in TiO_2 photocatalytic capabilities brought on by the noble metal NPs. In particular, AuNPs with the lowest size (13 nm) and intermediate Au content (0.5% w/w Au/TiO_2) displayed the maximum photoactivity, probably due to better charge separation and lower photogenerated electron–hole pair recombination. In addition, it is well known that more small NPs have a higher surface area, leading to an increase in the available surface-active sites for TiO_2-Au interactions with pollutants. This optimum formulation was then selected to investigate its self-cleaning and depolluting features using real pollutants, particularly, soot and NO. The results showed that better soot removal was observed for specimens coated with Au-TiO_2/SiO_2 nanocomposite. NO conversion was almost double for specimens coated with nanocomposite compared to their counterparts coated with only TiO_2/SiO_2, demonstrating that AuNPs similarly accelerated the photooxidation of this pollutant. Despite the interesting outcomes found in the study, no data about the biocide properties of the coatings or their durability were presented. The Cadiz group also prepared of Ag/modified-TiO_2 NPs using -SH or -NH ended alkoxysilanes to functionalize TiO_2 NPs for possible use as coating to protect heritage stone materials with visible self-cleaning activity and biocidal characteristics [154]. In order to reduce the potential harmful effects of AgNPs, a low Ag/TiO_2 ratio (1%) was used. The authors stated that reducing the discharge of Ag^+ into the environment would restrict the impact on non-target organisms such as aquatic fauna, soil bacteria, etc. In addition, excessive metal loading may increase the rate of electron–hole recombination, reduce the efficacy of the photocatalytic mechanism, and increase the creation of AgNP clusters during the synthesis process. The findings revealed that -NH functionalization produced the maximum stability, homogeneity, and visible range absorption. Moreover, the addition of -SH groups changed the size of the AgNPs and reduced their efficiency by producing Ag_2S and Ag/TiO_2 NPs with -SH functionalization, which exhibited no biocidal action. Next, in 2021, the Cádiz team in collaboration with Abdelmalek Essaadi University used Copper NPs to improve a TiO_2/SiO_2 nanocomposite [155]. In comparison to other noble metals, copper is more readily available on Earth and less expensive. Moreover, copper is accessible in different oxidation states (Cu^0, Cu^I, Cu^{II} and Cu^{III}). The authors declared that a photocatalyst containing 5% copper displayed the maximum degradation of both MB (95%) and soot (50%) within 1 h and 168 h of irradiation in a solar degradation reactor composed of a 2500 W xenon arc lamp equipped with an outdoor UV filter. Nevertheless, a higher Cu amount resulted in a reduction of nanocomposite performance. This may be attributed to defects in the Cu NPs, which could have promoted photogenerated electron–hole recombination and consequently led to decreased TiO_2 photocatalytic activity. In contrast to the results obtained in the self-cleaning test, air de-pollution findings carried out by investigating nanocomposites' performance through NOx reduction showed that their efficiency was proportional to the amount of copper. Indeed, samples coated with 15% Cu presented the highest NO oxidation activity, at 36.70%, while samples treated with pure TiO_2/SiO_2 showed only 25.56% oxidation activity. In order to investigate the durability of the coatings in real-life circumstances, samples were exposed in outdoor conditions for the first 12 months and following 17 months (February–February 2020; July–November 2020). Overall analysis of chromatic variations showed acceptable variation, with $\Delta E^* = 3.93$ and 3.28 for samples containing 5% and 15% Cu, respectively. In this study, chromatic variations were investigated only after the ageing test; more analyses are required to prove the durability of the coatings in real conditions. Other works have been performed to elaborate enhanced TiO_2 NPs as protective treatments in recent years, which are summarized in Table 2.

Table 2. Applications of enhanced TiO_2 NPs as coatings for the treatment of heritage stone materials (references are presented in a chronological order).

Nanomaterials Composition	Particles Size	Substrate	Obtained Results	Ref.
Ag-TiO_2	0.1–1 µm	Limestone slabs from quarry of Utrera (Seville, Spain)	• Ag/TiO_2: smaller particle sizes, more stable colloids, zeta potential < 20 mV, and antibacterial activity against *E. coli* • Nanocomposite showed good inhibition activity against biopatina formation compared to independently generated Ag or TiO_2 NPs	[156]
Au-SiO_2-TiO_2	10–30 nm	Limestone with an open porosity of around 12%	• The thickness of coatings ranged from 3 to 12 µm • ST12Au coating: highest self-cleaning efficiency by degrading methylene blue, acceptable chromatic variation ($\Delta E^* = 3.4 \pm 0.5$), adequate adhesion to the tested stone, and water repellence behaviour	[157]
Ag-TiO_2	94–234 nm	Carbonate stone mainly composed from calcite (95%–98%) and quartz (2%–5%) with open porosity of 10%	• Silver NPs were stabilized by citrate and prepared with TiO_2 to obtain nanocomposites • A noticeable chromatic variation was obtained; all coatings induced ($\Delta E^* > 5$) • Biopatina growth was significantly reduced by the developed nanocomposite	[158]

Table 2. *Cont.*

Nanomaterials Composition	Particles Size	Substrate	Obtained Results	Ref.
TiO_2/Au-SiO_2	20–25 nm	Fossiliferous limestone (calcite 98.5%, α-quartz 1.5%)	• The resultant nanocomposites produced crack-free surface coatings on limestone, effective adhesion, increased stone mechanical properties, and imparted hydrophobic and self-cleaning capabilities	[159]
Fe-TiO_2	13–21 nm	Limestone	• The low iron doping titania (0.05% and 0.10% w/w) had a favorable impact on the photocatalytic destruction of methyl orange under visible radiation	[160]
Au/N-TiO_2/SiO_2	5.2 nm	• Capri limestone composed from calcite (open porosity of 9%–12%) • Granite Grey Pearl (with open porosity < 1%) • HERPLAC® concrete (with an open porosity of 10%)	• The introduction of AuNPs has the drawback of causing significant color variations • Au/N-TiO_2/SiO_2 coatings can improve photocatalytic performance, but only when the conservation of the material aesthetic characteristics is not crucial	[161]
Ag-TiO_2	12.5 ± 4.3 nm	Fossiliferous limestone, extracted from Cabra (Cordoba, Spain), composed of calcite (~100%) and with a 6% porosity	• Superhydrophobic coatings were obtained with contact angle of 155° • Coatings were able to degrade 75% of MB dye after 100 min and reduced microbial growth on tested stone with inhibition up to 20% (*Saccharomyces cerevisiae*) and 70% (*Escherichia coli*)	[162]
N-TiO_2/SiO_2	50–200 nm	Portland cement mortar	• N-TiO_2/SiO_2 nanocomposites enhanced MB degradation compared to TiO_2/SiO_2 materials (MB removal of 85% and 78% after 1 h of irradiation, respectively).	[163]

ZnO NPs have been coupled with noble metals, particularly silver, to obtain enhanced nanocomposite coatings. It is well known that ZnO NPs have weak activity under visible light. Moreover, Ag nanoparticles are not stable when exposed to sunlight. These limitations can be eliminated when ZnO NPs are combined with Ag derivatives. Mu et al. (2021) [164] developed nanocomposite materials for use as antimicrobial coatings to protect heritage building materials, particularly limestone (natural rocks rich in $CaCO_3$). The antimicrobial activities of the prepared nanocomposites were tested against Gram-positive bacteria (*Bacillus subtilis*), Gram-negative bacteria (*Escherichia coli*), and fungi (*Aspergillus niger*). The results showed that the AgCl/ZnO combination exhibits less inhibitory activity against fungi than against bacteria. This might be explained by the sturdier cell walls fungi, which could serve as a defense against biocidal substances. However, additional research is required to assess the applicability of AgCl/ZnO to a larger variety of materials.

Fungi are among the most significant colonizers and biodegraders of stone substrates of all microbes. For this reason, there is an urgent need to develop antifungal materials [63]. In this regard, Fernández et al. (2017) [165] developed antifungal coatings with high potential for stone conservation. They prepared Zn-doped MgO ($Mg_{1-x}Zn_xO$, x = 0.096) NPs by a sol–gel process for use as an antifungal coating. Their research intended to combine the potential of ZnO NPs as antimicrobial agents with the high compatibility of MgO NPs with stone materials. The photocatalytic and antifungal activities of Zn-doped MgO NPs were compared with single ZnO and MgO NPs. The results showed that the doped nanomaterials reveal improved self-cleaning capabilities. In fact, 87% of methylene blue dye was destroyed using Zn-doped MgO NPs after one hour of UV exposure, while ZnO and MgO NPs alone only degraded 58% and 38%, respectively. Antifungal action was tested against *Aspergillus niger*, *Paraconiothyrium* sp., *Penicillium oxalicum*, and *Pestalotiopsis maculans* on two calcareous substrates (Calcitic and dolomitic stones) that are extensively used in the cultural heritage of Mexico and Spain. Such strains have demonstrated potential activity in the solubilization of calcium carbonate plates and limestone through the creation of oxalic acid [166]. The results showed that treatment using doped nanoparticles stopped invasive microbial growth on both tested calcareous stone materials, particularly against *Aspergillus niger* and *Penicillium oxalicum*.

Weththimuni et al. developed a ZrO_2-doped ZnO-PDMS nanocomposite (doped NPs 0.5% (w/w) in PDMS) which is a multi-functional and durable coating for protecting different types of stone e.g., Lecce stone (LS), Brick (B), and Marble (M) [96]. Nanocomposite protective coatings provided homogeneous distribution on the considered stone surfaces, as confirmed by optical microscope (Figure 11a–f) and SEM (Figure 11g–l) analyses. The authors mainly focused their study on evaluating the durability properties and self-cleaning effect of the newly prepared coating with respect to the well-known PDMS coating. They were assessed after exposure to two different ageing cycles: solar ageing (300 W OSRAM Ultravitalux light with an UV-A component (315–400 nm, 13.6 W) and UV-B component (280–315 nm, 3.0 W) for 1000 h) and humid chamber ageing (RH > 80%, T = 22 $\pm$ 3 °C, desiccator, 2 years). The preliminary results suggested that the PDMS (P) and ZrO_2-doped ZnO-PDMS (Zn-Zr-P) coatings did not significantly alter the original chromatic properties of any litho-types ($\Delta E^* < 5$), while water repellent behavior and vapor permeability were preserved at acceptable levels compared to their unaged counterparts [95,167]. However, each stone (Lecce stone, Brick, and Marble) showed its own behavior towards water and water vapor due to their different porosity properties. After assessing the durability of the coatings using two different ageing cycles, their performance was evaluated by the self-cleaning test. The newly developed coating showed a higher photocatalytic effect (self-cleaning effect) than plain polymer coating (PDMS) on all of the tested stone types based on its ability to discolor methylene blue dye (e.g., Figure 11: Y image) when exposed to UV light (Figure 11m,n graphs). The highest efficiency was reported in the case of marble treated with nanocomposite coating (the discoloration factor D^* was 72% in the case of ZrO_2-doped ZnO-PDMS-treated marble, while it was 56% in the case of plain PDMS). The doped NPs were able to increase the photocatalytic effect of the binder material (PDMS) to a good level and maintained their self-cleaning ability even after long-term ageing processes, indicating that the coating has good durability.

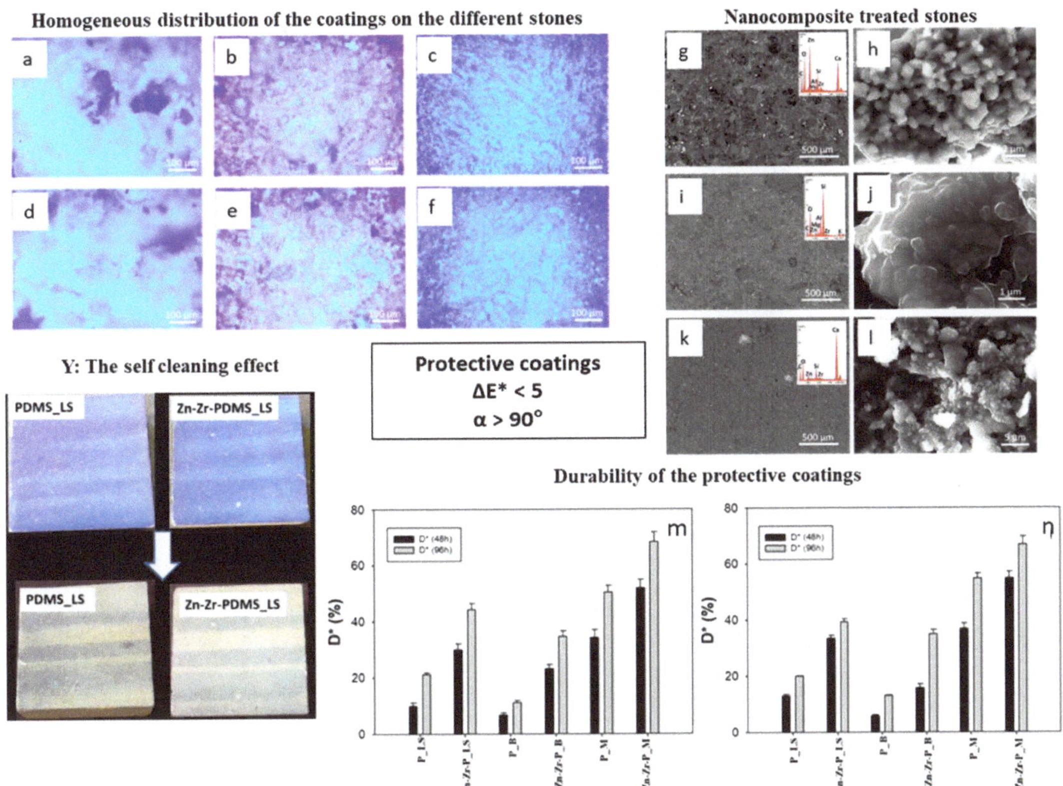

Figure 11. Optical microscope images (by UV light) of stones treated with two different coatings: (**a**) P_LS, (**b**) P_B, (**c**) P_M, (**d**) Zn-Zr-P_LS, (**e**) Zn-Zr-P_B, and (**f**) Zn-Zr-P_M. SEM-EDS analysis of nanocomposite-treated stones at two different magnifications: (**g,h**) Zn-Zr-P_LS, (**i,j**) Zn-Zr-P_B, and (**k,l**) Zn-Zr-P_M. The EDS spectra are the inset of the lowest magnification images (500 μm). Image Y: the self-cleaning effect of coated LS before and after the test, showing the MB discoloration percentage (D* (%)) of artificially aged samples after UV light exposure with (**m**) a humid chamber and (**n**) a solar lamp [95,96].

6. Risk of Toxicity and Preventive Measures for Use of Nanomaterials in Art Conservation

Handling products containing nanomaterials poses one of the biggest concerns when working on heritage conservation. However, the risk of nanomaterials does not exist only during the application of coatings on building surfaces. Actually, it starts when the synthesis procedure takes place, as operators are severely exposed when handling both the reagents and the resulting nanomaterials. Moreover, not only the operators but even people who are in the same work area are exposed to significant threats. Due to their small size, nanomaterials can pass into the human body, into animals, and even contaminate the environment.

The risks of nanomaterials in coatings during the application procedure depends on the application process. Nanomaterials dispersed in a solvent or mixture of solvents can be applied on building materials through three procedures: brushing, immersion, and spraying. In the case of application through immersion, the piece can be removed from the area and placed in a solution containing a mixture composed of nanomaterials for impregnation through capillary rise [31]. The application of coating through spraying processes can be considered the most hazardous procedure [168], as nanomaterials can enter into the body during the application process through the skin, mainly via the face, hands, and arms, in addition to possible access through the eyes, nose, and ears [31]. Simkó et al. (2010) [169] announced that there are various entrance points of nanomaterials into human body, with the most common being ingestion, aspiration by respiratory system, ear canals, tear ducts, and skin contact. Brushing processes can be considered less risky; however, nanomaterials can frequently persist on the brush, and contact between particles and the skin can take place accidentally. Additionally, the remaining amount can come into contact with laboratory workplace surfaces after the application process, where the risk of its spreading to other areas increases, and consequently could pose a threat to people, animals, and environment [31]. Therefore, it is crucial to implement the necessary safety precautions described later in this section.

Several factors are essential to take into consideration when evaluating nanomaterials toxicity, as presented in Figure 12, based on intensive research carried out on the ecotoxicity of nanomaterials conducted by international organizations as the European Commission [170]. It is known that the degree of cohesiveness between particles is fundamental to determining whether particles can be released into the environment. Agglomerates are defined as "weakly bound particles", while aggregates are "strongly bound particles". Thus, it is obvious that in the case of agglomerated nanomaterials the physical bond between particles can be easily broken and particles with tiny size can be simply released into the environment and human body. According to certain studies performed on the agglomeration of particles and their impact on health, agglomeration can more easily lead to diseases than aggregation [171]. On the other hand, particles exhibit numerous particular features depending on the synthesis technique, in particular morphology, which causes variations in their behavior and how they interact with their environment. Accordingly, toxicity may be higher in the ionic state, followed by spherical particles, and lower in the cases of cubic and prismatic particles [172]. However, other reports indicate that rod-like nanoparticles of iron oxide are more toxic than spherical ones [173]. Zhao et al. (2013) [174] stated that nano-hydroxyapatite with plate- and needle-shaped NPs produce a higher proportion of cell deaths compared to spherical and rod-shaped nanoparticles. Morphology has an impact on particle size, and consequently on surface area, which consequently affects the level of toxicity. It is well known that as the particle size decreases, the surface area increases, and consequently the level of interaction between NPs and their environment rises. The solubility of inorganic nanomaterials is another factor that needs to be considered. According to studies, it has been stated that soluble silver compounds are less hazardous than metallic silver and insoluble silver compounds [175]. The chemical composition of nanomaterials is one factor that strongly affects their toxicity. Compared to negative and neutral nanoparticles, positively charged NPs are more hazardous, as they are able to enter cells faster [176].

In general, basic nanomaterials act like gases, have quick diffusion capacities, and move over extended distances. When nanomaterials are released into the environment, they can come into contact with various organs of the human body and animals and react with different tissues or cells through a variety of access points. Studies on animals have revealed that certain NPs can pass through the blood–brain barrier, which generally defends the brain from toxins and contaminants in the bloodstream [177]. It has been reported that NPs inhaled by animals might result in inflammation of the lungs. In addition, NPs are able to move from the lungs to other organs and interfere with cell signaling [177]. Recently, Gomez-Villalba et al. (2023) [31] reported in detail on the side effects that may result from exposure to nanomaterials over time or continuously when performing ordinary conservation activities. They indicated that different organs could be affected due to nanomaterials accessing the body through different routes. For example, the nasal passage is distinguished from all pathways to the body. After entering the nasal channel, the particles move to the respiratory system and the lungs via the olfactory nerves (Figure 13a). Studies reported that diffusion, impaction, interception, and electrostatic attraction are the mechanisms causing deposition of particles in the pulmonary airways during the breath inspiratory phase [178]. In particular, it is important to know that how nanoparticle aerosols behave and how they affect human health depend on their electrical charge. The particle attraction on the surface of the respiratory tissue is caused by the presence of opposite charges between the particle and tissue, which can raise the amount of breathed-in particles that are deposited in the lungs. This process is influenced by the size of the particles. Indeed, the strength of this attraction is more efficient as particle size decreases. It has been stated that the Brownian deposition process predominates for atmospheric NPs when they are negatively charged and range in size from 6 nm to 30 nm [179]. Alveolar deposition increases dramatically in terms of surface area when the polarity of the nanoparticles changes, for instance, from 16 nm to 30 nm [179]. On the other hand, it is worth noting that nanoparticles that are collected at the nasal level are not only able to reach respiratory system and lungs, and could reach the brain through the olfactory nerves (Figure 13b). The eye is another organ by which nanoparticles can enter the brain, as shown in Figure 13c. Despite the existence of obstacles preventing materials from reaching the eyeball, particles with small size are able to create strong contact with the ocular surface [180]. Experts claim that the cornea is the primary location where nanomaterials could be deposited and persist for a long period, overcome the barriers of the ocular surface, and eventually reach the retina [181] and posterior segments of the eye.

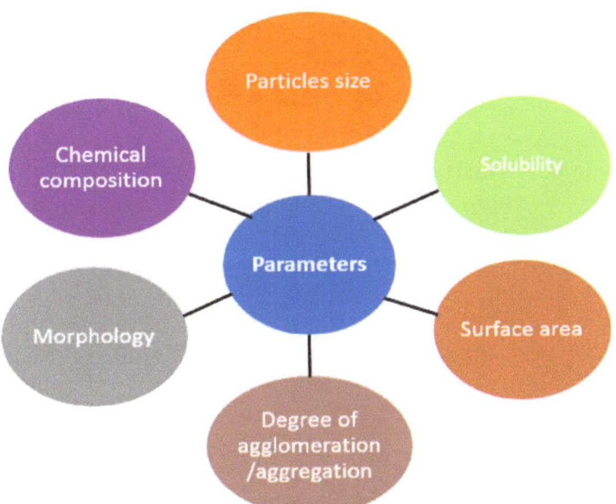

Figure 12. The main parameters that affect the toxicity of nanomaterials.

Prow et al. (2010) [181] stated that migration over the epithelial barrier is able to generate cytotoxicity and inflammatory reactions. Moreover, NPs have the ability to cause cellular damage and a broad immunological reactions, which has an impact on the optic nerve, retina, lens, and macula [181]. The impact of NPs on the eyes depends not only on their particles size but on their nature. Zhu et al. (2019) reported that an increase in retinopathies might result from exposure to ZnO NPs [180], while, Wu and Tang (2018) stated that Ag and TiO_2 NPs can enter the central nervous system and cause neuroinflammation by eye-to-brain routes [182]. The brain is not the only organ that NPs can reach; the kidneys are among the most influenced organs due to their high susceptibility to toxic metals and their contribution to the elimination of NPs from the blood supply (Figure 13d). Metal and metal oxide NPs have the ability to promote the production of ROS, which can lead to oxidative damage, activation of antioxidant enzymes, and cell death through apoptosis [183]. The accumulation of nanomaterials is influenced by their size and morphology, while their toxicity relies on their concentration and solubility [183]. The liver can be affected by nanomaterials as well; NPs can access it through gastrointestinal route to interconnect with the liver cells, altering their structure and even their function

(Figure 13e). Jayvadan and Champavat (2014) [184] reported a detailed investigation of the impacts of nanomaterials on the digestive system, the influence of the size, charge and solubility of particles, and the impact of the pH of the medium on their dispersion and cell-specific incorporation. The same authors stated that ingestion of TiO_2 NPs with size ranging from 25 to 80 nm can induce inflammation in stomach cells. Accumulation of ZnO NPs could result in intestinal and stomach inflammation and even intestinal blockages. In addition, depending on their size NPs can penetrate through the skin, hair follicles, pores, and wounds. Indeed, unlike particles with sizes larger than 30 nm, particles less than 10 nm can get inside human body and result in cell damage [185].

The negative effects of nanomaterials are not limited to the human body and animals; they can affect the environment as well. For example, silver NPs have been demonstrated to have harmful effects on plants, leading to chromosomal abnormalities [186]. Nanomaterials have been revealed to be effective against the natural enemies of mosquitos, posing a threat to public health by impairing the biological regulation of mosquito populations [187]. It has been reported that CuO NPs are harmful to the reproduction of *Enchytraeus crypticus* (i.e., the earthworm), which might modify soil processes, as earthworms play a crucial role in soil health [188]. Soil processes might be changed by impacts to the nitrogen cycle in plants due to TiO_2, Ag, and CuO NPs. Recently, a detailed study performed by Reyes-Estebanez et al. (2018) reported the ecotoxicological effect of engineered NPs on animals and plants [189].

In this review, we present the recent developments in the field of application of metal oxide nanomaterials for conservation of heritage building materials; thus, it is indispensable to present more details about the toxicity of these nanomaterials on the human body, animals, and the environment. Reports have investigated the impact of metal oxide nanomaterials, notably, TiO_2, ZnO, and CuO NPs [190–192]. As mentioned above, the main factors that affect the toxicity of metal oxide particles are their size, morphology, surface modification, solubility, and concentration. Oxidative stress and increased production of ROS are the main causes of metal oxide nanomaterial toxicity. Inflammation, cytotoxicity, DNA damage, chromosomal damage, and carcinogenic consequences as a result of oxidative stress are produced by the internalization of TiO_2 NPs into mammalian cells. Several studies have indicated that anatase phase NPs are more harmful than rutile, which reflects that the structure of TiO_2 polymorphs influences their toxicity. Numerous investigations have demonstrated that TiO_2 NPs were found in the gastrointestinal tract, lungs, spleen, liver, cardiac muscle, heart, and kidneys after oral exposure or inhalation [193]. Other studies have reported the risk of pulmonary harm when working with indoor paints containing TiO_2 nanopowders, which can induce inflammation and DNA damage [194]. Other studies have reported toxicity of TiO_2 NPs in plants [195], animals [196], and insects [197]. Several studies have demonstrated that ZnO NPs in a variety of forms, including rods and spheres, are likely to be harmful because of their capacity to enter the brain [198,199]. The harmful effects of ZnO nanoparticles due to ROS generation, oxidative stress, induction of apoptosis, inflammatory reactions, etc., have been demonstrated in numerous in vitro investigations [191]. Jia et al. (2017) [200] investigated the toxicity of ZnO NPs in the nervous system; their findings showed that both ZnO nanoparticles and Zn ions cause the production of ROS, which causes apoptosis and cytoskeleton disruption. Bioaccumulation ZnO NPs has been detected in the liver, gills, intestine, and brain of fish [201]. The researchers demonstrated that mature fish underwent neural and behavioral alterations as a result of exposure to ZnO nanoparticles [201]. Hepatocyte swelling has been detected in the livers of rats after various dosages of ZnO NPs [202]. Details of the impact of ZnO NPs on plants, soil, animals, and soil organisms can be found in [203]. Copper oxide NPs have the highest cytotoxicity and DNA damaging potential when compared to TiO_2 and ZnO NPs [204]. In vitro studies conducted on human cells exposed to CuO NPs have revealed their significant cytotoxicity, capacity to damage DNA, and ability to produce oxidative stress and cell death [205]. The capacity of particles to diffuse into tissues or cells, which is highly influenced by their surface charge, is necessary for the systemic dispersion of the particles [204]. After they enter cells, CuO NPs may interact with organelles to produce ROS, altering the normal cellular functions [206]. ROS stimulation of oxidative stress results in lipid peroxidation and damage to cellular defense mechanisms through the depletion of reduced glutathione [206]. CuO NPs are able to interact with the lung epithelium, resulting in inflammation [204]. The blood circulatory system carries NPs from the lungs to other parts of the body, where they accumulate in different bodily organs and have harmful effects at diverse locations [207]. According to the findings of in vivo research, oral administration of CuO NPs causes oxidative stress, increased ROS generation, inflammation, apoptosis, and histopathological abnormalities in a number of organs, including the liver, kidney, stomach, and bone marrow [208–210]. Other reports have presented the toxicity of CuO NPs on insects [211,212], animals [213], and plants [188,214]. Studies conducted on both in vivo and in vitro have revealed that NPs may be hazardous because they promote the formation of superoxide and other ROS, which results in an imbalance in the redox state and induces oxidative stress in cells [215]. It has been established that MgO NPs can penetrate through the skin, digestive system, and lungs, as well as accumulate in different tissues [216]. Other studies have demonstrated that MgO NPs are harmful to the human body, animals, and soil organisms when their concentration exceeds a certain level [216–218]. In this context, the toxicity of MgO NPs (10–15 nm) in Wistar rats was assessed in an in vivo setting [219]. The tested rats received varying dosages of MgO NPs through intraperitoneal injection to examine the effect of NP concentrations on their liver and kidney tissues. The findings demonstrated that when compared to the control group, high concentrations of MgO NPs (i.e., 250 and 500 g.mL^{-1}) dramatically increased white blood cells, red blood cells, hemoglobin, and hematocrit. Figure 14 reveals the different reported mechanisms of cell inactivation through metal oxide nanomaterials after the generation of reactive oxygen species.

According to the rules developed by various organizations, several precautions must be taken when working to conserve historical materials using nanomaterials. Nanomaterial application techniques must consider the maximum protection of workers against direct contact with NPs. In particular, the use of improper protective gloves when using brushing treatments increases the danger of skin contact with nanomaterials emitted into the environment. Thus, wearing gloves during brushing procedures is highly recommended. Moreover, in order to protect the skin gloves must be in accordance with the rules, extremely durable, not highly porous, and able to withstand contact with liquid solutions containing nanomaterials that may eventually react with the glove. A face shield, HEPA 14 filter mask, gloves, protective glasses, and suits are required for the operator when performing different application processes [31]. A glove box and pyramid portable glove bag are required to keep operators away from direct interaction with NPs. However, these are more useful when operators are working on specimens of small and medium size at the laboratory scale. Additionally, the laboratory needs to have a sufficient ventilation system and proper waste management containers. On the other hand, extra caution must be used when cleaning brushes, paintbrushes, and other tools, to prevent spilling their contents into the water pipes. As an alternative, they ought to be kept in cans that are clearly marked with information about the contents and risks [31]. Furthermore, selecting the most suitable application procedures depending on the nature and composition of coatings can decrease the risks of exposition to hazardous compounds. Figure 15 illustrates the main personal protection tools required for each application route and how the human body can be protected from nanomaterial penetration through the main access routes.

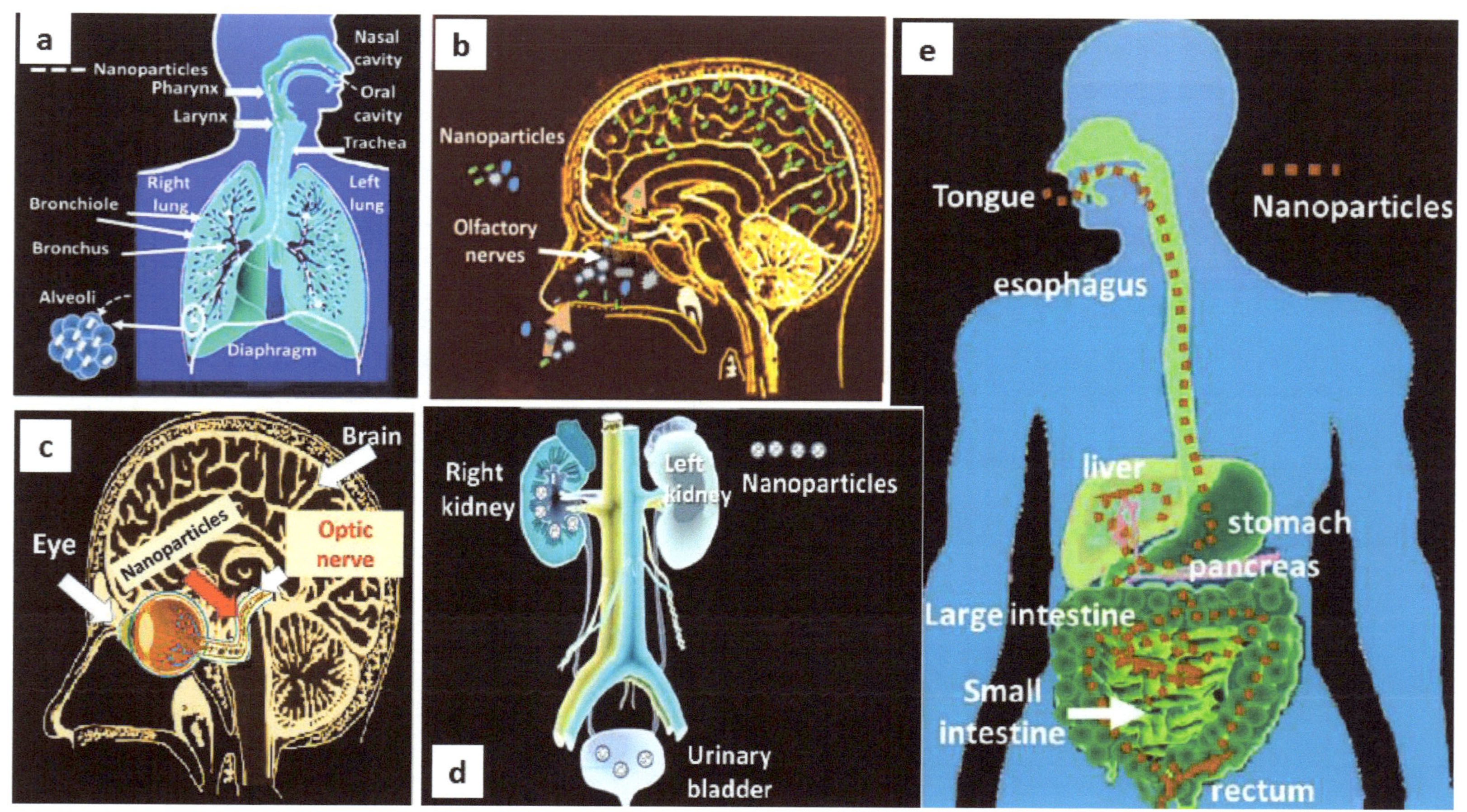

Figure 13. Access of nanomaterials (**a**) nasally through respiratory tract; (**b**) to the brain via the olfactory route; (**c**) to the brain through the tear ducts and eyes, then the optic nerve; (**d**) to the kidney and subsequent accumulation in the urinary track; and (**e**) through the gastrointestinal system [31].

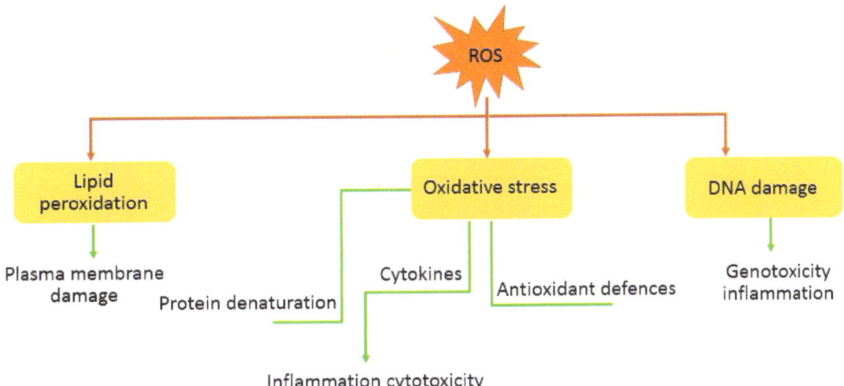

Figure 14. Different processes of cell disturbance due to metal oxide nanoparticles through the generation of ROS.

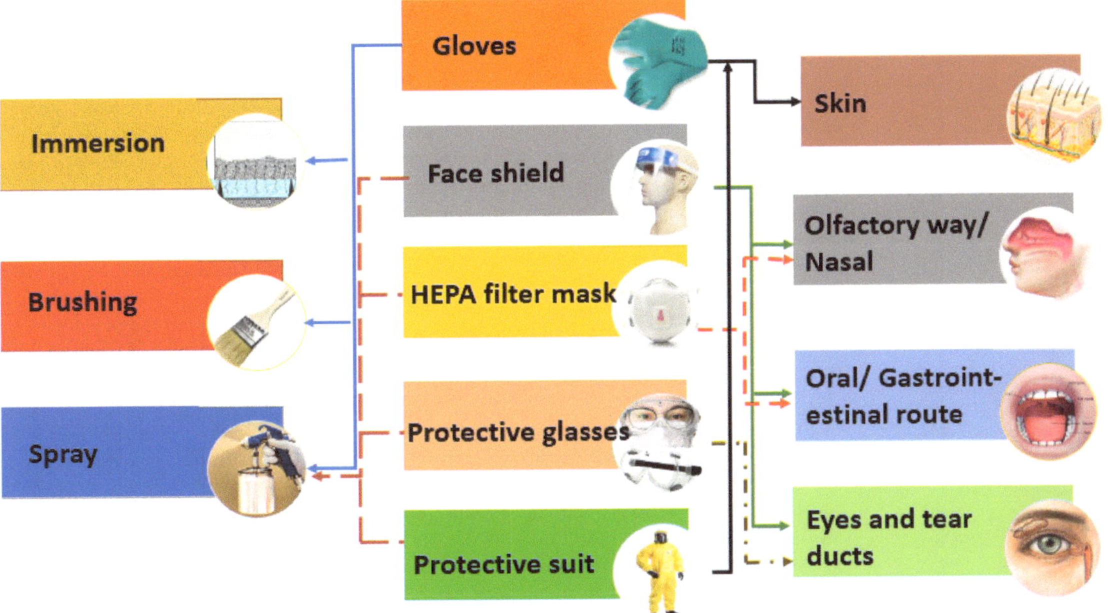

Figure 15. The main application methods of coatings, personal protection tools for each application type, and the utility of each protective tool to prevent nanomaterial penetration through the main access routes.

The access of nanomaterials in general and metal oxides in particular into the environment is feasible during synthesis procedures as well as during their application to heritage stone. The destiny of these nanomaterials after application has not yet been fully identified. Most studies have discussed the characteristics of coatings containing metal oxide NPs and their performance; however, almost no studies have dealt with the fate of these particles after their application, how they interact with environment, and how conservators might be able to remove coatings containing nanoparticles in order to perform renovations after several years. On the other hand, the control of treatments based on nanomaterials is almost always difficult for outdoor applications due to climate environments such as heavy rain, wind, flooding, etc. Alternatively, the use of low amounts of NPs dispersed and interconnected with a binder material, such as PDMS or any other binder, decreases the risk of toxicity. Indeed, metal oxide nanoparticles may be coated with polymeric suspension at low NP concentrations by forming chemical bonds between NPs and the binder, then links between the binder and the stone substrate, which decreases their release into environment or human body and moderate their harmfulness. Contrarily, preparing NP mixtures based on water or alcoholic suspensions increases the risk of NPs being dispersed outside, as water and alcoholic suspensions are easily evaporated over time and the NPs, without a bond or chemical link with the stone substrate, can easily become detached. Moreover, in general, combining NPs with organic fillers or surface functionalization can reduce their risk [96]. In fact, NPs used as aqueous dispersions or transformed to insoluble stable chemical compounds inside stone substrates (if possible) can decrease the risk of NPs being released into the environment [19]. This solution has several benefits due to its low cost and is safer for the environment as it is able to decrease the possibility of NPs accessing into the atmosphere. In addition, working with biocompatible nanomaterials, which have no negative impacts on the environment, has recently been considered as an alternative approach. The green route for synthesis of nanomaterials offers cost-effective, environmentally friendly, and non-toxic alternatives to traditional physical and chemical processes, and prevents operators from being exposed to hazardous chemical products.

7. Conclusions and Future Perspectives

Most of the cultural heritage materials of the globe are made of stone, which is deteriorating due to various causes. Several techniques have been developed for their conservation to preserve them for future generations. In particular, extensive research has been carried out on the development of coatings based on nanomaterial formulations with self-cleaning and/or antimicrobial activity for the protection of stone buildings. This review involves the recent advances in the development of metal oxide nanostructures as treatments for the cleaning, consolidation, and protection of natural stone in built heritage. Using metal oxide-based nanomaterial compounds has been considered an attractive approach by several research group working in this field due to the possible development of products with multifunctional features. In fact, coatings based on metal oxide nanomaterials can reduce water penetration thanks to their potential interaction with a wide range of hydrophobic binders, in addition to the so-called lotus effect. Moreover, their capacity to degrade pollutants such as stains [220], combat aerial pollutants such as nitrogen oxides gases (NO_x) [221], inhibit microbial growth, and promote resistance to ageing thanks to photocatalytic processes through the continuous generation of ROS when nanomaterials are exposed to a source of light has helped these kinds of nanomaterials to gain attention in this field. However, pure metal oxide NPs, particularly TiO_2 and ZnO nanomaterials, which are the most widely used metal oxides in this field, have two main limitations: weak solar energy absorption efficiency and fast recombination of photogenerated charge carriers. To overcome this issue, laboratory experiments have produced several promising outcomes, particularly for metal oxide nanomaterials coupled with metal ions/NPs. In the case of CuO and MgO, the use of hazardous materials, significant production expenses, and high energy consumption are their main drawbacks [222,223]. Green synthesis, which is based on using natural sources such as plants and microorganisms to create CuO or MgO nanomaterials, has developed as an option that is environmentally friendly with low costs and energy consumption.

Despite the extensive work which has been performed in the field of using metal oxide nanomaterials for the preservation of heritage stone artefacts, many existing works do not clearly mention the mechanical features of the coatings (e.g., peeling), their capacity to degrade a wide range of pollutants under normal solar light, their durability and resistance against ageing, or their microbial inhibitory capacity against a mixture of microorganisms. We expect that in the following years several research groups will work on developing multifunctional coatings with enhanced photocatalytic performance (i.e., active under natural solar light) with good resistance against ageing (at least 5 years) and which are able to support severe conditions (weather, climatic changes, pollution, etc.) and inhibit complex microbial mixture growth.

Because nanomaterials unfortunately have the potential to modify the environment and harm both human and animal health, it is necessary to work on new safe-by-design nanomaterials in order to reduce the environmental and toxic dangers, the consequences of release into the environment, and the required precautions. Furthermore, the potential for developing reversible treatments based on metal oxide nanomaterials is another issue that must be addressed in further research as a key request when discussing the preservation of stone artefacts.

Author Contributions: Conceptualization, M.B.C., M.L.W. and M.L.; methodology, M.B.C. and M.L.W.; validation, M.L.W., M.M., C.U. and M.L.; data curation, M.M., and C.U.; writing—original draft preparation, M.B.C.; writing—review and editing, M.L.W. and M.L.; visualization, C.U.; super-vision, C.U. and M.L. All authors have read and agreed to the published version of the manuscript.

Funding: This research received no external funding.

Institutional Review Board Statement: Not applicable.

Informed Consent Statement: Not applicable.

Data Availability Statement: The data presented in this research study are available in the present article.

Conflicts of Interest: The authors declare no conflict of interest.

References

1. Chobba, M.B.; Weththimuni, M.L.; Messaoud, M.; Urzi, C.; Maalej, R.; Licchelli, M. Silver Nanoparticles in the Cultural Heritage Conservation. In *Self-Assembly of Materials and Their Applications*; Rathnayake, H., Pathiraja, G., Sharmin, E., Eds.; IntechOpen: Rijeka, Croatia, 2023. [CrossRef]
2. Chobba, M.B.; Weththimuni, M.L.; Messaoud, M.; Bouaziz, J.; Licchelli, M. Enhanced Gd Doped TiO$_2$ NPs-PDMS Nanocomposites as Protective Coatings for Bio-Calcarenite Stone: Preliminarily Analysis. In *Design and Modeling of Mechanical Systems—V*; Walha, L., Jarraya, A., Djemal, F., Chouchane, M., Aifaoui, N., Chaari, F., Abdennadher, M., Benamara, A., Haddar, M., Eds.; Springer International Publishing: Cham, Switzerland, 2023; pp. 885–893. [CrossRef]
3. Licchelli, M.; Malagodi, M.; Weththimuni, M.; Zanchi, C. Anti-graffiti nanocomposite materials for surface protection of a very porous stone. *Appl. Phys. A* **2014**, *116*, 1525–1539. [CrossRef]
4. Weththimuni, M.; Canevari, C.; Legnani, A.; Licchelli, M.; Malagodi, M.; Ricca, M.; Zeffiro, A. Experimental characterization of oil-colophony varnishes: A preliminary study. *Int. J. Conserv. Sci.* **2016**, *7*, 813–826.
5. Angeli, M.; Bigas, J.-P.; Benavente, D.; Menéndez, B.; Hébert, R.; David, C. Salt crystallization in pores: Quantification and estimation of damage. *Environ. Geol.* **2007**, *52*, 205–213. [CrossRef]
6. Weththimuni, M.L.; Fiocco, G.; Milanese, C.; Spinella, A.; Saladino, M.L.; Malagodi, M.; Licchelli, M. Stradivari's Varnish Revisited: Feature Improvements Using Chemical Modification. *Polymers* **2023**, *15*, 3652. [CrossRef]
7. Spinella, A.; Malagodi, M.; Saladino, M.L.; Weththimuni, M.L.; Caponetti, E.; Licchelli, M. A step forward in disclosing the secret of stradivari's varnish by NMR spectroscopy. *J. Polym. Sci. Part A Polym. Chem.* **2017**, *55*, 3949–3954. [CrossRef]
8. Fichera, G.V.; Malagodi, M.; Cofrancesco, P.; Weththimuni, M.L.; Guglieri, C.; Olivi, L.; Ruffolo, S.; Licchelli, M. Study of the copper effect in iron-gall inks after artificial ageing. *Chem. Pap.* **2018**, *72*, 1905–1915. [CrossRef]
9. Baglioni, P.; Chelazzi, D. How Science Can Contribute to the Remedial Conservation of Cultural Heritage. *Chemistry* **2021**, *27*, 10798–10806. [CrossRef] [PubMed]
10. Rivera, L.E.C.; Ramos, A.P.; Sánchez, J.I.C.; Serrano, M.E.D. Origin and Control Strategies of Biofilms in the Cultural Heritage. In *Antimicrobials, Antibiotic Resistance, Antibiofilm Strategies and Activity Methods*; Kırmusaoğlu, S., Ed.; IntechOpen: Rijeka, Croatia, 2018. [CrossRef]
11. Weththimuni, M.L.; Licchelli, M. Heritage Conservation and Restoration: Surface Characterization, Cleaning and Treatments. *Coatings* **2023**, *13*, 457. [CrossRef]
12. Vinçotte, A.; Beauvoit, E.; Boyard, N.; Guilminot, E. Effect of solvent on PARALOID® B72 and B44 acrylic resins used as adhesives in conservation. *Herit. Sci.* **2019**, *7*, 42. [CrossRef]
13. Karapanagiotis, I.; Chatzigrigoriou, A.; Manoudis, P.N. Chapter 10—Waterborne superhydrophobic coatings for the conservation of the cultural heritage: A case study for the protection of mortar, ceramic, and wood. In *Handbook of Waterborne Coatings*; Zarras, P., Soucek, M.D., Tiwari, A., Eds.; Elsevier: Amsterdam, The Netherlands, 2020; pp. 229–247. [CrossRef]

14. Weththimuni, M.; Crivelli, F.; Galimberti, C.; Malagodi, M.; Licchelli, M. Evaluation of commercial consolidating agents on very porous biocalcarenite. *Int. J. Conserv. Sci.* **2020**, *11*, 251–260.

15. Artesani, A.; Di Turo, F.; Zucchelli, M.; Traviglia, A. Recent Advances in Protective Coatings for Cultural Heritage–An Overview. *Coatings* **2020**, *10*, 217. [CrossRef]

16. Wang, P.; Wei, W.; Li, Z.; Duan, W.; Han, H.; Xie, Q. A superhydrophobic fluorinated PDMS composite for wearable strain sensor with excellent mechanical robustness and liquid impalement resistance. *J. Mater. Chem. A* **2020**, *8*, 3509–3516. [CrossRef]

17. Licchelli, M.; Marzolla, S.J.; Poggi, A.; Zanchi, C. Crosslinked fluorinated polyurethanes for the protection of stone surfaces from graffiti. *J. Cult. Herit.* **2011**, *12*, 34–43. [CrossRef]

18. Licchelli, M.; Malagodi, M.; Weththimuni, M.L.; Zanchi, C. Water-repellent properties of fluoroelastomers on a very porous stone: Effect of the application procedure. *Prog. Org. Coat.* **2013**, *76*, 495–503. [CrossRef]

19. Weththimuni, M.L.; Licchelli, M.; Malagodi, M.; Rovella, N.; La Russa, M. Consolidation of bio-calcarenite stone by treatment based on diammonium hydrogenphosphate and calcium hydroxide nanoparticles. *Measurement* **2018**, *127*, 396–405. [CrossRef]

20. Yang, F.; Zhang, B.; Liu, Y.; Wei, G.; Zhang, H.; Chen, W.; Xu, Z. Biomimic conservation of weathered calcareous stones by apatite. *N. J. Chem.* **2011**, *35*, 887. [CrossRef]

21. Schifano, E.; Cavallini, D.; De Bellis, G.; Bracciale, M.P.; Felici, A.C.; Santarelli, M.L.; Sarto, M.S.; Uccelletti, D. Antibacterial Effect of Zinc Oxide-Based Nanomaterials on Environmental Biodeteriogens Affecting Historical Buildings. *Nanomaterials* **2020**, *10*, 335. [CrossRef] [PubMed]

22. Dei, L.; Salvadori, B. Nanotechnology in cultural heritage conservation: Nanometric slaked lime saves architectonic and artistic surfaces from decay. *J. Cult. Herit.* **2006**, *7*, 110–115. [CrossRef]

23. Facio, D.S.; Mosquera, M.J. Simple Strategy for Producing Superhydrophobic Nanocomposite Coatings In Situ on a Building Substrate. *ACS Appl. Mater. Interfaces* **2013**, *5*, 7517–7526. [CrossRef] [PubMed]

24. Pedna, A.; Pinho, L.; Frediani, P.; Mosquera, M.J. Obtaining SiO_2–fluorinated PLA bionanocomposites with application as reversible and highly-hydrophobic coatings of buildings. *Prog. Org. Coat.* **2016**, *90*, 91–100. [CrossRef]

25. Petcu, C.; Alexandrescu, E.; Bălan, A.; Tănase, M.A.; Cinteză, L.O. Synthesis and Characterisation of Organo-Modified Silica Nanostructured Films for the Water-Repellent Treatment of Historic Stone Buildings. *Coatings* **2020**, *10*, 1010. [CrossRef]

26. Xie, Z.; Duan, Z.; Zhao, Z.; Li, R.; Zhou, B.; Yang, D.; Hu, Y. Nano-materials enhanced protectants for natural stone surfaces. *Herit. Sci.* **2021**, *9*, 122. [CrossRef]

27. Sierra-Fernandez, A.; Gomez-Villalba, L.S.; Rabanal, M.E.; Fort, R. New nanomaterials for applications in conservation and restoration of stony materials: A review. *Mater. Construcción* **2017**, *67*, 107. [CrossRef]

28. Mishra, S.; Sundaram, B. A review of the photocatalysis process used for wastewater treatment. *Mater. Today Proc.* **2023**, *in press*. [CrossRef]

29. Ahmed, S.N.; Haider, W. Heterogeneous photocatalysis and its potential applications in water and wastewater treatment: A review. *Nanotechnology* **2018**, *29*, 342001. [CrossRef]

30. Chobba, M.B.; Messaoud, M.; Weththimuni, M.L.; Bouaziz, J.; Licchelli, M.; De Leo, F.; Urzì, C. Preparation and characterization of photocatalytic Gd-doped TiO_2 nanoparticles for water treatment. *Environ. Sci. Pollut. Res.* **2019**, *26*, 32734–32745. [CrossRef] [PubMed]

31. Gomez-Villalba, L.S.; Salcines, C.; Fort, R. Application of Inorganic Nanomaterials in Cultural Heritage Conservation, Risk of Toxicity, and Preventive Measures. *Nanomaterials* **2023**, *13*, 1454. [CrossRef]

32. Meng, H.; Katayama, Y.; Gu, J.-D. More wide occurrence and dominance of ammonia-oxidizing archaea than bacteria at three Angkor sandstone temples of Bayon, Phnom Krom and Wat Athvea in Cambodia. *Int. Biodeterior. Biodegrad.* **2017**, *117*, 78–88. [CrossRef]

33. Zammit, G.; Sánchez-Moral, S.; Albertano, P. Bacterially mediated mineralisation processes lead to biodeterioration of artworks in Maltese catacombs. *Sci. Total Environ.* **2011**, *409*, 2773–2782. [CrossRef]

34. Urzì, C.; Realini, M. Colour changes of Notos calcareous sandstone as related to its colonisation by microorganisms. *Int. Biodeterior. Biodegrad.* **1998**, *42*, 45–54. [CrossRef]

35. Cappitelli, F.; Principi, P.; Pedrazzani, R.; Toniolo, L.; Sorlini, C. Bacterial and fungal deterioration of the Milan Cathedral marble treated with protective synthetic resins. *Sci. Total Environ.* **2007**, *385*, 172–181. [CrossRef]

36. Rosado, T.; Reis, A.; Mirão, J.; Candeias, A.; Vandenabeele, P.; Caldeira, A.T. Pink! Why not? On the unusual colour of Évora Cathedral. *Int. Biodeterior. Biodegrad.* **2014**, *94*, 121–127. [CrossRef]

37. Schiavon, N.; De Caro, T.; Kiros, A.; Caldeira, A.T.; Parisi, I.E.; Riccucci, C.; Gigante, G.E. A multianalytical approach to investigate stone biodeterioration at a UNESCO world heritage site: The volcanic rock-hewn churches of Lalibela, Northern Ethiopia. *Appl. Phys. A* **2013**, *113*, 843–854. [CrossRef]

38. Ahmed, I.; Basharat, M.; Sousa, L.; Mughal, M.S. Evaluation of building and dimension stone using physico-mechanical and petrographic properties: A case study from the Kohistan and Ladakh batholith, Northern Pakistan. *Environ. Earth Sci.* **2021**, *80*, 759. [CrossRef]

39. Martínez-García, R.; de Rojas, M.S.; Jagadesh, P.; López-Gayarre, F.; Morán-Del-Pozo, J.M.; Juan-Valdes, A. Effect of pores on the mechanical and durability properties on high strength recycled fine aggregate mortar. *Case Stud. Constr. Mater.* **2022**, *16*, e01050. [CrossRef]

40. García-Del-Cura, M.; Benavente, D.; Martínez-Martínez, J.; Cueto, N. Sedimentary structures and physical properties of travertine and carbonate tufa building stone. *Constr. Build. Mater.* **2012**, *28*, 456–467. [CrossRef]

41. Wang, Y.; Li, L.; An, M.; Sun, Y.; Yu, Z.; Huang, H. Factors Influencing the Capillary Water Absorption Characteristics of Concrete and Their Relationship to Pore Structure. *Appl. Sci.* **2022**, *12*, 2211. [CrossRef]

42. Ruffolo, S.A.; La Russa, M.F.; Rovella, N.; Ricca, M. The Impact of Air Pollution on Stone Materials. *Environments* **2023**, *10*, 119. [CrossRef]

43. Graziani, G.; Sassoni, E.; Franzoni, E. Consolidation of porous carbonate stones by an innovative phosphate treatment: Mechanical strengthening and physical-microstructural compatibility in comparison with TEOS-based treatments. *Herit. Sci.* **2015**, *3*, 1. [CrossRef]

44. Licchelli, M.; Malagodi, M.; Weththimuni, M.; Zanchi, C. Nanoparticles for conservation of bio-calcarenite stone. *Appl. Phys. A* **2014**, *114*, 673–683. [CrossRef]

45. Jiang, X.; Mu, S.; Liu, J. Influence of chlorides and salt concentration on salt crystallization damage of cement-based materials. *J. Build. Eng.* **2022**, *61*, 105260. [CrossRef]

46. Menéndez, B. Estimation of salt mixture damage on built cultural heritage from environmental conditions using ECOS-RUNSALT model. *J. Cult. Herit.* **2017**, *24*, 22–30. [CrossRef]

47. Scrivano, S.; Gaggero, L. An experimental investigation into the salt-weathering susceptibility of building limestones. *Rock Mech. Rock Eng.* **2020**, *53*, 5329–5343. [CrossRef]

48. Urzì, C.; Krumbein, W.E. Microbiological Impacts on the Cultural Heritage. In *Durability and Change: The Science, Responsibility, and Cost of Sustaining Cultural Heritage—Chapter 10*; Krumbein, W.E., Brimblecombe, P., Cosgrove, D.E., Staniforth, S., Eds.; John Wiley and Sons Ltd.: Hoboken, NJ, USA, 1994. [CrossRef]

49. Zhang, X.; Hoff, I.; Saba, R.G. Response and Deterioration Mechanism of Bitumen under Acid Rain Erosion. *Materials* **2021**, *14*, 4911. [CrossRef]

50. Zhang, Y.; Gu, L.; Li, W.; Zhang, Q. Effect of acid rain on economic loss of concrete structures in Hangzhou, China. *Int. J. Low Carbon Technol.* **2019**, *14*, 89–94. [CrossRef]

51. Peng, N.; Hong, J.; Zhu, Y.; Dong, Y.; Sun, B.; Huang, J. Experimental Investigation of the Influence of Freeze–Thaw Mode on Damage Characteristics of Sandstone. *Appl. Sci.* **2022**, *12*, 12395. [CrossRef]

52. Huang, S.; Cai, Y.; Liu, Y.; Liu, G. Experimental and Theoretical Study on Frost Deformation and Damage of Red Sandstones with Different Water Contents. *Rock Mech. Rock Eng.* **2021**, *54*, 4163–4181. [CrossRef]

53. Zhang, H.; Meng, X.; Yang, G. A study on mechanical properties and damage model of rock subjected to freeze-thaw cycles and confining pressure. *Cold Reg. Sci. Technol.* **2020**, *174*, 103056. [CrossRef]

54. McNamara, C.; Mitchell, R. Microbial deterioration of historic stone. *Front. Ecol. Environ.* **2005**, *3*, 445–451. [CrossRef]

55. Cappitelli, F.; Cattò, C.; Villa, F. The Control of Cultural Heritage Microbial Deterioration. *Microorganisms* **2020**, *8*, 1542. [CrossRef] [PubMed]

56. Liu, X.; Koestler, R.J.; Warscheid, T.; Katayama, Y.; Gu, J.-D. Microbial deterioration and sustainable conservation of stone monuments and buildings. *Nat. Sustain.* **2020**, *3*, 991–1004. [CrossRef]

57. Pyzik, A.; Ciuchcinski, K.; Dziurzynski, M.; Dziewit, L. The Bad and the Good—Microorganisms in Cultural Heritage Environments—An Update on Biodeterioration and Biotreatment Approaches. *Materials* **2021**, *14*, 177. [CrossRef]

58. Ciferri, O. The role of microorganisms in the degradation of cultural heritage. *Stud. Conserv.* **2002**, *47*, 35–45. [CrossRef]

59. Lo Schiavo, S.; De Leo, F.; Urzì, C. Present and Future Perspectives for Biocides and Antifouling Products for Stone-Built Cultural Heritage: Ionic Liquids as a Challenging Alternative. *Appl. Sci.* **2020**, *10*, 6568. [CrossRef]

60. Miller, A.; Sanmartín, P.; Pereira-Pardo, L.; Dionísio, A.; Saiz-Jimenez, C.; Macedo, M.; Prieto, B. Bioreceptivity of building stones: A review. *Sci. Total Environ.* **2012**, *426*, 1–12. [CrossRef]

61. Roig, P.B.; Ros, J.L.R.; Estellés, R.M. Biocleaning of nitrate alterations on wall paintings by Pseudomonas stutzeri. *Int. Biodeterior. Biodegrad.* **2013**, *84*, 266–274. [CrossRef]

62. Abdelhafez, A.; El-Wekeel, F.M.; Ramadan, E.; Abed-Allah, A. Microbial deterioration of archaeological marble: Identification and treatment. *Ann. Agric. Sci.* **2012**, *57*, 137–144. [CrossRef]

63. De Leo, F.; Marchetta, A.; Urzì, C. Black Fungi on Stone-Built Heritage: Current Knowledge and Future Outlook. *Appl. Sci.* **2022**, *12*, 3969. [CrossRef]

64. Sterflinger, K.; Piñar, G. Microbial deterioration of cultural heritage and works of art—Tilting at windmills? *Appl. Microbiol. Biotechnol.* **2013**, *97*, 9637–9646. [CrossRef] [PubMed]

65. Krakova, L.; De Leo, F.; Bruno, L.; Pangallo, D.; Urzì, C. Complex bacterial diversity in the white biofilms of the Catacombs of St. Callixtus in Rome evidenced by different investigation strategies. *Environ. Microbiol.* **2015**, *17*, 1738–1752. [CrossRef] [PubMed]

66. Urzì, C.; Bruno, L.; De Leo, F. Biodeterioration of Paintings in Caves, Catacombs and other Hypogean Sites. In *Biodeterioration and Preservation in Art, Archaeology and Architecture*; Archetype Publications: London, UK, 2018; pp. 114–129.

67. Liang, X.; Meng, S.; He, Z.; Zeng, X.; Peng, T.; Huang, T.; Wang, J.; Gu, J.-D.; Hu, Z. Higher abundance of ammonia-oxidizing bacteria than ammonia-oxidizing archaea in biofilms and the microbial community composition of Kaiping Diaolou of China. *Int. Biodeterior. Biodegrad.* **2023**, *184*, 105647. [CrossRef]

68. Joseph, E. *Microorganisms in the Deterioration and Preservation of Cultural Heritage*; Springer Science and Business Media LLC.: Dordrecht, The Netherlands, 2021. [CrossRef]

69. Urzi, C. Microbial Deterioration of Rocks and Marble Monuments of the Mediterranean Basin: A Review. *Corros. Rev.* **2004**, *22*, 441–458. [CrossRef]

70. Lamp, J.L.; Marchant, D.R.; Mackay, S.L.; Head, J.W. Thermal stress weathering and the spalling of Antarctic rocks. *J. Geophys. Res. Earth Surf.* **2016**, *122*, 3–24. [CrossRef]

71. Sesana, E.; Gagnon, A.S.; Ciantelli, C.; Cassar, J.; Hughes, J.J. Climate change impacts on cultural heritage: A literature review. *WIREs Clim. Chang.* **2021**, *12*, e710. [CrossRef]

72. Kapsomenakis, J.; Douvis, C.; Poupkou, A.; Zerefos, S.; Solomos, S.; Stavraka, T.; Melis, N.S.; Kyriakidis, E.; Kremlis, G.; Zerefos, C. Climate change threats to cultural and natural heritage UNESCO sites in the Mediterranean. *Environ. Dev. Sustain.* **2023**, *25*, 14519–14544. [CrossRef]

73. Adamo, P.; Violante, P. Weathering of rocks and neogenesis of minerals associated with lichen activity. *Appl. Clay Sci.* **2000**, *16*, 229–256. [CrossRef]

74. Grøntoft, T.; Cassar, J. An assessment of the contribution of air pollution to the weathering of limestone heritage in Malta. *Environ. Earth Sci.* **2020**, *79*, 288. [CrossRef]

75. Huang, J.; Zheng, Y.; Li, H. Study of internal moisture condensation for the conservation of stone cultural heritage. *J. Cult. Herit.* **2022**, *56*, 1–9. [CrossRef]

76. Vyshkvarkova, E.; Sukhonos, O. Climate Change Impact on the Cultural Heritage Sites in the European Part of Russia over the Past 60 Years. *Climate* **2023**, *11*, 50. [CrossRef]

77. Masson-Delmotte, V.; Zhai, P.; Pirani, A.; Connors, S.L.; Péan, C.; Berger, S.; Caud, N.; Chen, Y.; Goldfarb, L.; Gomis, M.I.; et al. (Eds.) IPCC: Summary for Policymakers. In *Climate Change 2021: The Physical Science Basis*; Contribution of Working Group I to the Sixth Assessment Report of the Intergovernmental Panel on Climate Change; Cambridge University Press: Cambridge, UK, 2021; pp. 3–32.

78. UNESCO World Heritage Centre. *Climate Change and World Heritage: Report on Predicting and Managing the Impacts of Climate Change on World Heritage and Strategy to Assist States Parties to Implement Appropriate Management Responses*; UNESCO World Heritage Centre: Paris, France, 2007.

79. Ricca, M.; Le Pera, E.; Licchelli, M.; Macchia, A.; Malagodi, M.; Randazzo, L.; Rovella, N.; Ruffolo, S.A.; Weththimuni, M.L.; La Russa, M.F. The CRATI Project: New Insights on the Consolidation of Salt Weathered Stone and the Case Study of San Domenico Church in Cosenza (South Calabria, Italy). *Coatings* **2019**, *9*, 330. [CrossRef]

80. Ravankhah, M.; de Wit, R.; Argyriou, A.V.; Chliaoutakis, A.; Revez, M.J.; Birkmann, J.; Žuvela-Aloise, M.; Sarris, A.; Tzigounaki, A.; Giapitsoglou, K. Integrated Assessment of Natural Hazards, Including Climate Change's Influences, for Cultural Heritage Sites: The Case of the Historic Centre of Rethymno in Greece. *Int. J. Disaster Risk Sci.* **2019**, *10*, 343–361. [CrossRef]

81. Chavali, M.S.; Nikolova, M.P. Metal oxide nanoparticles and their applications in nanotechnology. *SN Appl. Sci.* **2019**, *1*, 607. [CrossRef]

82. Khan, M.M.; Adil, S.F.; Al-Mayouf, A. Metal oxides as photocatalysts. *J. Saudi Chem. Soc.* **2015**, *19*, 462–464. [CrossRef]

83. Carp, O.; Huisman, C.L.; Reller, A. Photoinduced reactivity of titanium dioxide. *Prog. Solid State Chem.* **2004**, *32*, 33–177. [CrossRef]

84. Fujishima, A.; Honda, K. Electrochemical Photolysis of Water at a Semiconductor Electrode. *Nature* **1972**, *238*, 37–38. [CrossRef] [PubMed]

85. Chobba, M.B.; Messaoud, M.; Bouaziz, J.; De Leo, F.; Urzì, C. The Effect of Heat Treatment on Photocatalytic Performance and Antibacterial Activity of TiO_2 Nanoparticles Prepared by Sol-Gel Method. In *Advances in Materials, Mechanics and Manufacturing*; Chaari, F., Barkallah, M., Bouguecha, A., Zouari, B., Khabou, M.T., Kchaou, M., Haddar, M., Eds.; Springer International Publishing: Cham, Switzerland, 2020; pp. 71–79. [CrossRef]

86. Abdelraheem, W.H.; Patil, M.K.; Nadagouda, M.N.; Dionysiou, D.D. Hydrothermal synthesis of photoactive nitrogen- and boron-codoped TiO_2 nanoparticles for the treatment of bisphenol A in wastewater: Synthesis, photocatalytic activity, degradation byproducts and reaction pathways. *Appl. Catal. B Environ.* **2018**, *241*, 598–611. [CrossRef]

87. Sunada, K.; Watanabe, T.; Hashimoto, K. Studies on photokilling of bacteria on TiO_2 thin film. *J. Photochem. Photobiol. A Chem.* **2003**, *156*, 227–233. [CrossRef]

88. Jin, C.; Tang, Y.; Yang, F.G.; Li, X.L.; Xu, S.; Fan, X.Y.; Huang, Y.Y.; Yang, Y.J. Cellular Toxicity of TiO_2 Nanoparticles in Anatase and Rutile Crystal Phase. *Biol. Trace Elem. Res.* **2011**, *141*, 3–15. [CrossRef] [PubMed]

89. Henningham, A.; Döhrmann, S.; Nizet, V.; Cole, J.N. Mechanisms of group A *Streptococcus* resistance to reactive oxygen species. *FEMS Microbiol. Rev.* **2015**, *39*, 488–508. [CrossRef]

90. Joost, U.; Juganson, K.; Visnapuu, M.; Mortimer, M.; Kahru, A.; Nõmmiste, E.; Joost, U.; Kisand, V.; Ivask, A. Photocatalytic antibacterial activity of nano-TiO_2 (anatase)-based thin films: Effects on Escherichia coli cells and fatty acids. *J. Photochem. Photobiol. B Biol.* **2015**, *142*, 178–185. [CrossRef] [PubMed]

91. Aminuzzaman, M.; Ying, L.P.; Goh, W.-S.; Watanabe, A. Green synthesis of zinc oxide nanoparticles using aqueous extract of Garcinia mangostana fruit pericarp and their photocatalytic activity. *Bull. Mater. Sci.* **2018**, *41*, 50. [CrossRef]

92. Raha, S. Ahmaruzzaman ZnO nanostructured materials and their potential applications: Progress, challenges and perspectives. *Nanoscale Adv.* **2022**, *4*, 1868–1925. [CrossRef] [PubMed]

93. Weththimuni, M.L.; Capsoni, D.; Malagodi, M.; Milanese, C.; Licchelli, M. Shellac/nanoparticles dispersions as protective materials for wood. *Appl. Phys. A* **2016**, *122*, 1058. [CrossRef]

94. Cinteză, L.O.; Tănase, M.A. Multifunctional ZnO Nanoparticle: Based Coatings for Cultural Heritage Preventive Conservation. In *Thin Films*; Ares, A.E., Ed.; IntechOpen: Rijeka, Croatia, 2020. [CrossRef]

95. Weththimuni, M.L.; Ben Chobba, M.; Tredici, I.; Licchelli, M. ZrO$_2$-doped ZnO-PDMS nanocomposites as protective coatings for the stone materials. *Acta IMEKO* **2022**, *11*, 5. [CrossRef]

96. Weththimuni, M.L.; Ben Chobba, M.; Sacchi, D.; Messaoud, M.; Licchelli, M. Durable Polymer Coatings: A Comparative Study of PDMS-Based Nanocomposites as Protective Coatings for Stone Materials. *Chemistry* **2022**, *4*, 60–76. [CrossRef]

97. Weththimuni, M.L.; Milanese, C.; Licchelli, M.; Malagodi, M. Improving the Protective Properties of Shellac-Based Varnishes by Functionalized Nanoparticles. *Coatings* **2021**, *11*, 419. [CrossRef]

98. Weththimuni, M.L.; Capsoni, D.; Malagodi, M.; Licchelli, M. Improving Wood Resistance to Decay by Nanostructured ZnO-Based Treatments. *J. Nanomater.* **2019**, *2019*, 6715756. [CrossRef]

99. Mokammel, M.A.; Islam, M.J.; Hasanuzzaman, M.; Hashmi, S. Nanoscale Materials for Self-Cleaning and Antibacterial Applications. In *Reference Module in Materials Science and Materials Engineering*; Elsevier: Amsterdam, The Netherlands, 2019. [CrossRef]

100. Brayner, R.; Ferrari-Iliou, R.; Brivois, N.; Djediat, S.; Benedetti, M.F.; Fiévet, F. Toxicological Impact Studies Based on *Escherichia coli* Bacteria in Ultrafine ZnO Nanoparticles Colloidal Medium. *Nano Lett.* **2006**, *6*, 866–870. [CrossRef] [PubMed]

101. Franklin, N.M.; Rogers, N.J.; Apte, S.C.; Batley, G.E.; Gadd, G.E.; Casey, P.S. Comparative Toxicity of Nanoparticulate ZnO, Bulk ZnO, and ZnCl$_2$ to a Freshwater Microalga (*Pseudokirchneriella subcapitata*): The Importance of Particle Solubility. *Environ. Sci. Technol.* **2007**, *41*, 8484–8490. [CrossRef]

102. Chen, X.; Xu, J.; Ji, B.; Fang, X.; Jin, K.; Qian, J. The role of nanotechnology-based approaches for clinical infectious diseases and public health. *Front. Bioeng. Biotechnol.* **2023**, *11*, 1146252. [CrossRef]

103. Bondarenko, O.; Ivask, A.; Käkinen, A.; Kahru, A. Sub-toxic effects of CuO nanoparticles on bacteria: Kinetics, role of Cu ions and possible mechanisms of action. *Environ. Pollut.* **2012**, *169*, 81–89. [CrossRef]

104. Busi, S.; Rajkumari, J. Chapter 15—Microbially synthesized nanoparticles as next generation antimicrobials: Scope and applications. In *Nanoparticles in Pharmacotherapy*; Grumezescu, A.M., Ed.; William Andrew Publishing: Norwich, NY, USA, 2019; pp. 485–524. [CrossRef]

105. Khan, M.I.; Akhtar, M.N.; Ashraf, N.; Najeeb, J.; Munir, H.; Awan, T.I.; Tahir, M.B.; Kabli, M.R. Green synthesis of magnesium oxide nanoparticles using Dalbergia sissoo extract for photocatalytic activity and antibacterial efficacy. *Appl. Nanosci.* **2020**, *10*, 2351–2364. [CrossRef]

106. Aziz, B.K.; Karim, M.A.H. Efficient catalytic photodegradation of methylene blue from medical lab wastewater using MgO nanoparticles synthesized by direct precipitation method. *React. Kinet. Mech. Catal.* **2019**, *128*, 1127–1139. [CrossRef]

107. Karthik, K.; Dhanuskodi, S.; Gobinath, C.; Prabukumar, S.; Sivaramakrishnan, S. Fabrication of MgO nanostructures and its efficient photocatalytic, antibacterial and anticancer performance. *J. Photochem. Photobiol. B Biol.* **2019**, *190*, 8–20. [CrossRef]

108. Dobrucka, R. Synthesis of MgO Nanoparticles Using Artemisia abrotanum Herba Extract and Their Antioxidant and Photocatalytic Properties. *Iran. J. Sci. Technol. Trans. A Sci.* **2018**, *42*, 547–555. [CrossRef]

109. Abinaya, S.; Kavitha, H.P.; Prakash, M.; Muthukrishnaraj, A. Green synthesis of magnesium oxide nanoparticles and its applications: A review. *Sustain. Chem. Pharm.* **2021**, *19*, 100368. [CrossRef]

110. Wetteland, C.L.; Nguyen, N.-Y.T.; Liu, H. Concentration-dependent behaviors of bone marrow derived mesenchymal stem cells and infectious bacteria toward magnesium oxide nanoparticles. *Acta Biomater.* **2016**, *35*, 341–356. [CrossRef] [PubMed]

111. Fruth, V.; Todan, L.; Codrea, C.I.; Poenaru, I.; Petrescu, S.; Aricov, L.; Ciobanu, M.; Jecu, L.; Ion, R.M.; Predoana, L. Multifunctional Composite Coatings Based on Photoactive Metal-Oxide Nanopowders (MgO/TiO$_2$) in Hydrophobic Polymer Matrix for Stone Heritage Conservation. *Nanomaterials* **2021**, *11*, 2586. [CrossRef]

112. He, Y.; Ingudam, S.; Reed, S.; Gehring, A.; Strobaugh, T.P., Jr.; Irwin, P. Study on the mechanism of antibacterial action of magnesium oxide nanoparticles against foodborne pathogens. *J. Nanobiotechnol.* **2016**, *14*, 54. [CrossRef] [PubMed]

113. Negrescu, A.M.; Killian, M.S.; Raghu, S.N.V.; Schmuki, P.; Mazare, A.; Cimpean, A. Metal Oxide Nanoparticles: Review of Synthesis, Characterization and Biological Effects. *J. Funct. Biomater.* **2022**, *13*, 274. [CrossRef]

114. D'orazio, L.; Grippo, A. A water dispersed Titanium dioxide/poly(carbonate urethane) nanocomposite for protecting cultural heritage: Preparation and properties. *Prog. Org. Coat.* **2015**, *79*, 1–7. [CrossRef]

115. La Russa, M.F.; Rovella, N.; de Buergo, M.A.; Belfiore, C.M.; Pezzino, A.; Crisci, G.M.; Ruffolo, S.A. Nano-TiO$_2$ coatings for cultural heritage protection: The role of the binder on hydrophobic and self-cleaning efficacy. *Prog. Org. Coat.* **2016**, *91*, 1–8. [CrossRef]

116. Eduok, U.; Faye, O.; Szpunar, J. Recent developments and applications of protective silicone coatings: A review of PDMS functional materials. *Prog. Org. Coat.* **2017**, *111*, 124–163. [CrossRef]

117. Kapridaki, C.; Maravelaki-Kalaitzaki, P. TiO$_2$–SiO$_2$–PDMS nano-composite hydrophobic coating with self-cleaning properties for marble protection. *Prog. Org. Coat.* **2013**, *76*, 400–410. [CrossRef]

118. Tavares, M.; Santos, A.; Santos, I.; Silva, M.; Bomio, M.; Longo, E.; Paskocimas, C.; Motta, F. TiO$_2$/PDMS nanocomposites for use on self-cleaning surfaces. *Surf. Coat. Technol.* **2014**, *239*, 16–19. [CrossRef]

119. Kapridaki, C.; Verganelaki, A.; Dimitriadou, P.; Maravelaki-Kalaitzaki, P. Conservation of Monuments by a Three-Layered Compatible Treatment of TEOS-Nano-Calcium Oxalate Consolidant and TEOS-PDMS-TiO$_2$ Hydrophobic/Photoactive Hybrid Nanomaterials. *Materials* **2018**, *11*, 684. [CrossRef]

120. Corcione, C.E.; Ingrosso, C.; Petronella, F.; Comparelli, R.; Striccoli, M.; Agostiano, A.; Frigione, M.; Curri, M.L. A designed UV–vis light curable coating nanocomposite based on colloidal TiO_2 NRs in a hybrid resin for stone protection. *Prog. Org. Coat.* **2018**, *122*, 290–301. [CrossRef]
121. Azadi, N.; Parsimehr, H.; Ershad-Langroudi, A. Cultural heritage protection via hybrid nanocomposite coating. *Plast. Rubber Compos.* **2020**, *49*, 414–424. [CrossRef]
122. Ruffolo, S.A.; La Russa, M.F. Nanostructured Coatings for Stone Protection: An Overview. *Front. Mater.* **2019**, *6*, 147. [CrossRef]
123. Peng, M.; Wang, L.; Guo, L.; Guo, J.; Zheng, L.; Yang, F.; Ma, Z.; Zhao, X. A Durable Nano-SiO_2-TiO_2/Dodecyltrimethoxysilane Superhydrophobic Coating for Stone Protection. *Coatings* **2022**, *12*, 1397. [CrossRef]
124. Goffredo, G.B.; Quagliarini, E.; Bondioli, F.; Munafò, P. TiO_2 nanocoatings for architectural heritage: Self-cleaning treatments on historical stone surfaces. *Proc. Inst. Mech. Eng. Part N J. Nanomater. Nanoeng. Nanosyst.* **2014**, *228*, 2–10. [CrossRef]
125. Kapridaki, C.; Pinho, L.; Mosquera, M.J.; Maravelaki-Kalaitzaki, P. Producing photoactive, transparent and hydrophobic SiO_2-crystalline TiO_2 nanocomposites at ambient conditions with application as self-cleaning coatings. *Appl. Catal. B Environ.* **2014**, *156–157*, 416–427. [CrossRef]
126. Bergamonti, L.; Alfieri, I.; Franzò, M.; Lorenzi, A.; Montenero, A.; Predieri, G.; Raganato, M.; Calia, A.; Lazzarini, L.; Bersani, D.; et al. Synthesis and characterization of nanocrystalline TiO_2 with application as photoactive coating on stones. *Environ. Sci. Pollut. Res.* **2014**, *21*, 13264–13277. [CrossRef]
127. Gherardi, F.; Roveri, M.; Goidanich, S.; Toniolo, L. Photocatalytic Nanocomposites for the Protection of European Architectural Heritage. *Materials* **2018**, *11*, 65. [CrossRef] [PubMed]
128. Crupi, V.; Fazio, B.; Gessini, A.; Kis, Z.; La Russa, M.F.; Majolino, D.; Masciovecchio, C.; Ricca, M.; Rossi, B.; Ruffolo, S.A.; et al. TiO_2-SiO_2–PDMS nanocomposite coating with self-cleaning effect for stone material: Finding the optimal amount of TiO_2. *Constr. Build. Mater.* **2018**, *166*, 464–471. [CrossRef]
129. Colangiuli, D.; Lettieri, M.; Masieri, M.; Calia, A. Field study in an urban environment of simultaneous self-cleaning and hydrophobic nanosized TiO_2-based coatings on stone for the protection of building surface. *Sci. Total Environ.* **2019**, *650*, 2919–2930. [CrossRef] [PubMed]
130. Wang, D.; Hou, P.; Stephan, D.; Huang, S.; Zhang, L.; Yang, P.; Cheng, X. SiO_2/TiO_2 composite powders deposited on cement-based materials: Rhodamine B removal and the bonding mechanism. *Constr. Build. Mater.* **2020**, *241*, 118124. [CrossRef]
131. Wang, D.; Geng, Z.; Hou, P.; Yang, P.; Cheng, X.; Huang, S. Rhodamine B Removal of TiO_2@SiO_2 Core-Shell Nanocomposites Coated to Buildings. *Crystals* **2020**, *10*, 80. [CrossRef]
132. Luna, M.; Delgado, J.; Romero, I.; Montini, T.; Gil, M.A.; Martínez-López, J.; Fornasiero, P.; Mosquera, M. Photocatalytic TiO_2 nanosheets-SiO_2 coatings on concrete and limestone: An enhancement of de-polluting and self-cleaning properties by nanoparticle design. *Constr. Build. Mater.* **2022**, *338*, 127349. [CrossRef]
133. Khannyra, S.; Luna, M.; Gil, M.A.; Addou, M.; Mosquera, M.J. Self-cleaning durability assessment of TiO_2/SiO_2 photocatalysts coated concrete: Effect of indoor and outdoor conditions on the photocatalytic activity. *Build. Environ.* **2022**, *211*, 108743. [CrossRef]
134. Xia, X.; Liu, J.; Liu, Y.; Lei, Z.; Han, Y.; Zheng, Z.; Yin, J. Preparation and Characterization of Biomimetic SiO_2-TiO_2-PDMS Composite Hydrophobic Coating with Self-Cleaning Properties for Wall Protection Applications. *Coatings* **2023**, *13*, 224. [CrossRef]
135. Ruffolo, S.A.; Macchia, A.; La Russa, M.F.; Mazza, L.; Urzì, C.; De Leo, F.; Barberio, M.; Crisci, G.M. Marine Antifouling for Underwater Archaeological Sites: TiO_2 and Ag-Doped TiO_2. *Int. J. Photoenergy* **2013**, *2013*, 251647. [CrossRef]
136. Roveri, M.; Gherardi, F.; Goidanich, S.; Gulotta, D.; Castelvetro, V.; Fischer, R.; Winandy, L.; Weber, J.; Toniolo, L. Self-cleaning and antifouling nanocomposites for stone protection: Properties and performances of stone-nanomaterial systems. *IOP Conf. Ser. Mater. Sci. Eng.* **2018**, *364*, 012070. [CrossRef]
137. Van der Werf, I.D.; Ditaranto, N.; Picca, R.A.; Sportelli, M.C.; Sabbatini, L. Development of a novel conservation treatment of stone monuments with bioactive nanocomposites. *Herit. Sci.* **2015**, *3*, 29. [CrossRef]
138. Aldosari, M.A.; Darwish, S.S.; Adam, M.A.; Elmarzugi, N.A.; Ahmed, S.M. Using ZnO nanoparticles in fungal inhibition and self-protection of exposed marble columns in historic sites. *Archaeol. Anthr. Sci.* **2019**, *11*, 3407–3422. [CrossRef]
139. Dakal, T.C.; Cameotra, S.S. Microbially induced deterioration of architectural heritages: Routes and mechanisms involved. *Environ. Sci. Eur.* **2012**, *24*, 36. [CrossRef]
140. Zanni, E.; Chandraiahgari, C.R.; De Bellis, G.; Montereali, M.R.; Armiento, G.; Ballirano, P.; Polimeni, A.; Sarto, M.S.; Uccelletti, D. Zinc Oxide Nanorods-Decorated Graphene Nanoplatelets: A Promising Antimicrobial Agent against the Cariogenic Bacterium *Streptococcus mutans*. *Nanomaterials* **2016**, *6*, 179. [CrossRef]
141. Zarzuela, R.; Carbú, M.; Gil, M.A.; Cantoral, J.M.; Mosquera, M.J. CuO/SiO_2 nanocomposites: A multifunctional coating for application on building stone. *Mater. Des.* **2017**, *114*, 364–372. [CrossRef]
142. Zarzuela, R.; Gil, M.A.; Carretero, J.; Carbú, M.; Cantoral, J.M.; Mosquera, M.J. Development of a novel engineered stone containing a CuO/SiO_2 nanocomposite matrix with biocidal properties. *Constr. Build. Mater.* **2021**, *303*, 124459. [CrossRef]
143. Ghasemi-Kahrizsangi, S.; Dehsheikh, H.G.; Boroujerdnia, M. MgO–CaO–Cr_2O_3 composition as a novel refractory brick: Use of Cr_2O_3 nanoparticles. *Boletín Soc. Española Cerámica Vidr.* **2017**, *56*, 83–89. [CrossRef]
144. Ali, T.; Ahmed, A.; Alam, U.; Uddin, I.; Tripathi, P.; Muneer, M. Enhanced photocatalytic and antibacterial activities of Ag-doped TiO_2 nanoparticles under visible light. *Mater. Chem. Phys.* **2018**, *212*, 325–335. [CrossRef]

145. Bahadur, J.; Agrawal, S.; Panwar, V.; Parveen, A.; Pal, K. Antibacterial properties of silver doped TiO_2 nanoparticles synthesized via sol-gel technique. *Macromol. Res.* **2016**, *24*, 488–493. [CrossRef]

146. Chobba, M.B.; Weththimuni, M.L.; Messaoud, M.; Urzi, C.; Bouaziz, J.; De Leo, F.; Licchelli, M. Ag-TiO_2/PDMS nanocomposite protective coatings: Synthesis, characterization, and use as a self-cleaning and antimicrobial agent. *Prog. Org. Coat.* **2021**, *158*, 106342. [CrossRef]

147. Ben Chobba, M.; Weththimuni, M.L.; Messaoud, M.; Sacchi, D.; Bouaziz, J.; De Leo, F.; Urzi, C.; Licchelli, M. Multifunctional and Durable Coatings for Stone Protection Based on Gd-Doped Nanocomposites. *Sustainability* **2021**, *13*, 11033. [CrossRef]

148. NORMAL 20/85, Interventi Conservativi: Progettazione Esecuzione e Valutazione Preventiva, Italian Standard, 1–20 Pages, CNR, Rome, Italy. 1985. Available online: https://moodle2.units.it/pluginfile.php/230506/mod_resource/content/1/Interventi%20 di%20conservazione.pdf (accessed on 26 December 2023).

149. Ben Chobba, M.; Weththimuni, M.L.; Messaoud, M.; Bouaziz, J.; Salhi, R.; De Leo, F.; Urzì, C.; Licchelli, M. Silver-Doped TiO_2-PDMS Nanocomposite as a Possible Coating for the Preservation of Serena Stone: Searching for Optimal Application Conditions. *Heritage* **2022**, *5*, 3411–3426. [CrossRef]

150. Awazu, K.; Fujimaki, M.; Rockstuhl, C.; Tominaga, J.; Murakami, H.; Ohki, Y.; Yoshida, N.; Watanabe, T. A Plasmonic Photocatalyst Consisting of Silver Nanoparticles Embedded in Titanium Dioxide. *J. Am. Chem. Soc.* **2008**, *130*, 1676–1680. [CrossRef] [PubMed]

151. Pinho, L.; Rojas, M.; Mosquera, M.J. Ag–SiO_2–TiO_2 nanocomposite coatings with enhanced photoactivity for self-cleaning application on building materials. *Appl. Catal. B Environ.* **2015**, *178*, 144–154. [CrossRef]

152. Luna, M.; Mosquera, M.J.; Vidal, H.; Gatica, J.M. Au-TiO_2/SiO_2 photocatalysts for building materials: Self-cleaning and de-polluting performance. *Build. Environ.* **2019**, *164*, 106347. [CrossRef]

153. Link, S.; El-Sayed, M.A. Spectral Properties and Relaxation Dynamics of Surface Plasmon Electronic Oscillations in Gold and Silver Nanodots and Nanorods. *J. Phys. Chem. B* **1999**, *103*, 8410–8426. [CrossRef]

154. Zarzuela, R.; Moreno-Garrido, I.; Gil, M.A.; Mosquera, M.J. Effects of surface functionalization with alkylalkoxysilanes on the structure, visible light photoactivity and biocidal performance of Ag-TiO_2 nanoparticles. *Powder Technol.* **2021**, *383*, 381–395. [CrossRef]

155. Khannyra, S.; Mosquera, M.; Addou, M.; Gil, M. Cu-TiO_2/SiO_2 photocatalysts for concrete-based building materials: Self-cleaning and air de-pollution performance. *Constr. Build. Mater.* **2021**, *313*, 125419. [CrossRef]

156. Becerra, J.; Zaderenko, A.; Ortiz, P. Silver/dioxide titanium nanocomposites as biocidal treatments on limestones. *Ge Conserv.* **2017**, *1*, 141–148.

157. Luna, M.; Delgado, J.J.; Gil, M.L.A.; Mosquera, M.J. TiO_2-SiO_2 Coatings with a Low Content of AuNPs for Producing Self-Cleaning Building Materials. *Nanomaterials* **2018**, *8*, 177. [CrossRef] [PubMed]

158. Becerra, J.; Zaderenko, A.; Sayagués, M.; Ortiz, R.; Ortiz, P. Synergy achieved in silver-TiO_2 nanocomposites for the inhibition of biofouling on limestone. *Build. Environ.* **2018**, *141*, 80–90. [CrossRef]

159. Truppi, A.; Luna, M.; Petronella, F.; Falcicchio, A.; Giannini, C.; Comparelli, R.; Mosquera, M.J. Photocatalytic Activity of TiO_2/AuNRs–SiO_2 Nanocomposites Applied to Building Materials. *Coatings* **2018**, *8*, 296. [CrossRef]

160. Kapridaki, C.; Xynidis, N.; Vazgiouraki, E.; Kallithrakas-Kontos, N.; Maravelaki-Kalaitzaki, P. Characterization of Photoactive Fe-TiO_2 Lime Coatings for Building Protection: The Role of Iron Content. *Materials* **2019**, *12*, 1847. [CrossRef]

161. Luna, M.; Gatica, J.M.; Vidal, H.; Mosquera, M.J. Use of Au/N-TiO_2/SiO_2 photocatalysts in building materials with NO depolluting activity. *J. Clean. Prod.* **2020**, *243*, 118633. [CrossRef]

162. Zarzuela, R.; Carbú, M.; Gil, A.; Cantoral, J.; Mosquera, M.J. Incorporation of functionalized Ag-TiO_2NPs to ormosil-based coatings as multifunctional biocide, superhydrophobic and photocatalytic surface treatments for porous ceramic materials. *Surf. Interfaces* **2021**, *25*, 101257. [CrossRef]

163. Khannyra, S.; Gil, M.L.A.; Addou, M.; Mosquera, M.J. Dye decomposition and air de-pollution performance of TiO_2/SiO_2 and N-TiO_2/SiO_2 photocatalysts coated on Portland cement mortar substates. *Environ. Sci. Pollut. Res.* **2022**, *29*, 63112–63125. [CrossRef]

164. Mu, B.; Ying, X.; Petropoulos, E.; He, S. Preparation of AgCl/ZnO nano-composite for effective antimicrobial protection of stone-made building elements. *Mater. Lett.* **2021**, *285*, 129143. [CrossRef]

165. Sierra-Fernandez, A.; De la Rosa-García, S.C.; Gomez-Villalba, L.S.; Gómez-Cornelio, S.; Rabanal, M.E.; Fort, R.; Quintana, P. Synthesis, Photocatalytic, and Antifungal Properties of MgO, ZnO and Zn/Mg Oxide Nanoparticles for the Protection of Calcareous Stone Heritage. *ACS Appl. Mater. Interfaces* **2017**, *9*, 24873–24886. [CrossRef]

166. Gómez-Cornelio, S.; Ortega-Morales, O.; Morón-Ríos, A.; Reyes-Estebanez, M.; De la Rosa-García, S. Changes in fungal community composition of biofilms on limestone across a chronosequence in Campeche, Mexico. *Acta Bot. Mex.* **2016**, *117*, 59–77. [CrossRef]

167. Weththimuni, M.; Chobba, M.B.; Tredici, I.; Licchelli, M. Polydimethylsiloxane (PDMS)/ZrO2-Doped ZnO Nanocomposites as Protective Coatings for Stone Materials. In Proceedings of the 2020 IMEKO TC-4 International Conference on Metrology for Archaeology and Cultural Heritage, Tento, Italy, 16 October 2020.

168. Yang, Y.-F.; Wang, W.-M.; Chen, C.-Y.; Lu, T.-H.; Liao, C.-M. Assessing human exposure risk and lung disease burden posed by airborne silver nanoparticles emitted by consumer spray products. *Int. J. Nanomed.* **2019**, *14*, 1687–1703. [CrossRef]

169. Simko, M.; Fiedeler, U.; Gazsó, A.; Nentwich, M. How Nanoparticles Enter the Human Body and Their Effects There. 2011. Available online: http://hw.oeaw.ac.at/nanotrust-dossier (accessed on 26 December 2023).

170. European Commission. Guidance on the Protection of the Health and Safety of Workers from the Potential Risks Related to the Nanomaterials at Work. Guidance for Employers and Health and Safety Practitioners. 2013. Available online: https://ec.europa.eu/social/BlobServlet?docId=13087&langId=en%252 (accessed on 26 December 2023).

171. Bian, Y.; Kim, K.; Ngo, T.; Kim, I.; Bae, O.-N.; Lim, K.-M.; Chung, J.-H. Silver nanoparticles promote procoagulant activity of red blood cells: A potential risk of thrombosis in susceptible population. *Part. Fibre Toxicol.* **2019**, *16*, 9. [CrossRef] [PubMed]

172. Auclair, J.; Gagné, F. Shape-Dependent Toxicity of Silver Nanoparticles on Freshwater Cnidarians. *Nanomaterials* **2022**, *12*, 3107. [CrossRef] [PubMed]

173. Lee, J.H.; Ju, J.E.; Kim, B.I.; Pak, P.J.; Choi, E.; Lee, H.; Chung, N. Rod-shaped iron oxide nanoparticles are more toxic than sphere-shaped nanoparticles to murine macrophage cells. *Environ. Toxicol. Chem.* **2014**, *33*, 2759–2766. [CrossRef] [PubMed]

174. Zhao, X.; Ng, S.; Heng, B.C.; Guo, J.; Ma, L.; Tan, T.T.Y.; Ng, K.W.; Loo, S.C.J. Cytotoxicity of hydroxyapatite nanoparticles is shape and cell dependent. *Arch. Toxicol.* **2013**, *87*, 1037–1052. [CrossRef] [PubMed]

175. Panyala, N.R.; Peña-Méndez, E.M.; Havel, J. Silver or silver nanoparticles: A hazardous threat to the environment and human health? *J. Appl. Biomed.* **2008**, *6*, 117–129. [CrossRef]

176. Sukhanova, A.; Bozrova, S.; Sokolov, P.; Berestovoy, M.; Karaulov, A.; Nabiev, I. Dependence of Nanoparticle Toxicity on Their Physical and Chemical Properties. *Nanoscale Res. Lett.* **2018**, *13*, 44. [CrossRef] [PubMed]

177. Thomas, S.P.; Al-Mutairi, E.M.; De, S.K. Impact of Nanomaterials on Health and Environment. *Arab. J. Sci. Eng.* **2013**, *38*, 457–477. [CrossRef]

178. Löndahl, J.; Möller, W.; Pagels, J.H.; Kreyling, W.G.; Swietlicki, E.; Schmid, O. Measurement Techniques for Respiratory Tract Deposition of Airborne Nanoparticles: A Critical Review. *J. Aerosol Med. Pulm. Drug Deliv.* **2014**, *27*, 229–254. [CrossRef]

179. Sonwani, S.; Madaan, S.; Arora, J.; Suryanarayan, S.; Rangra, D.; Mongia, N.; Vats, T.; Saxena, P. Inhalation Exposure to Atmospheric Nanoparticles and Its Associated Impacts on Human Health: A Review. *Front. Sustain. Cities* **2021**, *3*, 690444. [CrossRef]

180. Zhu, S.; Gong, L.; Li, Y.; Xu, H.; Gu, Z.; Zhao, Y. Safety Assessment of Nanomaterials to Eyes: An Important but Neglected Issue. *Adv. Sci.* **2019**, *6*, 1802289. [CrossRef] [PubMed]

181. Prow, T.w.; Bhutto, I.; Kim, S.Y.; Grebe, R.; Merges, C.; McLeod, D.S.; Uno, K.; Mennon, M.; Rodriguez, L.; Leong, K.; et al. Ocular nanoparticle toxicity and transfection of the retina and retinal pigment epithelium. *Nanomed. Nanotechnol. Biol. Med.* **2008**, *4*, 340–349. [CrossRef] [PubMed]

182. Wu, T.; Tang, M.; Grigoriadis, N.; Lagoudaki, R.; Tascos, N.; Milonas, I.; Lopez-Campos, J.L.; Calero-Acuña, C.; Lopez-Ramirez, C.; Abad-Arranz, M.; et al. The inflammatory response to silver and titanium dioxide nanoparticles in the central nervous system. *Nanomedicine* **2017**, *13*, 233–249. [CrossRef] [PubMed]

183. Makhdoumi, P.; Karimi, H.; Khazaei, M. Review on Metal-Based Nanoparticles: Role of Reactive Oxygen Species in Renal Toxicity. *Chem. Res. Toxicol.* **2020**, *33*, 2503–2514. [CrossRef] [PubMed]

184. Jayvadan, P.; Champavat, V. Toxicity of Nanomaterials on the Gastrointestinal Tract. In *Biointeractions of Nanomaterials*; CRC Press: Boca Raton, FL, USA, 2014; pp. 259–284. [CrossRef]

185. Gautam, A.; Singh, D.; Vijayaraghavan, R. Dermal exposure of nanoparticles: An understanding. *J. Cell Tissue Res.* **2011**, *11*, 2703–2708.

186. Rafique, T.; Naseem, S.; Usmani, T.H.; Bashir, E.; Khan, F.A.; Bhanger, M.I. Geochemical factors controlling the occurrence of high fluoride groundwater in the Nagar Parkar area, Sindh, Pakistan. *J. Hazard. Mater.* **2009**, *171*, 424–430. [CrossRef] [PubMed]

187. Kumar, P.M.; Murugan, K.; Madhiyazhagan, P.; Kovendan, K.; Amerasan, D.; Chandramohan, B.; Dinesh, D.; Suresh, U.; Nicoletti, M.; Alsalhi, M.S.; et al. Biosynthesis, characterization, and acute toxicity of *Berberis tinctoria*-fabricated silver nanoparticles against the Asian tiger mosquito, *Aedes albopictus*, and the mosquito predators *Toxorhynchites splendens* and *Mesocyclops thermocyclopoides*. *Parasitol. Res.* **2016**, *115*, 751–759. [CrossRef]

188. Nair, P.M.G.; Kim, S.-H.; Chung, I.M. Copper oxide nanoparticle toxicity in mung bean (*Vigna radiata* L.) seedlings: Physiological and molecular level responses of in vitro grown plants. *Acta Physiol. Plant.* **2014**, *36*, 2947–2958. [CrossRef]

189. Reyes-Estebanez, M.; Ortega-Morales, B.O.; Chan-Bacab, M.; Granados-Echegoyen, C.; Camacho-Chab, J.C.; Pereañez-Sacarias, J.E.; Gaylarde, C. Antimicrobial engineered nanoparticles in the built cultural heritage context and their ecotoxicological impact on animals and plants: A brief review. *Herit. Sci.* **2018**, *6*, 52. [CrossRef]

190. Suman, T.Y.; Li, W.-G.; Pei, D.-S. Chapter 5—Toxicity of metal oxide nanoparticles. In *Nanotoxicity*; Rajendran, S., Mukherjee, A., Nguyen, T.A., Godugu, C., Shukla, R.K., Eds.; Elsevier: Amsterdam, The Netherlands, 2020; pp. 107–123. [CrossRef]

191. Girigoswami, K. Toxicity of Metal Oxide Nanoparticles. In *Cellular and Molecular Toxicology of Nanoparticles*; Saquib, Q., Faisal, M., Al-Khedhairy, A.A., Alatar, A.A., Eds.; Springer International Publishing: Cham, Switzerland, 2018; pp. 99–122. [CrossRef]

192. Sengul, A.B.; Asmatulu, E. Toxicity of metal and metal oxide nanoparticles: A review. *Environ. Chem. Lett.* **2020**, *18*, 1659–1683. [CrossRef]

193. Shabbir, S.; Kulyar, M.F.-E.; Bhutta, Z.A.; Boruah, P.; Asif, M. Toxicological Consequences of Titanium Dioxide Nanoparticles (TiO$_2$NPs) and Their Jeopardy to Human Population. *BioNanoScience* **2021**, *11*, 621–632. [CrossRef] [PubMed]

194. Saber, A.T.; Jacobsen, N.R.; Mortensen, A.; Szarek, J.; Jackson, P.; Madsen, A.M.; Jensen, K.; Koponen, I.K.; Brunborg, G.; Gützkow, K.B.; et al. Nanotitanium dioxide toxicity in mouse lung is reduced in sanding dust from paint. *Part. Fibre Toxicol.* **2012**, *9*, 4. [CrossRef] [PubMed]

195. Cox, A.; Venkatachalam, P.; Sahi, S.; Sharma, N. Silver and titanium dioxide nanoparticle toxicity in plants: A review of current research. *Plant Physiol. Biochem.* **2016**, *107*, 147–163. [CrossRef] [PubMed]

196. Shakeel, M.; Jabeen, F.; Shabbir, S.; Asghar, M.S.; Khan, M.S.; Chaudhry, A.S. Toxicity of Nano-Titanium Dioxide (TiO$_2$-NP) Through Various Routes of Exposure: A Review. *Biol. Trace Element Res.* **2016**, *172*, 1–36. [CrossRef]

197. Hilal, S.M.; Mohmed, A.S.; Barry, N.M.; Ibrahim, M.H. Entomotoxicity of TiO$_2$ and ZnO Nanoparticles Against Adults Tribolium Castaneum (Herbest) (Coleoptera: Tenebrionidae). *IOP Conf. Ser. Earth Environ. Sci.* **2021**, *910*, 012088. [CrossRef]

198. Amara, S.; Ben Slama, I.; Omri, K.; EL Ghoul, J.; EL Mir, L.; Ben Rhouma, K.; Abdelmelek, H.; Sakly, M. Effects of nanoparticle zinc oxide on emotional behavior and trace elements homeostasis in rat brain. *Toxicol. Ind. Health* **2013**, *31*, 1202–1209. [CrossRef]

199. Shrivastava, R.; Raza, S.; Yadav, A.; Kushwaha, P.; Flora, S.J.S. Effects of sub-acute exposure to TiO$_2$, ZnO and Al$_2$O$_3$ nanoparticles on oxidative stress and histological changes in mouse liver and brain. *Drug Chem. Toxicol.* **2014**, *37*, 336–347. [CrossRef]

200. Liu, J.; Kang, Y.; Yin, S.; Song, B.; Wei, L.; Chen, L.; Shao, L. Zinc oxide nanoparticles induce toxic responses in human neuroblastoma SHSY5Y cells in a size-dependent manner. *Int. J. Nanomed.* **2017**, *12*, 8085–8099. [CrossRef]

201. Chen, T.-H.; Lin, C.-C.; Meng, P.-J. Zinc oxide nanoparticles alter hatching and larval locomotor activity in zebrafish (*Danio rerio*). *J. Hazard. Mater.* **2014**, *277*, 134–140. [CrossRef]

202. Mansouri, E.; Khorsandi, L.; Orazizadeh, M.; Jozi, Z. Dose-dependent hepatotoxity effects of Zinc oxide nanoparticles. *Nanomed. J.* **2015**, *2*, 273–282. [CrossRef]

203. Rajput, V.D.; Minkina, T.M.; Behal, A.; Sushkova, S.N.; Mandzhieva, S.; Singh, R.; Gorovtsov, A.; Tsitsuashvili, V.S.; Purvis, W.O.; Ghazaryan, K.A.; et al. Effects of zinc-oxide nanoparticles on soil, plants, animals and soil organisms: A review. *Environ. Nanotechnol. Monit. Manag.* **2018**, *9*, 76–84. [CrossRef]

204. Tulinska, J.; Mikusova, M.L.; Liskova, A.; Busova, M.; Masanova, V.; Uhnakova, I.; Rollerova, E.; Alacova, R.; Krivosikova, Z.; Wsolova, L.; et al. Copper Oxide Nanoparticles Stimulate the Immune Response and Decrease Antioxidant Defense in Mice After Six-Week Inhalation. *Front. Immunol.* **2022**, *13*, 874253. [CrossRef]

205. He, H.; Zou, Z.; Wang, B.; Xu, G.; Chen, C.; Qin, X.; Yu, C.; Zhang, J. Copper Oxide Nanoparticles Induce Oxidative DNA Damage and Cell Death via Copper Ion-Mediated P38 MAPK Activation in Vascular Endothelial Cells. *Int. J. Nanomed.* **2020**, *15*, 3291–3302. [CrossRef]

206. Ameh, T.; Sayes, C.M. The potential exposure and hazards of copper nanoparticles: A review. *Environ. Toxicol. Pharmacol.* **2019**, *71*, 103220. [CrossRef]

207. Adefarati, T.; Bansal, R. Integration of renewable distributed generators into the distribution system: A review. *IET Renew. Power Gener.* **2016**, *10*, 873–884. [CrossRef]

208. Bugata, L.S.P.; Venkata, P.P.; Gundu, A.R.; Fazlur, R.M.; Reddy, U.A.; Kumar, J.M.; Mekala, V.R.; Bojja, S.; Mahboob, M. Acute and subacute oral toxicity of copper oxide nanoparticles in female albino Wistar rats. *J. Appl. Toxicol.* **2019**, *39*, 702–716. [CrossRef] [PubMed]

209. Abdel-Azeem, A.M.; Abdel-Rehiem, E.S.; Farghali, A.A.; Khidr, F.K.; Abdul-Hamid, M. Ameliorative role of nanocurcumin against the toxicological effects of novel forms of Cuo as nanopesticides: A comparative study. *Environ. Sci. Pollut. Res.* **2023**, *30*, 26270–26291. [CrossRef]

210. De Jong, W.H.; De Rijk, E.; Bonetto, A.; Wohlleben, W.; Stone, V.; Brunelli, A.; Badetti, E.; Marcomini, A.; Gosens, I.; Cassee, F.R. Toxicity of copper oxide and basic copper carbonate nanoparticles after short-term oral exposure in rats. *Nanotoxicology* **2018**, *13*, 50–72. [CrossRef]

211. Gomes, S.I.; Murphy, M.; Nielsen, M.T.; Kristiansen, S.M.; Amorim, M.J.; Scott-Fordsmand, J.J. Cu-nanoparticles ecotoxicity— Explored and explained? *Chemosphere* **2015**, *139*, 240–245. [CrossRef]

212. Carmona, E.R.; Inostroza-Blancheteau, C.; Obando, V.; Rubio, L.; Marcos, R. Genotoxicity of copper oxide nanoparticles in Drosophila melanogaster. *Mutat. Res. Genet. Toxicol. Environ. Mutagen.* **2015**, *791*, 1–11. [CrossRef]

213. Gosens, I.; Cassee, F.R.; Zanella, M.; Manodori, L.; Brunelli, A.; Costa, A.L.; Bokkers, B.G.H.; De Jong, W.H.; Brown, D.; Hristozov, D.; et al. Organ burden and pulmonary toxicity of nano-sized copper (II) oxide particles after short-term inhalation exposure. *Nanotoxicology* **2016**, *10*, 1084–1095. [CrossRef]

214. Liu, J.; Dhungana, B.; Cobb, G.P. Environmental behavior, potential phytotoxicity, and accumulation of copper oxide nanoparticles and arsenic in rice plants: Phytotoxity of copper oxide nanoparticles and arsenic. *Environ. Toxicol. Chem.* **2017**, *37*, 11–20. [CrossRef]

215. Dizaj, S.M.; Lotfipour, F.; Barzegar-Jalali, M.; Zarrintan, M.H.; Adibkia, K. Antimicrobial activity of the metals and metal oxide nanoparticles. *Mater. Sci. Eng. C* **2014**, *44*, 278–284. [CrossRef]

216. Rempel, S.; Ogliari, A.J.; Bonfim, E.; Duarte, G.W.; Riella, H.G.; Silva, L.L.; Mello, J.M.M.; Baretta, C.R.D.M.; Fiori, M.A. Toxicity effects of magnesium oxide nanoparticles: A brief report. *Matéria* **2020**, *25*, e-1287. [CrossRef]

217. Mangalampalli, B.; Dumala, N.; Grover, P. Acute oral toxicity study of magnesium oxide nanoparticles and microparticles in female albino Wistar rats. *Regul. Toxicol. Pharmacol.* **2017**, *90*, 170–184. [CrossRef] [PubMed]

218. Wang, Z.-L.; Zhang, X.; Fan, G.-J.; Que, Y.; Xue, F.; Liu, Y.-H. Toxicity Effects and Mechanisms of MgO Nanoparticles on the Oomycete Pathogen *Phytophthora infestans* and Its Host *Solanum tuberosum*. *Toxics* **2022**, *10*, 553. [CrossRef] [PubMed]

219. Mazaheri, N.; Naghsh, N.; Karimi, A.; Salavati, H. In vivo Toxicity Investigation of Magnesium Oxide Nanoparticles in Rat for Environmental and Biomedical Applications. *Iran. J. Biotechnol.* **2019**, *17*, e1543. [CrossRef] [PubMed]

220. Vadiraj, K.T.; Shivaraju, H.P. Metal Oxide-Based Nanomaterials for the Treatment of Industrial Dyes and Colorants. In *Advanced Oxidation Processes in Dye-Containing Wastewater*; Springer Nature: Singapore, 2022; pp. 233–251. [CrossRef]
221. Cruz-Yusta, M.; Sánchez, M.; Sánchez, L. Metal Oxide Nanomaterials for Nitrogen Oxides Removal in Urban Environments. In *Tailored Functional Oxide Nanomaterials: From Design to Multi-Purpose Applications*; Wiley: Hoboken, NJ, USA, 2022; pp. 229–276. [CrossRef]
222. Labanni, A.; Nasir, M.; Arief, S. Research progress and prospect of copper oxide nanoparticles with controllable nanostructure, morphology, and function via green synthesis. *Mater. Today Sustain.* **2023**, *24*, 100526. [CrossRef]
223. Farani, M.R.; Farsadrooh, M.; Zare, I.; Gholami, A.; Akhavan, O. Green Synthesis of Magnesium Oxide Nanoparticles and Nanocomposites for Photocatalytic Antimicrobial, Antibiofilm and Antifungal Applications. *Catalysts* **2023**, *13*, 642. [CrossRef]

Review

Innovative Methodologies for the Conservation of Cultural Heritage against Biodeterioration: A Review

Martina Cirone [1,2], Alberto Figoli [2], Francesco Galiano [2], Mauro Francesco La Russa [1], Andrea Macchia [1], Raffaella Mancuso [3], Michela Ricca [1], Natalia Rovella [2,*], Maria Taverniti [4] and Silvestro Antonio Ruffolo [1]

1   Department of Biology, Ecology and Earth Sciences, University of Calabria, Via P. Bucci Cubo 12 B, 87036 Arcavacata di Rende, Italy; martina.cirone@unical.it (M.C.); mauro.larussa@unical.it (M.F.L.R.); andrea.macchia@unical.it (A.M.); michela.ricca@unical.it (M.R.); silvestro.ruffolo@unical.it (S.A.R.)
2   Institute on Membrane Technology, CNR-ITM, Via P. Bucci 17/C, CS, 87036 Rende, Italy; a.figoli@itm.cnr.it (A.F.); f.galiano@itm.cnr.it (F.G.)
3   Laboratory of Industrial and Synthetic Organic Chemistry (LISOC), Department of Chemistry and Chemical Technologies, University of Calabria, Via Pietro Bucci 12/C, CS, 87036 Arcavacata di Rende, Italy; raffaella.mancuso@unical.it
4   Institute of Informatics and Telematics, CNR-IIT, Via P. Bucci 17/C, CS, 87036 Rende, Italy; maria.taverniti@cnr.it
*   Correspondence: n.rovella@itm.cnr.it

Abstract: The use of traditional biocidal products in cultural heritage has suffered a slowdown due to the risks related to human health and the environment. Thus, many studies have been carried out with the aim of testing innovative and environmentally friendly alternatives. In this framework, this review attempts to provide an overview of some novel potential products with biocidal action, tested to counteract the process of degradation of paper and stone materials due to microbial activity, keeping in mind the sustainability criteria. In particular, we have focused our attention on the testing of nanotechnologies, essential oils, DES (deep eutectic solvents) with low toxicity, and colloidal substances for conservation purposes.

Keywords: biodegradation; biocide; paper restoration; stone restoration; essential oils; DES; polymers; nanotechnologies

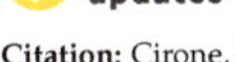

Citation: Cirone, M.; Figoli, A.; Galiano, F.; La Russa, M.F.; Macchia, A.; Mancuso, R.; Ricca, M.; Rovella, N.; Taverniti, M.; Ruffolo, S.A. Innovative Methodologies for the Conservation of Cultural Heritage against Biodeterioration: A Review. *Coatings* **2023**, *13*, 1986. https://doi.org/10.3390/coatings13121986

Academic Editor: Qiao Dong

Received: 11 October 2023
Revised: 16 November 2023
Accepted: 20 November 2023
Published: 22 November 2023

1. Introduction

The cultural heritage is largely affected by biodeterioration, a phenomenon manifested commonly by microbial growth on surfaces causing aesthetic, physical, and chemical damage to a cultural object [1]. The prevention and the control of the diffusion of biodeteriogens represent a global necessity in the conservation of the cultural heritage [2]. The intensity of the biodeterioration processes depends on several factors, such as the environmental conditions to which the materials are exposed and the biological species.

Biodeteriogens consist of microorganisms [3,4], such as bacteria and fungi, capable of using a substrate to support their growth and reproduction [5]; for this reason, environmental factors such as the presence of water, temperature, natural or artificial light radiation, and the characteristics of the material, such as porosity, represent crucial aspects to be considered. These biofilms can cause both mechanical and chemical transformations, respectively, due to the microdecohesions and metabolic processes of microorganisms, as well as their secretions, combined with air pollutants, which accumulate in biofilms and corrode the substrate [6]. These effects can occur both on the surfaces and on the inner areas of the material, producing its long-term decay and loss of strength and durability over time [7]. Every material can be affected more or less intensely; among these, stone and cellulosic are particularly sensitive. In museum, archive, or library environments, paper can be colonized by numerous microorganisms that can cause serious damage to valuable documents. Therefore, the preservation of historical paper-based artifacts against deterioration due to the presence of bacteria and fungi is important in order to protect the cultural heritage [8]. Similarly, the stones used in the monuments or in decorative elements represent an ideal environment for the colonization of a wide variety of microorganisms, and some physical properties, such as porosity and surface roughness, make the stone susceptible to biological colonization, which can lead to aesthetic and/or physical and chemical damage [9].

Therefore, restoration interventions are necessary in order to hinder the development of biodeteriogens. Mainly, the techniques include both physical and chemical methods [6], and the latter ones will be described and discussed in this review. Even though traditional procedures with biocidal products act on microorganisms efficiently, they are harmful to the environment and the operator and therefore do not represent a sustainable alternative [10]. In fact, most biocidal products containing compounds such as acetone, ethanol, toluene, ethyl acetate, quaternary ammonium salts, etc., are toxic substances or pollutants that do not degrade easily and can remain in the environment for a long time [11], hence the need to develop innovative, sustainable, and safe products [12,13].

When it comes to biocides, reference is made to many chemicals that can kill unwanted organisms [2]; these products inhibit the action of microorganisms by the presence of certain bioactive substances. The term biocidal product does not always have a negative connotation; its dangerousness depends on the circumstances of exposure and the levels of toxicity. However, the extensive use and repeated applications of biocides could lead to the evolution of other microorganisms and can create resistant and harmful species on cultural heritage objects [14]. In this regard, an interesting challenge that also represents a gap in this topic is the long-term monitoring of the evolution of the biofilms and the microorganisms at substrate/air interface after the application of biocidal products. In fact, sometimes, the interactions between organisms, the environment, and products can provide unpredictable effects such as the formation of new secondary biofilms, which is a fact worth investigating. In addition, it is crucial to not underestimate and to investigate biodeterioration due to the indirect action of compounds originating from the previous activity of the microorganisms.

Some biocides may produce undesirable effects that cause physical damage, such as color change, structure and permeability damage, or even chemical damage such as mineral solubilization and pH change [15,16]. Studies for the development of natural products aim to create alternatives that will, over time, replace conventional biocides, which are respectful of the environment and the material while, at the same time, being economical and effective [2,14].

Thus, this review highlights the complexity of the biodeterioration issue, account for the biodiversity, the different bioreceptivities, and adaptability of microorganisms to environmental changes. Moreover, it is worth noting the capacity of the microorganisms to develop resistance versus traditional biocidal materials; this requires the definition of ever more efficient strategies able to mitigate the effects of biodeterioration while respecting the environment, the health of the operators, and, obviously, the integrity of the artwork. For this reason, the review means to identify the status of progress in conservation practices applied to cellulosic and stone materials, comparing traditional techniques with the most recent products, clarifying the different performances, and identifying these new solutions as real, sustainable, and efficient alternatives.

Microbial agents on paper can cause damage due to chromatic changes but also due to structural alterations to both basic and additive components due to hydrolysis and oxidation phenomena, while, among the forms of degradation that affect stone materials, there are blemishes, patinas, pitting, crusts, biological patina, etc.

In what follows, alternative and ecological approaches will be presented, such as the use of nanotechnologies, essential oils, deep eutectic solvents (DES), and colloidal substances.

2. Cellulosic Materials: Paper

Paper consists mainly of cellulose, a naturally occurring polysaccharide. Cellulose is represented by its empirical formula, $(C_6H_{10}O_5)_n$, where "n" indicates the degree of polymerization, indicating the number of basic units (monomers) that make up its structure [17,18].

The degree of polymerization (DP) constitutes a primary parameter that reflects the number of glucose monomeric units joined together in the structure of cellulose. To determine the degree of polymerization of paper samples, viscosity measurements are employed. These measurements involve dissolving the samples in a 50% CED (cupryleethylenediamine) solution, following the international standard ASTM 04 243 from 1999 [19]. The calculation of the degree of polymerization is based on the empirical equation $DP = K[\eta]^{\alpha}$, where η represents intrinsic viscosity, K is equal to 1.5, and α is equal to 1 [20]. This value can vary depending on the plant origin of cellulose (Table 1) and has a significant impact on the overall strength of the cellulose molecule. Non-uniformity in paper quality is observed due to discrepancies in DP values. For instance, spruce wood exhibits cellulose with an approximate DP of 600, which corresponds to a less resistant fiber compared to hemp or flax, which have a DP exceeding 2000. The DP value is frequently employed to define the characteristics of a paper sample and evaluate the preservation degree of ancient documents. In such contexts, the average DP is considered, representing the measure of the mean degree of polymerization of all cellulose molecules present in the analyzed sample. In general, high-quality paper tends to have an average DP around 1000, while extremely fragile or deteriorated paper may exhibit an average DP lower than 100. A significant decline in the average DP value denotes a series of ruptures in cellulose chains, indicating severe deterioration of the paper material [21].

Table 1. Degree of polymerization of the different materials from which cellulose is obtained [21].

Origin of the Paper	Degree of Polymerization (DP)
Fir	600
Pine	650
Straw	1600
Raw Cotton	2100
Jute	2200
Hemp	2250
Linen	2400

Cellulose comprises numerous glucose molecules (Figure 1) that bond together to create macromolecules organized in a linear chain. These macromolecules then cluster via hydrogen bonds, giving rise to individual microfibrils. The assembly of multiple microfibrils leads to the formation of fibrils, and, in the end, these fibrils unite to constitute fibers [22].

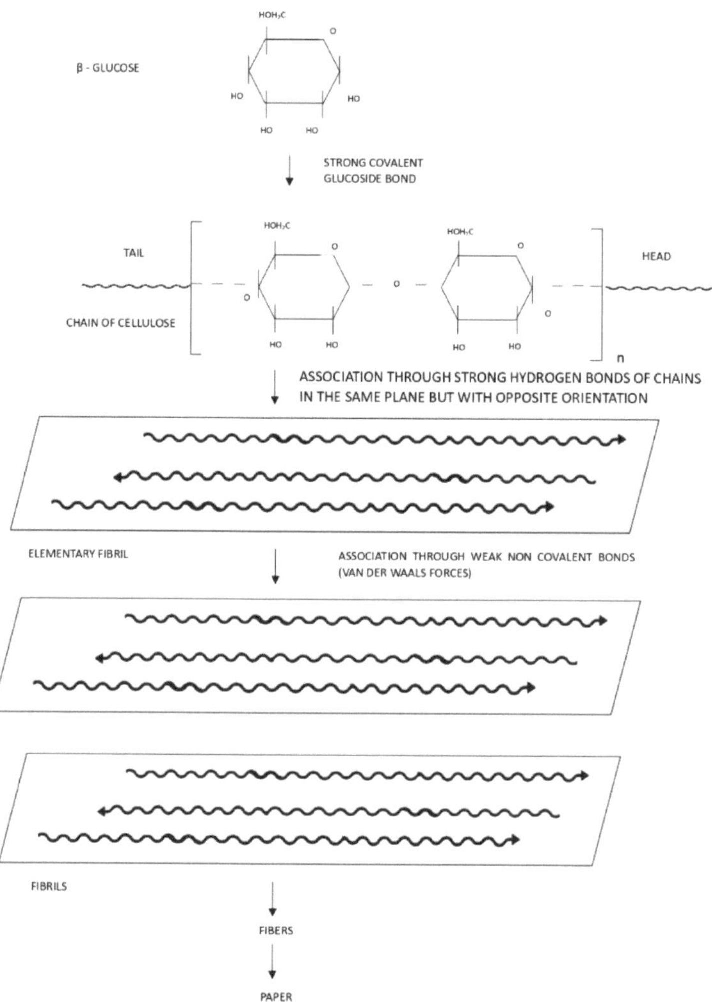

Figure 1. Representation of the hierarchical levels leading from the glucose molecule to the paper material (modified from [23]).

Cellulose is predominantly sourced from various tree species, including poplar, eucalyptus, pine, red fir, beech, and others. For centuries, other materials, such as cotton rags, hemp, and flax, have also been traditional sources for cellulose pulp production. However, wood serves as the primary raw material for cellulose production, requiring treatment in order to separate cellulose fibers from other components, such as lignin, which is an intricate three-dimensional polymer composed of aromatic alcohols connected by various carbon–carbon bonds—the molecule is notably resilient, with only a select few fungi and bacteria possessing the capability to break it down—and hemicelluloses, encompassing pectic substances and diverse heteropolysaccharides, which are polysaccharides with limited solubility that are closely linked to cellulose. Exoenzymes known as hemicellulases, produced by most bacteria and fungi, facilitate their hydrolysis [24].

Aside from cellulose, lignin, and hemicellulose, paper includes additional elements tailored to bestow specific properties essential for applications. These supplementary components encompass binders, introduced to restrict ink absorption by the paper; mineral fillers, usually white, functioning as pore fillers to yield a smoother, whiter surface, while also acting as opacifiers; and dyes, incorporated to ensure even distribution of color throughout the paper's thickness [25].

2.1. Paper Biodeterioration

Museums constitute the main institutions providing knowledge and cultural, political, economic, scientific, and historical information within communities. In the context of a museum, library, or archival environment, the presence of pests can lead to significant damage to materials of invaluable and irreplaceable value. Paper, like other materials, undergoes a degradation process over time [26]. However, the rate of deterioration can be accelerated by various factors, both internal, such as acidity, the presence of metal ions, lignin, or degradation products, and external, such as heat, humidity, ultraviolet light, oxygen, pollutants, or biodeteriogens. Biodeteriogens are organisms characterized by their ability to use a substrate as a source of sustenance for their growth and reproduction [5], and biodeterioration is defined as any undesired form of alteration of a material caused by the vital activities of such organisms. This phenomenon occurs as the microclimatic conditions become favorable for the proliferation of microorganisms that act chemically, physically, or mechanically [27]. Such attacks pose a formidable obstacle to the preservation of Cultural Heritage. To prevent the progressive destruction of our Cultural Heritage, restorers, conservators, and researchers have developed various methodologies to combat chemical and biological enemies. In the case of a biological attack, it is possible to act in two ways: by avoiding contact between aggressive bioagents and the objects to be preserved, or by eliminating such bioagents [28].

2.1.1. Factors Leading to the Biodeterioration of Paper

Paper artifacts primarily consist of an organic matrix that provides a favorable environment for the proliferation of biological agents responsible for their degradation. In fact, the degradation of paper directly affects the structure of its fibers, weakening the cellulose chains. In particular, microorganisms such as fungi, which are the main biodeteriogens of paper, use cellulose as a nutrient source, transforming it into simpler and easily assimilable molecules [29]. Among the degradation agents, fungi represent the most common and widespread group, as many species demonstrate a high capacity for adaptation to different environmental conditions they encounter [25].

Over two thousand fungal species are known to cause damage to both wood and cellulose, primarily through the physical penetration of their hyphae and the production of primary and secondary metabolites [30–32]. These metabolites include fungal pigments and various extracellular hydrolytic enzymes such as cellulases, amylases, and proteases; these enzymes—in particular, the cellulases—make the depolymerization of the cellulose possible, and the microorganism will be able to use the glucose monomer as nutrient. [31]. The main microorganisms responsible for biodeterioration encompass fungi and bacteria. Common species encountered in this context include the following: *Alternaria, Bacillus, Chaetomium globosum, Trichoderma viride, Penicillium chrysogenum, Cladosporium herbarum, Aspergillus niger, Stachybotrys atra, Trichoderma koningii, Chaetomium elatum Pseudomonas, Staphylococcus, Micromonospora,* and *Virgibacillus* [32,33].

Once fungal colonization takes place, the degraded cellulose material provides an enriched substrate for bacterial growth [34]. It has been reported that various bacterial species colonize and damage paper, but bacteria of the genus Bacillus are considered to be among the main causes of paper deterioration [35]. However, it is important to note that these bacteria may be present on the surfaces of paper documents as contaminants from dust and may not play a significant role as causal agents of degradation [36].

These agents are primarily heterotrophic organisms that feed on organic substances containing carbon and nitrogen, and they mostly require oxygen for their aerobic metabolism. Therefore, such biological agents initiate the process of the biodeterioration of paper materials, which involves a combination of physical and chemical processes triggered by the growth of organisms that, mainly transported through the air, settle on the surfaces of art objects, causing their alteration [25].

In addition to microorganisms, insects also contribute to the process of the biodeterioration of paper. Regarding insects that pose a threat to paper conservation, it is important to mention some common families of beetles, such as silverfish and wood-boring beetles, as well as those that develop and live within paper materials, such as termites.

Annually, a significant number of volumes are subject to attacks by insects, microorganisms, and, to a lesser extent, rodents. These attacks exhibit variations in terms of characteristics and magnitudes across different geographical areas, in relation to the specific climatic conditions of each area, the present microflora, and the architectural peculiarities of the buildings housing library collections, which do not always adhere to the standards considered essential by the scientific community for conservation. Biological alterations are largely favored by inadequate environmental conditions and, when they occur, often involve entire sections of libraries; furthermore, they can be exacerbated by extraordinary events such as floods, earthquakes, and collapses, assuming even more significant proportions in such circumstances.

In addition to the presence of biological deposits on a substrate, biodeterioration phenomena are influenced by a range of complex factors, including environmental and climatic conditions that promote the growth of biological organisms. Within libraries, biodeterioration processes are favored by conditions such as high temperatures, elevated humidity levels, and inadequate ventilation.

To ensure the optimal preservation of paper, it is crucial to maintain specific parameters of temperature and relative humidity. The ideal range for paper conservation is generally considered to be between 16 °C and 20 °C for temperature and between 40% and 60% for relative humidity. These values aim to provide a stable and controlled environment that minimizes the risks of biodeterioration and effectively preserves the paper material.

Paper, being a hygroscopic material, is susceptible to the influence of moisture content, which can promote the development of microorganisms and insects. Like all hygroscopic substances, paper can absorb and release water in both liquid and vapor states. When the water content exceeds 10%, favorable conditions for spore germination can occur, especially in environments with temperatures around 20 °C and relative humidity of 80%. To prevent the proliferation of microorganisms, it is essential to maintain the temperature below 18 °C and the relative humidity below 65% (Table 2). However, these values are not sufficient to prevent the survival of certain insect species, which can thrive even in environments characterized by low relative humidity (40%–60%) and very low temperatures (below 0 °C). Therefore, to effectively counter the risk of insect infestation, additional preventive measures need to be implemented, in addition to humidity and temperature control, in order to preserve the integrity of the paper material [29].

Table 2. Components of the microclimate: the lower the thermal values, the higher the hygrometric values at which microorganisms develop [29].

Fungal Species	Temperature	Minimum Relative Humidity for Microorganism Development
Penicillium Chrysogenum	10 °C	83.5%
	15 °C	77%
	25 °C	72.5%
Aspergillus Flavus	12 °C	95%
	16 °C	90%
	30 °C	81%
Aspergillus Tamarii	15 °C	90%
	20 °C	85%
	30 °C	79%

Bacteria can have deteriorating effects on paper; however, due to their relatively higher moisture requirements, it is more likely for fungi to proliferate in environments such as libraries, archives, and museums. The environmental conditions typically found in such settings are more conducive to fungal growth compared to bacterial growth [26]. The organic components of books and archival documents readily absorb water, making these materials susceptible to microbial attacks. Consequently, the biodeterioration of paper is primarily attributed to the action of microorganisms that exploit its constituent components, such as cellulose, hemicellulose, lignin, etc., and transform them into simpler and easily assimilable molecules as a source of nutrients.

Another significant factor is the pH value, which plays a crucial role in the chemical stability of paper supports. pH influences the hydrolysis processes that lead to the breaking of bonds within cellulose, thereby affecting the fragility of the paper. Acidic treatments can compromise the polymeric structure of cellulose, and the rate of such deterioration depends significantly on the pH value. Microorganisms can adapt to pH values ranging from 3 to 9, although the intensity of their metabolic activity is most pronounced within a narrow range that varies from species to species. For most fungi, this range falls between 5 and 7 [29].

2.1.2. Consequences of Paper Biodeterioration

Several environmental and biological factors have been identified as responsible for the degradation process affecting archival materials, manifesting through phenomena such as chromatic alterations, discoloration, and loss of paper structure [37,38].

As previously mentioned, fungal attack is a common occurrence, although the damage caused by microorganisms often combines different forms of deterioration. These damages, which can vary in terms of appearance and severity, include chromatic alterations (known as "foxing") resulting from the formation of stains with various shapes and colors, structural alterations of the underlying components of the materials leading to material fragility and potential destruction, and alterations to essential additive components such as binders, pigments, and inks. Although these components are not the primary elements of the material, their deterioration leads to the loss of support characteristics.

The damage inflicted on paper by microbial agents can manifest in various forms and severity, primarily attributable to the following:

- Chromatic modifications resulting from pigments and exudates, associated with cellular structures such as fungal mycelium and spores, giving rise to the appearance of stains with a wide range of chromatic characteristics, shapes, and sizes.
- Structural modifications of the fundamental components of the materials, induced by specific enzymes such as cellulases, proteases, lipases, etc., are evident through the intrinsic fragility of the materials themselves, which can lead to their complete disintegration. This type of damage tends to occur in advanced stages of infection and is widely recognized as the most devastating.
- Additionally, modifications occur in the crucial additive components of the materials, as in certain cases, microorganisms proliferate by utilizing specific substances that, although present in the material, do not constitute its main element (such as adhesives, plasticizers, antioxidants, etc.). This phenomenon leads to the loss of the substrate's unique characteristics, sometimes rendering it unusable [25,29].

Therefore, it is of paramount importance to prioritize prevention and mitigation of degradation, as well as the risk of infections and infestations, using specifically designed biocidal products for paper conservation. These products are developed with the aim of controlling microbiological deterioration and countering the deposition of microorganisms on the surfaces of materials without causing lasting damage to the paper and paper supports. Furthermore, it is essential that such biocides are safe for the environment and for humans [39]. Moreover, it has to be taken into account that exposure to microorganisms, fungal spores, and their related metabolites can cause significant adverse effects on the health of people working in libraries, as well as that of visitors, particularly in terms of skin and respiratory system [40].

2.2. Innovative Methods for Paper Preservation

Various physical and chemical strategies have been employed for paper conservation, such as gamma radiation and the use of chemical compounds, such as orthophenylphenol, calcium propionate, ethanol, ethylene oxide, formaldehyde, etc. However, some methods have several limitations, including temporary effects, high costs, and the utilization of toxic chemicals. For example, ethylene oxide, which, in addition to causing an increase in the susceptibility of objects to future microbial attacks, is classified as carcinogenic; formaldehyde, which, at low relative humidity, undergoes polymerization and precipitates on materials, thus having a low penetration power (as well as is being carcinogenic); and pentachlorophenol, which is highly toxic and carcinogenic [26]. Consequently, there is an urgent need to develop an alternative sustainable and environmentally friendly strategy (Table 3) that achieves long-term efficiency, cost-effectiveness, and safety [41].

Table 3. Summary of innovative methods for paper conservation.

Innovative Method/Activity	Reference Work
Innovative deacidification material/procedure	[8] [42]
Supercritical carbon dioxide (SCCO$_2$)	[43]
Microbial inhibition	[41] [44]
Bait for insects/parasites	[45,46]
Use of DES	[47–51]

The objective of conservation is to mitigate degradation processes and extend the lifespan of the artifact. Numerous strategies are available for paper conservation, targeting various aspects of deterioration, and which are further tailored according to specific objectives and implementation constraints. Paper conservation activities are classified into distinct categories, which can be considered as approximate representations of the generic phases involved in a paper conservation treatment: intervention preparation, pest control and disinfection/sterilization, surface dry cleaning, wet washing, deacidification, paper repair, and consolidation/strengthening. Documentation of the treatment is concurrently carried out throughout all these stages [52].

2.2.1. Technologies against Fungal and Bacterial Attacks

The overwhelming majority of antifungal strategies employed to prevent and/or counteract fungal biodeterioration in the preservation of paper-based artifacts can range from restricting fungal access to water to the application of chemical agents in gaseous or liquid form, as well as the use of physical methods such as extreme temperatures, radiation, or currents [26]. A frequently employed physical method is represented by gamma irradiation. However, gamma irradiation is responsible for a reduction in the mechanical strength of paper, as well as its acidification and subsequent yellowing. Additionally, it is important to note that chemical agents can also cause damage to paper; for example, the use of a 70% (v/v) aqueous ethanol solution can result in loss of gloss and damage to book parts [43].

In general, physical intervention methods exhibit long-term characteristics as their antimicrobial action is immediate and leaves no residues. Conversely, most chemical compounds, including those in gaseous form such as ethanol, leave residues that can prolong the antimicrobial effect for a limited period. A simple and harmless approach to inhibit fungal growth involves restricting access to water, thereby reducing water activity on the substrate. An appropriate antifungal method for materials should possess a broad spectrum of activity, good chemical stability, low cost, be non-toxic to humans, and have no negative effects on the treated material [26].

Among biocidal treatment research, a first potential tested product is represented by supercritical carbon dioxide ($SCCO_2$) in combination with an ethanol additive used for disinfecting ancient paper artifacts affected by fungal contamination [43]. Supercritical carbon dioxide ($SCCO_2$) refers to the physicochemical state assumed by CO_2 when confined in an environment characterized by temperature and pressure conditions exceeding the critical point [53]. This solution offers advantages such as eco-compatibility, cost-effectiveness, and wide availability, as well as non-flammability, without compromising the mechanical strength of cellulose fibers. Among the numerous fungi detected in a set of 294 samples of ancient documents affected by fungal growth, *Aspergillus niger* exhibited the highest frequency (36%), followed by *Aspergillus flavus* (20.7%), *Eurotium amstelodami* (15.5%), *Acremonium* spp. (14.9%), *Aspergillus versicolor* (4.4%), *Penicillium chrysogenum* (4.2%), *Cladosporium subuliforme* (3.4%), *Rhizopus* spp. (0.6%), and *Epicoccum nigrum* (0.3%). The results of this study indicate that fungal contamination of the samples was reduced through the application of supercritical carbon dioxide ($SCCO_2$) containing concentrations of 4% and 8% ethanol. The comparative analysis did not reveal significant differences in efficacy between the use of the two ethanol concentrations (4% and 8%) in combination with $SCCO_2$ [43]. Therefore, it was stated by the authors that this procedure could represent a highly promising option for the preservation and control of fungi in ancient documentary artifacts in the future.

In recent years, significant progress has been made in the application of nanomaterials as deacidification agents for the preservation of paper artifacts. The presence of bacteria and fungi on paper materials plays a crucial role in the acidification process. The extracellular secretions of these microorganisms contribute to increased acidity within the cellulose structure, ultimately leading to paper degradation. Therefore, to indirectly reinforce cellulose, the removal of bacterial and fungal colonies from the artifact can also be anticipated, preventing acidification. Considering this, the use of antimicrobial agents can be an effective strategy to prevent microbial growth. Additionally, nanomaterials can be employed to enhance antimicrobial activity. In relation to the latter, a study [8] involved the preparation of calcium/chitosan nanoparticles (Ca/CS NPs) to explore their potential as a novel approach for the preservation of paper documents. The aim of this study was to improve the efficacy of Ca/CS NPs for paper preservation. Tests were conducted on specific fungal and bacterial strains, including *Micrococcus luteus* (ATCC 15307), *Bacillus megaterium* (ATCC 14581), *Bacillus subtilis* (ATCC 6051), *Aspergillus niger* (CRM-16404), and *Aspergillus fumigatus* (MYA-4916). The synthesis of chitosan nanoparticles (CS NPs) was performed for the first time via dropwise addition of sodium tripolyphosphate (TPP) to the chitosan solution. By mixing calcium hydroxide with empty CS NPs, Ca/CS NPs were also obtained. Subsequently, the Ca/CS NPs were applied to the paper samples using a spray method. The spraying process was conducted at room temperature and at 20 cm from the paper samples.

The deacidification effect of the nanoparticles was studied through periodic pH measurements. The initial pH level of the paper samples was measured as pH = 4.71, and after spraying the Ca/CS NPs onto the samples, the pH level increased to pH = 6.17. This pH level was maintained for nearly 10 days and then gradually decreased in the following days. The pH level increase and stability were attributed to the removal of microorganisms facilitated by the antimicrobial effect of the nanoparticles. It can be inferred that the incorporation of calcium with CS NPs enhances antifungal activity. SEM images of the Ca/CS NPs on the paper samples revealed their presence both on the surface and interfaces, forming a protective coating against microorganisms.

The Ca/CS NPs exhibit superior antimicrobial effects against specific strains of bacteria and fungi commonly found on paper documents compared to empty CS NPs. Therefore, these nanoparticles showed increased antifungal and antibacterial potency with the addition of calcium, resulting in a stronger inhibitory effect on specific gram-positive bacteria such as *Bacillus megaterium* and *Bacillus subtilis*. Furthermore, this antimicrobial activity contributed to the pH stability of paper-based artifacts. Based on the results, it can be deduced that these nanoparticles have the potential to be preferred as conservation materials for paper-based artifacts.

In the field of nanotechnology, an assessment was conducted on the preservative effect of paper models treated with biosynthesized zinc oxide (ZnO) and silver (Ag_2O) nanoparticles (NPs) against strains of *Bacillus subtilis* (B3) and *Penicillium chrysogenum* (F9) [41]. Isolation, identification, and characterization of microorganisms involved in the deterioration of archaeological manuscripts dating back to the 17th century (1677 A.D.) were performed, leading to the identification of 11 bacterial species and 15 fungal species. Among the obtained bacterial and fungal strains, B3 and F9 strains showed the highest cellulolytic activity, and their antimicrobial effect was evaluated using ZnO and Ag_2O NPs.

Scanning electron microscopy (SEM) analysis revealed that a reduced concentration of nanoparticles, used in the context of this study, exhibits a protective effect on the cellulose fibers against bacterial deterioration, through its inhibitory influence was evident on the growth pattern of the B3 strain but not on the F9 strain. On the other hand, higher concentrations of NPs could provide protection against fungal deterioration of the cellulose fibers. Additionally, observation through SEM confirmed the presence of fillers, added to enhance paper properties such as opacity, whiteness, brightness, and printability. The constituent elements and chemical characterization of the filler material in the investigated archaeological manuscript were analyzed using energy-dispersive X-ray spectroscopy (EDX). The highest percentage was found for carbon, followed by oxygen. Other trace elements included silicon and aluminum. The analyses confirmed the use of kaolinite as a filler material, providing a soft characteristic to the paper. Furthermore, it was observed that calcium carbonate was utilized as a filler material to increase the opacity of the manuscript.

The application of Ag_2O nanoparticles at concentrations of 1.0 or 2.0 mM demonstrated a remarkable preservation effect on the paper models, achieving complete microbial inhibition against bacteria and fungi, respectively. Additionally, the NP-treated paper models exhibited slight chromatic changes. Therefore, the deterioration of manuscripts caused by bacterial action can be effectively treated with a low concentration (1.0 mM) of NPs, while for fungal inhibition and the long-term protection of archaeological documents from microbial attack, the use of a high concentration of NPs (2.0 mM) is recommended.

The newly developed functionalized polyamidoamines (PAAs) have been introduced for the sustainable protection of wood and paper [42]. These eco-friendly polymers exhibit a broad-spectrum protective action and have biostatic/biocide properties against various organisms responsible for biotic degradation. PAAs are synthetic linear polymers containing amide and tertiary amine groups that are evenly distributed along the polymer chain. They can be easily incorporated into aqueous media within lignocellulosic artifacts, where they primarily exert a deacidifying effect against bacteria, fungi, molds, and insects.

The objectives of treating paper with PAAs include deacidification, followed by protection against biotic degradation and consolidation. Preliminary studies have been conducted by immersing paper samples in PAAOH, which contains alcoholic groups, and it is soluble in aqueous environments. The deacidification process has proven to be highly effective; raising the pH from 5.4 to 7.4. PAAOH demonstrates promising biostatic activity against some molds responsible for paper degradation.

The collected data unequivocally demonstrate that treatments with PAAs effectively improve both the dynamic–mechanical properties and strength. It can thus be stated that such a treatment emerges as a valid approach to safeguarding the paper support of aged manuscripts from weakening caused by chemical and physical aging phenomena.

Essential oils have also been employed as antifungal agents against the growth of certain fungi. Their antifungal potency has been observed with three natural oils extracted from the peels of *Citrus reticulata*, the leaves of *C. aurantifolia*, and *Linum usitatissimum* (flaxseed), which were used as antifungal agents against the growth of *Aspergillus flavus* and *Penicillium chrysogenum* [44].

To assess the effect of the oil on paper, a sample of historical manuscript paper was considered. The samples were placed in Petri dishes containing cotton saturated with these oils, with no direct contact between the oil and the paper. Therefore, the process occurred through oil sublimation without impregnation.

The oils exhibited varying levels of antifungal activity against the studied fungi (*Aspergillus flavus* and *Penicillium chrysogenum*). Generally, the inhibitory effect of the oils increased proportionally with concentration, with maximum inhibition achieved at a final concentration of 2000 µL/ml. Thermogravimetric analysis revealed that the paper samples exhibited an initial weight loss of 2.86% at approximately 105 °C, primarily attributed to the evaporation of any absorbed moisture.

The oils played a crucial role in reducing color fading values in the historical paper, with *Linum usitatissimum* (flaxseed oil) exhibiting the most effective protection against color fading. Thermogravimetric analysis further demonstrated that flaxseed oil outperformed the other two oils, exhibiting superior inhibitory activity on historical documents and producing positive effects on them. Additionally, flaxseed oil provided the best protection against color change.

2.2.2. Strategies against Insect Infestation

The preservation of collections in museums, libraries, and archives, which comprise organic artifacts such as wood, linen, wool, and others, faces a significant risk from a diverse array of harmful insects jeopardizing their long-term conservation [45]. Non-chemical methods offer effective elimination of insect pests in museum, library, and historic building settings, prioritizing object preservation, comprehensive eradication of insects at all stages, and ensuring the well-being of both the environment and museum staff [54]. Physical treatments, such as freezing [55,56], controlled heating [57,58], microwave radiation exposure, and gamma irradiation, are employed. However, it is crucial to highlight that not all materials can be suitably treated using these methodologies. For delicate objects, anoxic treatments [59,60] are preferred, creating a low-oxygen atmosphere through nitrogen or argon-based approaches.

There is no universally perfect treatment method, and the selection of the most appropriate approach should consider factors such as time constraints, financial resources, available means, and the specific type of pests and materials requiring treatment [48].

An investigation was conducted to explore and identify the biodiversity of harmful insects infesting the Manuscript Library of the Coptic Museum (Egypt) [45]. Over a period of one year, from October 2018 to September 2019, monthly sampling was carried out using adhesive traps with a non-toxic sticky substance. The adhesive traps were strategically placed on the floor surface (at the corners of the library), on window and door edges, and behind fabrics and manuscripts to capture insects walking on them.

In total, 1047 specimens belonging to nine species were identified and categorized into five orders (Hymenoptera, Coleoptera, Thysanura, Psocoptera, Blattodea) and six families (Formicidae, Ptinidae, Dermestidae, Lepismatidae, Liposcelidae, Blattidae) within the studied library context [45]. The most abundant and consistently present species throughout all months was Monomorium pharaonis. It was found that the presence of insects in the library was higher during the summer months, with a range of 108 to 222 individuals, compared to the winter months, when 4 to 23 individuals were trapped. This seasonal difference could be attributed to the hot and humid climate, which favors the proliferation of pests.

The parasites infesting paper materials include both pests associated with molds and high humidity, such as *Liposcelis bostrychophila* (Booklice), which feed on microscopic molds that develop on paper and starch-based glue used in binding, often indicating the presence of harmful moisture and mold issues, as well as generic pests such as *Thermobia domestica* (Silverfish) that can damage paper, bindings, starch-based glue, and other cellulose materials. Additionally, common pests such as cockroaches, ants, and other insects can attack and cause problems within museums.

In conclusion, the baitless adhesive traps used proved to be more effective in collecting a wide range of harmful insects within museums compared to baited ones. To prevent the presence of moisture-related pests, measures need to be taken to adjust climatic conditions and reduce the presence of dust, microscopic fungi, and other organic substances. It is crucial to promptly address infested artifacts present in the library to prevent further spreading and damage.

The *Ctenolepisma longicaudatum*, also known as the "long-tailed silverfish", represents an intradomestic parasitic agent of significant importance for libraries, archives, and museums. In a study [46], the impact of using an insecticidal gel bait containing indoxacarb as the active ingredient on populations of *C. longicaudatum* in three libraries, seven archives, and seven museums located in Norway and Austria was examined. To assess the effectiveness of bait application, the activity of the parasites was monitored using adhesive traps. This is a long-acting control strategy, which demonstrated a 90% reduction within a period of 3–4 months, requiring a minimal amount of bait to achieve complete eradication of the individuals. Significant decreases in parasite populations were observed in all sites where a few drops of bait were applied inside the buildings. In the context of this study, the maximum amount of bait used in a single location was 0.02 g per m^2.

Paper is universally recognized as a source of carbohydrates from which insects derive nourishment and can survive under abiotic conditions. However, to ensure their survival, *C. longicaudatum* also requires the intake of proteins, which are not present in the paper itself but can be obtained from dead insects, including those of the same species. It is important to note that these insects die because of ingesting baits containing sugars as stimulant agents, leading to primary poisoning. Once deceased, they become poisoned protein sources for other insects, thereby causing secondary poisoning.

Therefore, the use of bait has proven effective in preventing infestations within museums, representing an advantageous and low-risk approach that can be adopted as a conservation strategy in libraries, archives, and museums. The treatment cost, evaluated based on the amount of bait used and the hours of labor involved, has been found to be reasonable. The use of gel bait can provide a cost-effective alternative to reduce the presence of the parasite community; furthermore, the method has a limited impact on the health of operators and visitors in the treated areas, as well as on the objects present.

2.2.3. DES (Deep Eutectic Solvent)

In the field of cultural heritage, degradation processes are widespread, and in the context of restoration and preservation of such materials, there is a gradual shift towards a "green" chemistry approach that focuses on the use of non-toxic solvents and reagents. In this perspective, deep eutectic solvents (DES) have been proposed as an alternative to Ionic Liquids (IL) [61]. DES can be described as a mixture of acids and bases or as a mixture of hydrogen bond donors and acceptors (HBD and HBA, respectively). When employed in the treatment of lignocellulosic biomass, DES can also serve as a pre-treatment medium to enhance the fiber structure.

In order to facilitate the production of micro fibrillated cellulose, a study was conducted to investigate the effects of DES on induced fibrillation processes [47]. Cellulose derived from eucalyptus and cotton pulp was utilized to evaluate these effects. The study examined the influence of three different DES systems (acidic, neutral, and basic) on the chemical and physical characteristics, including internal and external fibrillation, of cellulose fibers obtained from the aforementioned materials. The primary objective of the research was to assess the capacity of these solvents, when used as pre-treatment, to induce fiber fibrillation and achieve micro fibrillation without introducing significant modifications to the cellulose structure.

The obtained results highlight that the treatment with acid and neutral DES has modified the chemical composition of eucalyptus fibers. This treatment induced both internal and external fibrillation without significantly altering the chemical composition, crystallinity, and polymerization degree of the fibers. Furthermore, a notable improvement in the mechanical properties of the produced papers was observed, attributed to an increased degree of adhesion and preservation of fiber length. Therefore, the DES treatment can be regarded as a gentle and environmentally friendly "chemical" refinement process that could be employed as a pre-treatment method for cellulose micro-fibrillation.

In a previous study, the utilization of microwave-assisted deep eutectic solvent (MV-DES) treatment, combined with ultrasound treatment, was employed for the preparation of lignin-containing cellulose nanofibers (LCNF) [48]. Subsequently, the effectiveness of LCNF was evaluated through the application of reinforcement agents and UV-absorbing substances on polyanionic cellulose (PAC) films.

The resulting LCNF exhibited an intricate matrix, which can be attributed to the agglomerating role played by lignin among the cellulose nanofibers. The incorporation of LCNF led to a significant improvement in the stability of the film-forming suspension of polyanionic cellulose (PAC). Therefore, this study provides an environmentally friendly methodology to produce biomass derived LCNF. The addition of 5% LCNF to PAC films demonstrated a high capacity for UV protection, suggesting the broad applicability of the MV-DES method in the preparation of such nanofibers. The stability of the LCNF/PAC composite film suspensions exhibited excellent mechanical properties and an adjustable UV protection capacity based on the quantity and type of LCNF used.

Another study presents a novel approach for the direct modification of cellulose using deep eutectic solvents, providing a sustainable strategy for the modification of other materials [49]. The surface of cellulose particles was subjected to carboxylation in a deep eutectic solvent containing carboxylic acid (oxalic acid), acting as a hydrogen bond donor (HBD), while choline chloride was employed as an acceptor (HBA). The effects of key factors on cellulose carboxylation were investigated based on the carboxyl content.

The results of the analysis revealed that the type of carboxylic acid employed, the size of cellulose particles, and the pre-treatment methods significantly impact the efficiency of cellulose carboxylation process. The carboxylation reaction did not induce any alteration in the internal crystalline structure of cellulose; however, it led to the destruction of the surface crystalline structure, resulting in a remarkable enhancement in the dispersibility and hydrophilic properties of the product. In order to assess the potential practical applications of carboxymethyl cellulose modified using DES, a systematic study was conducted to investigate its adsorption capacity using methylene blue. Cellulose with a highly fine particle size exhibits a high susceptibility to degradation and carbonization when heated in DES containing oxalic acid, resulting in a significant reduction in the yield of carboxymethyl cellulose. This study revealed that cellulose carboxylation is more efficient in the presence of carboxylic acids characterized by high acidity in the DES, as well as a lower molecular size.

Another study presented a novel approach for investigating solvents for cellulose [50], focusing on the application of three DES for cellulose dissolution. Three new solvents were synthesized from combinations of urea/caprolactam, caprolactam/acetamide, and urea/acetamide, which are organic solid compounds with respective molar ratios of 1:3, 1:1, and 1:2. Their physical characteristics, including melting point, conductivity, and solubility, were studied.

The melting point represents a fundamental parameter for evaluating the behavior of a deep eutectic solvent, and in this study, it was observed that this parameter depended on the molar ratio of the two compounds. Specifically, it was found that the solvent composed of urea/caprolactam with a molar ratio of 1:3 had a melting point of 30 °C, the caprolactam/acetamide solvent with a molar ratio of 1:1 had a melting point of 18 °C, while the urea/acetamide solvent with a molar ratio of 1:2 exhibited a melting point of 48 °C.

Furthermore, taking the caprolactam/urea deep eutectic solvent as an example, it was observed that during the heating of the mixing system, the intramolecular hydrogen bonding of each precursor compound was disrupted, leading to the formation of new intermolecular hydrogen bonds between one urea molecule and multiple caprolactam molecules. This reaction resulted in a decrease in various physical factors, such as hydrogen bond energy, molecular arrangement degree, and electrostatic attraction between hydrogen and the organic molecule.

These weak ionic interactions contributed to the reduction of the crystalline lattice energy of the salts, resulting in salts with a low melting point. The conductivity of the deep eutectic solvent synthesized with the optimal molar ratio was approximately in the range of 10^{-5}–10^{-4} S/m. Currently, one of the distinguishing properties of deep eutectic solvents compared to ionic liquids is their low conductivity, particularly for the deep eutectic solvent synthesized from urea and caprolactam, which was only approximately 3.2×10^{-5} S/m at 38 °C. At the same temperature, the conductivity of the three systems followed the following order: urea/caprolactam < caprolactam/acetamide < urea/acetamide. With increasing temperature, the conductivity increases.

It has been observed that the three solvents exhibit the ability to dissolve cellulose in the following order: urea/caprolactam > caprolactam/acetamide > urea/acetamide. The solubility of the deep eutectic solvent for cellulose is influenced by the quantity and intensity of hydrogen bonds formed with cellulose. In other words, a greater formation of hydrogen bonds with cellulose corresponds to a higher solubility of cellulose.

In conclusion, the three deep eutectic solvents exhibit a low melting point and low conductivity. Additionally, it has been found that the solubility of cellulose is limited. The studies have examined the application of DES on cellulose materials. However, studies have also been conducted on the use of these new molecules as preservation agents for stone materials as an alternative to traditional biocides that are toxic to the environment and operators.

A study was conducted on a mosaic located in the Archaeological Park of Ostia Antica to evaluate the effectiveness of five DES with different compositions containing choline chloride, ethylene glycol, malonic acid, glycerol, oxalic acid, and urea. These solvents were compared to the traditional biocide Preventol RL50 [51].

The DES employed in this research for the removal of biological patina are of pure nature and synthesized from natural precursors. Therefore, they are considered safe for both the environment and the operator, and they do not interact with the stone layer but exert solely antibacterial activity on the biofilm. It has been observed that DES 2 and 4, composed of malonic acid and oxalic acid, respectively, exhibit a lower pH and a higher biocidal effect, thereby inhibiting bacterial growth.

DES 5 (ChCl/U) has demonstrated significant efficacy at a slightly higher pH (7.2 ± 0.5). Subsequently, the effectiveness of DES was compared with that of Preventol RI50, and despite achieving positive performance, these new solvents do not reach the standards achieved by the traditional biocide. However, there is a substantial difference between the two products; in fact, DES exhibits stability, non-volatility, eco-compatibility, and does not require dilution with solvents as they can be used in their pure form as slightly viscous liquids at room temperature. The results obtained from this study highlight how the use of DES can represent a new environmentally friendly strategy for cleaning and preserving materials in the context of cultural heritage.

3. Stone Materials

Stone materials assume a predominant significance within the extensive panorama of our historical and artistic heritage, embodying a category of works that undergoes thorough and comprehensive investigation within the academic context of conservation and restoration.

The term "stone materials" refers both to natural substrates, encompassing any variety of rocks utilized in the architectural field, and to artificial ones, which are materials resulting from the manipulation and transformation of naturally sourced raw materials such as mortars or ceramics. Within the natural stone materials, the possibility of classification based on their geological origin is highlighted: sedimentary rocks, such as limestone and sandstone, metamorphic rocks, including marble, and igneous rocks, such as granite, are distinguished. In each of the aforementioned categories, there exist further subdivisions or classifications that can be associated with chemical composition, color, porosity, utilization, and other related attributes.

Regarding artificial stone materials such as mortars and ceramics, they can be defined as a combination of inorganic or organic binders, predominantly fine-grained aggregates, water, and possible additions of organic and/or inorganic additives (or a mixture consisting solely of binder and water), in proportions necessary to confer the mixture with a suitable fresh state for processing, as well as appropriate physical properties in the hardened state, such as porosity, water permeability, and so forth, in addition to mechanical characteristics including compressive strength, tensile strength, flexural strength, adhesion, aesthetic appearance, and durability [62].

3.1. Degradation of Stone Material

Stone materials undergo a range of natural and non-natural processes that inevitably lead to a progressive alteration of their initial characteristics. In many cases, these processes occur without causing significant harm to the preservation of the material itself, while in others, they can manifest as genuine forms of degradation. These processes that can compromise the very structure of the asset can be linked to both natural factors (e.g., rainfall, temperature fluctuations, etc.) and anthropogenic factors such as pollution. It is worth pointing out that the mineropetrographic, physical, and chemical features of the stone substrate represent the first discriminant influencing the intensity and the typology of biodeterioration affecting the material.

Every kind of natural and artificial stone should be treated deeply and separately based on the different response to biodeterioration. However, the authors mean to provide a general overview on the progress of the strategies against this phenomenon, especially valorizing the "green" and most sustainable ones.

The in-depth analysis and understanding of the factors responsible for degradation, therefore emerge as a crucial element to implement preventive measures or, at the very least, delay the deteriorative processes, especially when dealing with materials exposed to external environmental conditions [63]. The implementation of strategies aimed at modifying moisture levels or mitigating the impact of pollutant agents proves to be inherently complex and requires meticulous application and management.

The modifications affecting the material are manifested through more or less significant variations in the stone matrices, giving rise to phenomena of alteration (material alteration that does not necessarily imply a deterioration of its chemical and physical properties from a conservation perspective) and/or degradation (alteration that always entails a deterioration of the original material's chemical and physical characteristics) [64].

The magnitude of these effects will vary depending on the specific characteristics of the material itself, such as composition, origin, structure, and texture, as well as the combination of external agents, including environmental, physical, chemical, and biological factors.

Physical degradation occurs when external forces act upon the stone material, causing surface decohesion and a reduction in its mechanical strength, which can lead to the loss of material fragments in extreme cases [63]. The main factors responsible for physical degradation include water, salt crystallization, wind action, and thermal stress [65]. The result of these physical phenomena that affect the stone are the phenomena of alveolization [66,67], efflorescence [68], etc. Chemical degradation assumes a prominent role in stone materials due to its ability to induce chemical alterations in the original material, resulting in the formation of byproducts characterized by high solubility in many cases. The factors contributing to chemical degradation can stem from both natural and anthropogenic causes. Natural factors may be associated with marine aerosol, volcanic eruptions, and similar circumstances. As for anthropogenic chemical degradation factors, it is pertinent to mention the presence of atmospheric gases and particulate matter, including carbon compounds (CO_x), sulfur compounds (SO_x), and nitrogen compounds (NO_x). Acid precipitation, along with atmospheric particulate deposition, represent primary degradation factors that significantly impact the preservation of stone materials exposed to the outdoor environment [11]. Among the forms of physical degradation are the formation of black crusts [69,70], chromatic alteration [71], etc. Biological degradation is also observed, which is associated with the activity of microorganisms and organisms that attack the surface of stone materials. These agents include bacteria, algae, and higher organisms (i.e., plants) and often induce damages both aesthetically and chemical–physical damage that manifest itself through forms of degradation such as biological deposits [72] and biological patina [73].

It is essential, therefore, to identify the main causes of degradation and assess the feasibility of interventions aimed at mitigating or resolving them.

3.1.1. Stone Material Biodeterioration

Rocks, whether in natural geological formations or in stone monuments, serve as favorable environments that harbor a diverse array of microorganisms. The pervasive presence and metabolic activities of microorganisms contribute significantly to the degradation process experienced by stone structures [4,74–76].

Biodeterioration is a phenomenon characterized by the modification of an artifact's original state, resulting from the metabolic activity of communities of organisms and microorganisms. These communities, comprising algae, bacteria, fungi, and lichens, collectively referred to as biodeteriogens, contribute to the process of alteration.

Biodeterioration is a subsequent phase that ensues after previous forms of alteration, establishing conducive conditions for subsequent colonization by living organisms. Stone surfaces undergo a sequence of transformative processes following their initial installation, involving diminished luster, heightened roughness, crack formation, organic matter accumulation, and increased moisture levels. These phenomena facilitate the colonization and proliferation of biodeteriogens.

Biodeterioration refers to a phenomenon in which the presence of microorganisms or organisms triggers physical and chemical processes that can utilize the substrate for both nutritional purposes (heterotrophs) and structural support (autotrophs).

Within the specific context of stone artifacts, which predominantly consist of inorganic materials, the initial colonization is mainly driven by autotrophic or photoautotrophic organisms, including a diverse array of bacteria, cyanobacteria, algae, and lichens. This is followed by the subsequent diffusion of heterotrophic organisms, encompassing various bacteria and fungi.

The adherence of microorganisms to the stone substrate occurs through the production of extracellular polymeric substances (EPS), which form a complex layer known as a biofilm [25]. These microbial aggregates form communities in which cells are enclosed within a self-generated gelatinous matrix composed of EPS [77]. This matrix, primarily consisting of water, encompasses diverse constituents, such as polysaccharides, proteins, and extracellular DNA, playing a vital role in substrate adhesion and the sustenance of the microbial community [3,78].

After the initial attachment, there is the establishment of a microbial community comprising diverse bacterial strains, progressing through subsequent stages of growth and maturation (Figure 2).

Within this progression, novel microbial species, including heterotrophic microorganisms, integrate into the community, utilizing organic matter in the form of deceased cells and secondary products as a source of nutritional sustenance.

The occurrence of biodeteriogens on surfaces is contingent upon multiple factors, encompassing the inherent composition of the structures, the degree of light exposure, the moisture levels, and the surrounding microclimate. These factors require meticulous preliminary analysis before making any decisions regarding the preservation of the object being examined.

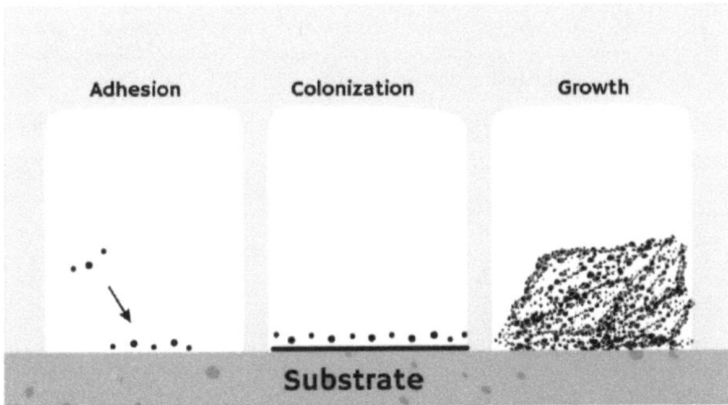

Figure 2. Stages of formation of a biofilm (modified from [25]).

The impact of these exogenous degradation agents is intensified by the bio-receptivity phenomenon, which pertains to the distinct response exhibited by the artifact, and it is influenced by a range of endogenous factors. These encompass the rock type, its origin, and the processing techniques employed, as well as its mineralogical and structural properties, including porosity, permeability, capillarity, and surface roughness. Furthermore, distinct microclimatic conditions associated with moisture, solar irradiation, wind, pollution, and, particularly, the presence of water, have a further impact on the bio-receptivity of the material.

Specifically, the hygroscopic nature of the material, intimately associated with its porosity, assumes a crucial role in the biodeterioration process. The amplification in relative humidity percentage within the external environment, coupled with an augmented water absorption capacity of the material, engenders more propitious conditions conducive to the proliferation of surface microflora.

Biofilms can also act as reservoirs for atmospheric pollutants, thereby exacerbating the processes of chemical substrate corrosion [6].

The process of biocorrosion, as an additional factor impacting biodeterioration, is initiated by the microbial secretion of both inorganic acids (such as nitric acid, sulfuric acid, carbonic acid, and nitrous acid) and organic acids. This secretion ultimately results in the solubilization and weakening of the stone matrix.

3.1.2. Consequences of Stone Biodeterioration

Microbial activity in biodegradation processes can lead to diverse forms of damage to stone monuments, encompassing aesthetic manifestations such as the development of surface crusts and patinas, mechanical repercussions resulting in significant fractures, and chemical modifications induced by metabolite production.

There exists a multitude of biological alterations that can manifest in various forms, such as patina, crust, discoloration, stains, and pitting, among others (Table 4). Biological patinas exhibit variations in color, thickness, and microstructure. These formations have the potential to comprise biofilms or evolve into authentic crusts, irrespective of the material type and environmental conditions. These "biopatinas" have the potential to generate mechanical actions resulting in the microdecohesion of substrates, as well as to initiate metabolic processes capable of inducing chemical transformations in materials.

Table 4. Summary of degradation patterns induced by biological growths ([64] and modified from [79]).

Biodeteriogenic Agent	Description
Bacteria	Black crusts, black patinas, exfoliation, pulverization, colour change, stains
Fungi	Staining, exfoliation, pitting
Algae	Patinas and films of various colors and consistency
Lichens	Scaling, mottling, pitting
Superior plants	Grass, shrubs, and woody species induce fractures, collapses, detachment of material

The presence of biofilm formations can lead to the accumulation of atmospheric pollutants, thus intensifying the processes of chemical corrosion on substrates. Additionally, the occurrence of stains causes superficial chromatic alterations to the artwork, with microorganisms being implicated in certain instances. The occurrence of pitting signifies a manifestation of degradation resulting from the activity of microorganisms with the capacity to penetrate the microcavities found in the stone structure. This infiltration process leads to the generation of numerous invisible openings characterized by their proximity and high density.

3.2. New Technologies for Stone Restoration

Currently, there are multiple strategies available for the consolidation and disinfection of such artifacts; however, numerous issues arise, primarily related to compatibility with the original materials and the durability of restoration interventions.

Every restoration intervention on stone artifacts requires an initial phase of diagnosis of their conservation status, which includes the analysis of composition and deterioration. Subsequently, a cleaning phase is carried out to remove any superficial deposits such as dirt and alteration patinas. Finally, a consolidation phase is implemented to restore the structural characteristics of the artwork.

The widely adopted restoration techniques for the removal of biological patinas from stone surfaces involve the use of both physical and chemical approaches. Among the physical approaches, mechanical brushes are employed to remove the biofilm; however, it is important to consider that they can potentially damage the material itself, thereby increasing its susceptibility to biodegradation [80]. Traditional chemical methods, on the other hand, involve the use of biocides such as benzalkonium chloride, which can reduce the presence of the microbial community. However, this biocide is known to be potentially toxic to operators and harmful to the environment, especially when used in confined spaces.

Consequently, there emerges a necessity to conceive pioneering, ecologically sustainable, and secure solutions [12,13] to mitigate these impacts. As a result, research efforts have recently intensified towards the exploration of alternative eco-compatible tools [9]. In Table 5, we have summarized the innovative methods proposed against biodeterioration. They are based on a chemical action, since all the active components have a biocidal and/or bacterial inhibition feature. The main difference between biocidal and bacterial inhibition effect is that the first term refers to the ability to kill the microorganism, while the second one describes the ability to avoid the biological growth. The reviewed methods are divided in two classes: those methods related to the removal of the biofilm and those mainly focused on bacterial inhibition. Biofilm removal consists of the killing of the microorganisms and their removal from the surface; these methods cover all the conservation phases related to the cleaning of the stone. A bacterial inhibition method should be adopted after a cleaning phase since it is aimed to protect the surface against further colonization over time.

Table 5. Summary of the innovative methods for stone reviewed.

Innovative Method	Activity	Reference Work
Sodium alginate hydrogel and hypochlorite ions	Biofilm removal	[11] Gabriele et al., 2021
Solvent gel of dimethyl sulfoxide		[81] Toreno et al., 2018
Lavandula angustifolia and *Thymus vulgaris* essential oil encapsulated in alginate hydrogel	Bacterial inhibition	[6] Ranaldi et al., 2022
Nanocapsules with essential oil of *Origanum vulgare* and *Thymus capitatus*		[82] Romano et al., 2020
Glyco-alkaloids extracted from *Solanum nigrum*, filtered without cells of the fungus *Trichoderma harzianum* and the bacterium *Burkholderia gladioli*		[9] Sasso et al., 2016
Silver nanoparticles		[10] Carillo-Gonzàlez et al., 2016

An alternative approach to the use of hypochlorite as a biocidal agent on stone substrates has been identified in order to mitigate the negative impact on such materials and preserve the health of the operators [11].

A hydrogel formulated with sodium alginate as an inert matrix and hypochlorite ions as biocidal agents has been tested to remove biological patinas from samples of Lecce limestone. It is one of the predominant materials used in the construction of historical buildings in the Baroque style located in southern Italy. This material generates considerable interest in the field of stone conservation studies, as it presents significant degradation issues both aesthetically and structurally, primarily associated with biofouling and chemical pollution.

Three samples of Lecce stone were selected as the control group, kept under conditions of intact preservation, while the other samples were subjected to induced biodeterioration or different treatments. Subsequently, an analysis was conducted to identify the phototrophic microorganisms constituting the present biofilms. It was found that these biofilms were predominantly composed of filamentous cyanobacteria and green microalgae. Optimizations were carried out on two different formulations of biocidal hydrogels to determine the best combinations of sodium alginate and calcium hypochlorite quantities to achieve a hydrogel with suitable consistency and maximum effectiveness.

The samples of Lecce limestone were subjected to specific biodeterioration processes, varying the duration to obtain three different levels of biodeterioration. Additional samples were subjected to a cleaning process using an aqueous solution containing 2.5% calcium hypochlorite. It was observed that in order to achieve complete removal of microorganisms from the stone, it would be necessary to use an oxidizing agent quantity approximately 10 times higher than the concentration present in the biocidal hydrogel with a higher amount of calcium hypochlorite. The presence of the biocide incorporated within the cross-linked structure of the hydrogel allows for the use of a reduced concentration of oxidizing agent.

The obtained results highlighted the effectiveness of the two optimized hydrogels in removing microorganisms from the stone surface, also demonstrating how the quantity of oxidizing agent to be incorporated in the hydrogel depends on the extent of the biocolonization process. This sanitization approach has made it possible to use hypochlorite as a biocidal agent, as potential side effects and health risks associated with operators are mitigated by its retention within the gel matrix. Thanks to the gel's adherence to the stone surface, which ensures prolonged exposure times, an extremely low concentration of oxidizing agent could be used to eliminate the biofilm.

Another environmentally friendly method proposed for the removal of biofilms and lichens from surfaces is represented by a solvent gel containing dimethyl sulfoxide (DMSO) applied on marble artifacts affected by biological patinas [81].

The gel was applied to the surfaces of the marble blocks and subsequently removed after a 24-h period, followed by cleaning with distilled water. This protocol was repeated from two to six times. Surprisingly, excellent results were achieved after only two treatments with the gel.

Preliminary tests conducted on a non-colonized marble sample have demonstrated that treatment with the DMSO gel solvent does not significantly alter the physical parameters of the substrate. In fact, measurements taken using a profilometer and colorimeter before and after six applications did not show significant variations. Therefore, the DMSO gel treatment did not alter the natural color of the marble, and no significant variations in the surface profile of the stone were observed after all the applied treatments. There were no significant variations in terms of porosity, pore radius, saturation, and adsorption parameters, indicating that the marble substrate remains substantially unchanged after the cleaning treatments with the gel.

Before the gel treatments, it is possible to observe the widespread presence of a biofilm. However, after the first application of the gel, the biofilm appears partially detached from the substrate. At the end of the treatments, the rocky surfaces are completely clean and free of residues.

The comparative study conducted between the DMSO-based solvent gel and the currently used biocides revealed that, after the same number of treatments, the gel proved to be the most efficient. The number of gel applications required may be related to the thickness of the biofilm and its penetration into the substrate. In fact, the best results were achieved through repeated gel applications. Subsequently, a culture test conducted one year later did not detect the presence of colonies, confirming the effectiveness of the treatment.

In the context of cultural heritage restoration, the observation regarding the use of essential oils (EO) as natural biocidal agents has emerged as an effective alternative to conventional chemical compounds. Evidence supporting this observation has been provided by a scientific study [6] demonstrating the efficacy of *Lavandula angustifolia Mill.* and *Thymus vulgaris* L. essential oils at a concentration of 5% in combating cyanobacterial biofilms. During the comparative analysis, it was observed that the 5% concentration of *Thymus vulgaris* essential oil (EO) exhibited greater inhibitory efficacy against the photosynthetic activity of cyanobacterial biofilms over an extended period. Samples treated with thyme essential oil displayed significant discoloration without evidence of microbial growth, whereas only one of the samples treated with lavender essential oil exhibited such discoloration.

To develop new non-invasive methodologies for the conservation of stone monuments, various concentrations of thyme essential oil (EO) were experimented with. These concentrations were encapsulated within an alginate hydrogel (EO-HG) and applied for different time periods on cyanobacterial biofilms.

All the examined concentrations of *Thymus vulgaris* essential oil encapsulated in an alginate hydrogel (*T. vulgaris* EO HG) demonstrated the ability to completely inhibit the photosynthetic activity of the biofilms.

As a result, the utilization of *T. vulgaris* EO-HG, even at very low concentrations, demonstrated a potent inhibitory power against microorganisms. Consequently, the utilization of Thymus vulgaris essential oil in an alginate hydrogel is considered a recommended formulation as it allows for the use of extremely limited amounts of biocide while maintaining high antimicrobial efficacy against cyanobacteria and reducing the volatility of terpenic components.

In the study [82], the use of nanocapsules (NC) containing essential oils of *Origanum vulgare* (Or-EO) and *Thymus capitatus* (Th-EO) is described to counteract the bacterial proliferation of two microorganisms (*Escherichia coli* and *Kokuria rhizophila*) on marble surfaces.

Plant essences, in the form of essential oils (EO), could represent an effective alternative to conventional chemical biocides. However, characteristics such as color, high volatility, insolubility in water, and sensitivity to oxygen, light, and heat may pose challenges in the use of these essential oils for the preservation of cultural heritage.

The process of nanoencapsulation of essential oils can be an effective strategy for creating new non-toxic and eco-friendly nanomaterials to be used in the field of cultural heritage. This technique allows for the protection of compounds, such as EO, from adverse environmental conditions, ensuring color masking, improved solubility, and reduced volatility. Furthermore, it enables controlled release of EO even at different temperatures.

No significant changes in the color, structure, or chemical composition of the stone were observed after the application of nano-encapsulated essential oil (EO-NC). After 6 h of contact, both Or-NC and Th-NC were able to reduce the bacterial concentration by three to four orders of magnitude for both microorganisms. After 24 h of contact, no viable bacterial cells were observed.

The nanostructured systems (EO-NC) have shown the ability to inhibit bacterial growth on the stone previously treated with a bacterial inoculum. It has been found that the nanocapsules loaded with thyme essential oil are more efficient compared to the nanocapsules loaded with oregano essential oil.

In the field of research, glycoalkaloids (GA) are also being studied as possible natural biocides, which are produced by plants of the *Solanaceae* family. In [9], they were tested against a phototrophic culture that develops on a substrate of limestone rock, filtered without cells of the fungus *Trichoderma harzianum* and the *bacterium Burkholderia gladioli*. To promote the development of a photosynthetic biofilm on the surface of a limestone rock, a phototrophic culture composed of different species was employed. The main components of the inoculum consisted of the genera *Chlorella* and *Stichococcus*. Sixteen blocks of white bioclastic limestone were used as the lithotype, onto which 200 mL of a multispecies phototrophic culture were applied on their respective upper surfaces.

The cell-free filtrates of *T. harzianum* and *B. gladioli*, along with the glycoalkaloids, exhibit limited biocidal activity that is primarily focused on filamentous phototrophic microorganisms in general. The treatment was not successful in effectively inhibiting the growth of photosynthetic biomass on stone surfaces. The glycoalkaloids appeared to reduce microbial growth during the initial month of incubation, but after 30 days of incubation, they promoted biofilm expansion. This also demonstrates the ineffectiveness of glycoalkaloids as biocides, especially against unicellular phototrophs, as these natural compounds can act as nutrient sources for microbial growth instead of controlling biological colonization. Among the three tested natural biocides, only the cell-free filtrate of the fungus *Trichoderma* proved to be effective against the photosynthetic community developed on limestone blocks during a 45-day incubation period. Additionally, the application of the treatments caused chromatic variations, manifesting as darkening and the yellowing of the blocks.

In summary, the application of cell-free filters and plant extracts on stone samples has been shown to suppress the growth of filamentous phototrophic microorganisms, but at the same time, it promoted the growth of unicellular phototrophic microorganisms. Therefore, the use of this biocide has proven to be ineffective.

Nanotechnology, which involves the production of nanomaterials using biological or natural components (green synthesis), is among the emerging alternatives in this context. It offers numerous advantages over chemical methods and has a low environmental impact. Silver nanoparticles (AgNP) have been synthesized using a green synthesis method and have been evaluated as potential antimicrobial inhibitors [10].

Two plant species, *Foeniculum vulgare* and *Tecoma stans*, were utilized in the synthesis of nanoparticles using aqueous leaf extracts and two extraction procedures. The potential of AgNP was evaluated as preventive/corrective treatments to safeguard materials (stucco, basalt, and limestone) from biodeterioration. AgNP exhibit unique physicochemical properties compared to individual particles, thereby conferring increased reactivity and high efficacy against a wide range of microorganisms.

A total of twenty-three bacterial species and fourteen fungal species were isolated from colored stains, patinas, and biofilms present on surfaces. AgNP produced by *Foeniculum vulgare* exhibited high efficacy in suppressing bacterial growth, surpassing that of nanoparticles produced by *Tecoma stans*. Studies have reported a correlation between bacterial growth and the increase in AgNP concentration. In particular, it was observed that a concentration of 200 µL of AgNP produced by both *Foeniculum vulgare* and *Tecoma stans* did not result in any significant reduction of bacterial growth. However, it was found that dosages of 300 µL consistently yielded high percentages of bacterial inhibition (85% and 77%, respectively). Furthermore, higher dosages (500 µL) demonstrated effective reduction of bacterial growth (71% and 64%).

A significant enhancement of the antibacterial effect was observed when the size of the nanoparticles (NP) was reduced from 100 to 20 nm. Additionally, it was observed that bacteria exhibit lower sensitivity to AgNP compared to fungi. The three types of materials show a similar degree of susceptibility to microbial colonization; however, limestone exhibits a higher tendency for alteration compared to other stone varieties due to processes such as material dissolution, salt crystallization, and biological activity. However, the use of AgNP contributes to enhancing the protection of the stone material. It is worth highlighting that during and after the treatments with AgNP, no surface alterations were observed in the three tested materials. The employment of AgNP as a preventive or corrective treatment to reduce microbial colonization in three different stone varieties used in historical masonry structures has shown remarkable success.

4. Conclusions and New Perspectives

From the examined literature it has been seen that the effectiveness of the compound or method used against biodeterioration varies greatly in the function of various factors, such as the mechanism of action, the concentration used, and the period of action.

Nowadays, the international scientific community dedicates considerable attention to researching and experimenting with innovative natural biocides that are both long-lasting, non-invasive, and eco-friendly.

Among the innovative methods concerning the biodeterioration of paper, functionalized nanoparticles and essential oils are the most effective methods of inhibiting microbial growth. Essential oils also have the benefit of being ecofriendly; however, some issue related to their color changing over time can arise. DES, in addition to having an inhibitory effect when combined with Preventol RL50, has been found to be effective in improving the mechanical properties of paper, which also affect biodeterioration. This is an example of the many possibilities of intervention that can be formulated and combined in a suitable way to test, case-by-case, different compounds of various origins.

This review conducted a critical assessment of the evidence regarding the biocidal properties of natural substances in the context of stone material preservation. In some cases, encouraging results have been observed when testing several natural substances, such as various essential oils and other non-harmful compounds, as biocidal agents on stone materials. Moreover, for stone materials, essential oils seem to be promising, although long-term observations should be carried out in order to properly understand the behavior. The performance of nanoparticles used as bacterial inhibitors has to be balanced with the recently emerging issues related to the dispersion of nanoparticles in the environment.

The issue is widely open considering the great heterogeneity of the artworks needing to be preserved, as well as the complexity of the degradation phenomena and the factors influencing them. Moreover, it is worth underlining the great variety of biodeteriogens, their strong adaptation skills with respect to the surrounding environment, and, consequently, their increasing resistance to intervention techniques over time. All this context means that the research around biodeterioration, restoration efforts, and the conservation of cultural heritage is constantly evolving.

In the future, it would be reasonable to improve the classification of the microorganisms present on paper and stone objects, investigating their metabolic functions and their close connection with the environment. Moreover, knowledge of the processes regulating biofilm composition, growth, and activities could contribute to identifying or formulating new biocontrol agents able to be, at once, efficient against biodeteriogens and preservative of the safety of other organisms, "from the micro to the macro", such as the operators and end-users of cultural assets. Finally, it is worth noting that an experimental approach, with a case-to-case evaluation of the materials involved (whether stone or paper), the surrounding environmental conditions, and probable past restorative interventions, represents a most effective and always-current strategy.

Author Contributions: Conceptualization, M.F.L.R.; Writing—original draft preparation, M.C., F.G., N.R. and S.A.R.; writing—review and editing, M.F.L.R., A.M., R.M., M.R., N.R., S.A.R. and M.T.; supervision, A.F., M.F.L.R., and S.A.R. All authors have read and agreed to the published version of the manuscript.

Funding: This research received no external funding.

Institutional Review Board Statement: Not applicable.

Informed Consent Statement: Not applicable.

Data Availability Statement: Data are contained within the article.

Conflicts of Interest: The authors declare no conflict of interest.

References

1. Farooq, M.; Hassan, M.; Gull, F. Mycobial deterioration of stone monuments of Dharmarajika Taxila. *JMAN* **2015**, *2*, 29–33. [CrossRef]
2. Fidanza, M.; Caneva, G. Natural biocides for the conservation of stone cultural heritage: A review. *J. Cult. Herit.* **2019**, *38*, 271–286. [CrossRef]
3. Di Martino, P. What about biofilms on the surface of stone monuments? *Open Conf. Proc. J.* **2016**, *7*, 14–28. [CrossRef]
4. Gaylarde, C.; Morton, L. Deteriogenic biofilms on buildings and their control: A review. *Biofouling J. Bioadhesion Biofilm Res.* **1999**, *14*, 59–74. [CrossRef]
5. Pinzari, F.; Pasquariello, G.; de Mico, A. Biodeterioration of Paper: A SEM Study of Fungal Spoilage Reproduced Under Controlled Conditions. *Macromol. Symp.* **2006**, *238*, 57–66. [CrossRef]
6. Ranaldi, R.; Rugnini, R.; Gabriele, F.; Spreti, N.; Casieri, C.; di Marco, G.; Gismondi, A.; Bruno, L. Plant essential oils suspended into hydrogel: Development of an easy-to-use protocol for the restoration of stone cultural heritage. *Int. Biodeter. Biodegr.* **2022**, *172*, 105436. [CrossRef]
7. Ashraf, M.; Mohamed, K.K. Biological nanosilver particles for the protection of archaeological stones against microbial colonization. *Int. Biodeter. Biodegr.* **2014**, *94*, 31–37.
8. Can Egil, A.; Ozdemir, B.; Gunduz, S.K.; Altikatoglu-Yapaoz, M.; Budama-Kilinc, Y.; Mostafavi, E. Chitosan/calcium nanoparticles as advanced antimicrobial coating for paper documents. *Int. J. Biol. Macromol.* **2022**, *215*, 521–530. [CrossRef]
9. Sasso, S.; Miller, S.; Rogerio-Cadelera, M.; Cubero, B.; Coutinho, M.; Scrano, L. Potential of natural biocides for biocontrolling phototrophic colonization on limestone. *Int. Biodeter. Biodegr.* **2016**, *107*, 102–110. [CrossRef]
10. Carillo-Gonzàlez, R.; Martìnez-Gòmes, M.A.; Gonzàlez-Chavèz, M.D.C.A.; Hernàndez, J.C.M. Inhibition of microorganisms involved in deterioration of an archaeological site by silver nanoparticles produced by a green synthesis method. *Sci. Total Environ.* **2016**, *565*, 872–881. [CrossRef]
11. Gabriele, F.; Tortora, M.; Bruno, L.; Casieri, C.; Chiarini, M.; Germani, R.; Spreti, N. Alginate-biocide hydrogel for the removal of biofilms from calcareous stone artworks. *J. Cult. Herit.* **2021**, *49*, 106–114. [CrossRef]
12. Urzì, C.; de Leo, F.; Krakova, L.; Pangallo, D.; Bruno, L. Effects of biocide treatments on the biofilm community in Domitilla's catacombs in Rome. *Sci. Total Environ.* **2016**, *572*, 252–262. [CrossRef]
13. Palla, F.; Bruno, M.; Mercurio, F.; Tantillo, A.; Rotolo, V. Essential Oils as Natural Biocides in Conservation of Cultural Heritage. *Molecules* **2020**, *25*, 730. [CrossRef]
14. Pinna, D.; Salvadori, B.; Galeotti, M. Monitoring the performance of innovative and traditional biocides mixed with consolidants and water-repellents for the prevention of biological growth on stone. *Sci. Total Environ.* **2012**, *423*, 132–141. [CrossRef] [PubMed]
15. Fonseca, A.; Pina, F.; Macedo, M.; Leal, N. Anatase as an alternative application for preventing biodeterioration of mortars: Evaluation and comparison with other biocides. *Int. Biodeter. Biodegr.* **2010**, *64*, 388–396. [CrossRef]
16. Gomez-Villaba, L.; Lòpez-Arce, P.; Fort, R.; Alvarez De Buergo, M. Structural stability of a colloidal solution of $Ca(OH)_2$ nanocrystals exposed to high relative humidity conditions. *Appl. Phys. A* **2011**, *104*, 1249–1254. [CrossRef]

17. Ceragioli, G. Caratteristiche delle fibre e proprietà della carta. In *Proceedings of the Ciclo di Conferenze Agli Studenti dell'Istituto Tecnico di Fabriano, Ente Nazionale per la Cellulosa e la Carta, Rome, Italy, 1970–1971*; Ente Nazionale per la Cellulosa e la Carta: Rome, Italy, 1975; pp. 161–172.
18. Culicchi, P. Generalità sulle cellulose. In *Ciclo di conferenze agli studenti dell'Istituto Tecnico di Fabriano, Ente nazionale per la Cellulosa e la Carta, Rome, Italy, 1974–1975*; Ente Nazionale per la Cellulosa e la Carta: Rome, Italy, 1978; pp. 137–146.
19. *Carta Unitex, Norma UNI 8282*; AA. VV. Cellulosa in Soluzioni Diluite-Determinazione Dell'indice Della Viscosità Limite-Metodo Che Usa Una Soluzione di Cupriletilendiammina (CED). UNI—Ente Italiano di Normazione: Milan, Italy, 1982.
20. Giorgi, R.; Chelazzi, D.; Baglioni, P. Il ruolo degli inchiostri metallo-gallici nei processi degradativi di manoscritti cartacei. In Proceedings of the V Congresso Nazionale IGIIC—Lo Stato dell'Arte 5, Palazzo Cittanova-Palazzo Trecchi, Cremona, Italy, 4–6 October 2007; Nardini Editore: Firenze, Italy, 2003; pp. 273–280.
21. Aulascienze.scuola.zanichelli.it. Available online: https://aulascienze.scuola.zanichelli.it/blog-scienze/come-te-lo-spiego-scienze/la-chimica-della-carta#leggi (accessed on 27 June 2023).
22. Pagano, D. Restauro e Conservazione Delle Opere su Carta. Master's Thesis, Politecnico di Torino, Torino, Italy, 2019.
23. Banik, G.; Cremonesi, P.; de la Chapelle, A.; Montalbano, L. *Nuove Metodologie Nel Restauro Del Materiale Cartaceo*; Il Prato, Collana i Talenti: Padova, Italy, 2003; pp. 5–32.
24. Cialei, V. Restauro di Superfici Cartacee Biodeteriorate: Batteri Pulitori e Nuovi Metodi Enzimatici Integrati. Doctoral Thesis, Sapienza Università di Roma, Rome, Italy, 2008.
25. Caneva, G.; Nugari, M.; Salvadori, O. *La Biologia Vegetale per i Beni Culturali*; Nardini Editore: Firenze, Italy, 2005; pp. 65–70.
26. Sequiera, S.; Cabrita, E.J.; Macedo, M.F. Antifungals on paper conservation: An overview. *Int. Biodeter. Biodegr.* **2012**, *74*, 67–86. [CrossRef]
27. Eckhardt, F. Mechanisms of microbial degradations of minerals in sandstone monuments, mediaeval frescoes and plasters. In Proceedings of the Vth International Congress on Biodeterioration and Conservation of Stone, Lausanne, Switzerland, 25–27 September 1985; Volume 5, pp. 643–652.
28. Campanella, L.; Angeloni, R.; Cibin, F.; Dell'Aglio, E.; Grimaldi, F.; Reale, R.; Vitali, M. Olio essenziale capsulato in sfere di gel per la tutela del patrimonio culturale cellulosico. *Naz. Prod. Ris.* **2021**, *35*, 116–123. [CrossRef]
29. Gallo, F. *Il Biodeterioramento di Libri e Documenti*; Centro di Studi per la Conservazione Della Carta ICCROM: Rome, Italy, 1992.
30. Bennett, J.W.; Klich, M.; Moselio, S. Encyclopedia of Microbiology. In *Mycotoxins*; Academic Press: Oxford, UK, 2009; pp. 559–565.
31. Sterflinger, K. Fungi: Their role in deterioration of cultural heritage. *Fungal Biol. Rev.* **2010**, *24*, 47–55. [CrossRef]
32. Zyska, B. Fungi isolated from library materials: A review of the literature. *Int. Biodeter. Biodegr.* **1997**, *40*, 43–51. [CrossRef]
33. Kraková, L.; Šoltys, K.; Otlewska, A.; Pietrzak, K.; Purkrtová, S.; Savická, D.; Pangallo, D. Comparison of methods for identification of microbial communities in book collections: Culture-dependent (sequencing and MALDI-TOF MS) and culture-independent (Illumina MiSeq). *Int. Biodeter. Biodegr.* **2018**, *131*, 51–59. [CrossRef]
34. Michaelsen, A.; Pinar, G.; Pinzari, F. Molecular and Microscopical Investigation of the Microflora Inhabiting a Deteriorated Italian Manuscript Dated from the Thirteenth Century. *Microb. Ecol.* **2010**, *60*, 69–80. [CrossRef]
35. Olubanke, M.A. review of biological deterioration of library materials and possible control strategies in the tropics. *Libr. Rev.* **2010**, *24*, 47–55.
36. Jacob, S.M.; Bhagwat, A.M.; Kelkar-Mane, V. Bacillus species as an intrinsic controller of fungal deterioration of archival documents. *Int. Biodeter. Biodegr.* **2015**, *104*, 46–52. [CrossRef]
37. Baty, J.; Maitland, C.; Minter, W.; Hubbe, M.; Jordan-Mowery, S. Deacidification for the conservation and preservation of paper-based works: A review. *BioResources* **2010**, *5*, 1955–2023. [CrossRef]
38. Espejo, T.; Duran, A.; Lopez-Montes, A.; Blanc, R. Microscopic and spectroscopic techniques for the study of paper supports and textile used in the binding of hispano-arabic manuscripts from Al-Andalus: A transition model in the 15th century. *J. Cult. Herit.* **2010**, *11*, 50–58. [CrossRef]
39. Karbowska-Berent, J.; Gorniak, B.; Czajkowska-Wagner, L.; Rafalska, K.; Jarmiłko, J.; Koziekec, T. The initial disinfection of paper-based historic items—Observations on some simple suggested methods. *Int. Biodeter. Biodegr.* **2018**, *131*, 60–66. [CrossRef]
40. Sterflinger, S.; Pinzari, F. The revenge of time: Fungal deterioration of cultural heritage with particular reference to books, paper and parchment. *Environ. Microbiol.* **2012**, *14*, 559–566. [CrossRef]
41. Fouda, A.; Abdel-Maksoud, G.; Abdel-Rahman, M.A.; Salem, S.S.; Hassan, S.E.; El-Sadany, M.A. Eco-friendly approach utilizing green synthesized nanoparticles for paper conservation against microbes involved in biodeterioration of archaeological manuscript. *Int. Biodeter. Biodegr.* **2019**, *142*, 160–169. [CrossRef]
42. Bergamonti, L.; Graiff, C.; Isca, C.; Predieri, G.; Lottici, P.; di Maggio, R.; Palanti, S.; Maistrello, L.; Montanari, M. Trattamenti sostenibili per la protezione e il consolidamento di legno e carta. *Chem. Green Chem.* **2017**, *5*, 9–17.
43. Teixeira, F.S.; dos Reis, T.A.; Sgubin, L.; Thomè, L.E.; Bei, I.W.; Clemencio, R.E.; Correa, B.; Salvadori, M.C. Disinfection of ancient paper contaminated with fungi using supercritical carbon dioxide. *J. Cult. Herit.* **2018**, *30*, 110–116. [CrossRef]
44. Mansour, M.M.A.; Salem, M.Z.M.; Hassam, R.R.A.; Ali, H.M. Antifungal potential of three natural oils and their effects o the thermogravimetric and chromatic behaviors when applied to historical paper and various commercial paper sheets. *Nat. Oils TGA* **2021**, *16*, 492–514. [CrossRef]
45. El-Hassan, G.M.M.A.; Alfarraj, S.; Alharbi, S.A.; Atiya, N.H. Survey of insect pests in the manuscripts library of Coptic museum in Egypt. *Saudi J. Biol. Sci.* **2021**, *28*, 5061–5064. [CrossRef] [PubMed]

46. Rukke, B.A.; Querner, P.; Hage, M.; Steinert, M.; Kaldager, M.; Somhovd, A.; Dominiak, P.; Garrido, M.; Hansson, T.; Aak, A. Insecticidal gel bait for the decimation of Ctenolepisma longicaudatum (Zygentoma: Lepismatidae) populations in libraries, museums, and archives. *J. Cult. Herit.* **2023**, *59*, 255–263. [CrossRef]

47. Mnasri, A.; Dhaouadi, H.; Khiari, R.; Halila, S.; Mauret, E. Effect of Deep Eutectic Solvents on cellulosic fibres and paper properties Green «chemical» refining. *Carbohydr. Polym.* **2022**, *292*, 119606. [CrossRef]

48. Liu, C.; Li, M.; Chen, W.; Huang, R.; Hong, S.; Wu, Q.; Mei, C. Production of lignin-containing cellulose nanofibers using deep eutectic solvents for UV-absorbing polymer reinforcement. *Carbohydr. Polym.* **2020**, *246*, 116548. [CrossRef]

49. Cao, X.; Liu, M.; Bi, W.; Lin, J.; Chen, D.; Da, Y. Direct carboxylation of cellulose in deep eutectic solvent and its adsorption behavior of methylene blue. *Carbohydr. Polym. Technol. Appl.* **2022**, *4*, 100222. [CrossRef]

50. Zhou, E.; Liu, H. A Novel Deep Eutectic Solvents Synthesized by Solid Organic Compounds and Its Application on Dissolution for Cellulose. *Asian J. Chem.* **2014**, *26*, 3626–3630. [CrossRef]

51. Macchia, A.; Strangis, R.; de Angelis, S.; Cersosimo, M.; Docci, A.; Ricca, M.; Gabirele, B.; Mancuso, R.; La Russa, M.F. Deep Eutectic Solvents (DESs): Preliminary Results for Their Use Such as Biocides in the Building Cultural Heritage. *Materials* **2022**, *15*, 4005. [CrossRef] [PubMed]

52. Zervos, S.; Alexopoulou, I. Paper conservation methods: A literature review. *Cellulose* **2015**, *22*, 2859–2897. [CrossRef]

53. Wang, Y.J.; Tan, W.; Liu, C.Y.; Fang, Y.X. Deacidification of Paper in Supercritical Carbon Dioxide (CO_2SCF) Solvent System with Magnesium Acetate and Calcium Hydroxide. *Adv. Mater. Res.* **2011**, *347–353*, 504–507.

54. Querner, P. Insect Pests and Integrated Pest Management in Museums, Libraries and Historic Buildings. *Insects* **2015**, *6*, 595–607. [CrossRef] [PubMed]

55. Florian, M. Freezing for museum pest eradication. *Collect. Forum* **1990**, *6*, 1–7.

56. Berzolla, A.; Reguzzi, M.C.; Chiappini, E. Preliminary observations on the use of low temperatures in the cultural. *JEAR* **2011**, *43*, 191–196.

57. TStrang, T.J.K. A review of published temperatures for the control of pest insects in museums. *Collect. Forum* **1992**, *8*, 41–67.

58. Pinniger, D.B. Saving our treasures—Controlling Museum pests with temperature extremes. *Pestic. Outlook* **2003**, *14*, 10–11. [CrossRef]

59. Gilberg, M. Inert atmosphere fumigation of museum objects. *Stud Conserv* **1989**, *34*, 80–84.

60. Berzolla, A.; Reguzzi, M.; Chiappini, E. Controlled atmospheres against insect pests in museums: A review and some considerations. *JEAR* **2011**, *43*, 197–204. [CrossRef]

61. Abbott, A.P.; Capper, G.; Davies, D.L.; Rasheed, R.K.; Tambyrajah, V. Novel solvent properties of choline chloride/urea mixtures. *Chem. Comm.* **2003**, *1*, 70–71. [CrossRef]

62. *NorMal UNI 10924*; Beni Culturali—Malte per Elementi Costruttivi e Decorativi—Classificazione e Terminologia. UNI—Ente Italiano di Normazione: Milan, Italy, 2001.

63. Siegesmund, S.; Weiss, T.; Vollbrecht, A. Natural Stone, Weatcering Pcenomena, Conservation Strategies and Case Studies. *Geol. Soc.* **2002**, *205*, 1–7.

64. *NorMal UNI 11182*; Beni Culturali—Materiali Lapidei Naturali ed Artificiali—Descrizione Della Forma di Alterazione—Termini e Definizioni. UNI—Ente Italiano di Normazione: Milan, Italy, 2006.

65. Sage, J.D. Thermal microfracturing of marble. In *Engineering Geology of Ancient Works, Monuments and Historical Sites*; Marinos, P.G., Koukis, G.C., Eds.; Balkema: Rotterdam, The Netherlands, 1988; Volume 2, pp. 1013–1018.

66. Anania, L.; Badalà, A.; Barone, G.; Belfiore, C.M.; Calabrò, C.; La Russa, M.F.; Mazzoleni, P.; Pezzino, A. The stones in monumental masonry buildings of the "Val di Noto" area: New data on the relationships between petrographic characters and physical-mechanical properties. *Constr. Build Mater.* **2012**, *33*, 122–132. [CrossRef]

67. Punturo, R.; Russo, L.G.; Lo Giudice, A.; Mazzoleni, P.; Pezzino, A. Building stone employed in the historical monuments of Eastern Sicily (Italy). An example: The ancient city centre of Catania. *Environ. Geol.* **2006**, *50*, 156–169. [CrossRef]

68. Malaga-Starzec, K.; Panas, I.; Lindqvist, J.E.; Lindqvist, O. Efflorescence on thin sections of calcareous stones. *J. Cult. Herit.* **2003**, *4*, 313–318. [CrossRef]

69. Comite, V.; Miani, A.; Ricca, M.; La Russa, M.F.; Pulimeno, M.; Fermo, P. The impact of atmospheric pollution on outdoor cultural heritage: An analytic methodology for the characterization of the carbonaceous fraction in black crusts present on stone surfaces. *Environ. Res.* **2021**, *201*, 111565. [CrossRef]

70. La Russa, M.F.; Belfiore, C.M.; Comite, V.; Barca, D.; Bonazza, A.; Ruffolo, S.; Crisci, G.; Pezzino, A. Geochemical study of black crusts as a diagnostic tool in cultural heritage. *Appl. Phys. A* **2013**, *113*, 1151–1162. [CrossRef]

71. Cons, E.; Appolonia, L.; Galinetto, P.; Riccardi, M.; Tarantino, S.; Zema, M. Chromatic Alteration of Roman Heritage in Aosta (Italy). *Procedia Chem.* **2013**, *8*, 78–82. [CrossRef]

72. La Russa, M.F.; Ruffolo, S.; Malagodi, M.; Barca, D.; Cirrincione, R.; Pezzino, A.; Crisci, G.; Miriello, D. Petrographic, biological, and chemical techniques used to characterize two tombs in the Protestant Cemetery of Rome (Italy). *Appl. Phys. A* **2010**, *100*, 865–872. [CrossRef]

73. Urzì, C.; Realini, C. Colour changes of Noto's calcareous sandstone as related. *Int. Biodeter. Biodegr.* **1998**, *42*, 45–54. [CrossRef]

74. Dornieden, T.; Gorbushina, A.; Krumbein, W. Biodecay of cultural heritage as a space/time-related ecological situation—An evaluation of a series of studies. *Int. Biodeter. Biodegr.* **2000**, *46*, 261–270. [CrossRef]

75. Miller, A.Z.; Rogerio-Candelera, M.A.; Laiz, L.; Wierzchos, J.; Ascaso, C.; Sequeira Braga, M.A.; Hernandez-Marinè, M.; Maurício, A.A.; Dionísio, A.A.; Macedo, A.; et al. Laboratory-induced endolithic growth in calcarenites: Biodeteriorating potential assessment. *Microb. Ecol.* **2010**, *60*, 55–68. [CrossRef]
76. Sterflinger, K.; Pinar, G. Microbial deterioration of cultural heritage and works of art—Tilting at windmills? *Appl. Microbiol. Biotechnol.* **2013**, *97*, 9637–9646. [CrossRef]
77. Flemming, H.; Wingender, J.; Szewzyk, U.; Steinberg, P.; Rice, S.; Kjelleberg, S. Biofilms: An emergent form of bacterial life. *Nat. Rev. Microbiol* **2016**, *14*, 563–575. [CrossRef]
78. Rossi, F.; Micheletti, E.; Bruno, L.; Adhikary, S.; Albertano, P.; de Philippis, R. Characteristics and role of the exocellular polysaccharides produced by five cyanobacteria isolated from phototrophic biofilms growing on stone monuments. *J. Bioadhesion Biofilm Res.* **2012**, *28*, 215–224. [CrossRef]
79. Pastore, T. Utilizzo di Malte Biologiche in ambito Cultural Heritage. Master's Thesis, Università Cà Foscari, Venezia, Italy, 2020.
80. Bruno, L.; Valle, V. Effect of white and monochromatic lights on cyanobacteria and biofilms from Roman Catacombs. *Int. Biodeter. Biodegr.* **2017**, *123*, 286–295. [CrossRef]
81. Toreno, G.; Isola, D.; Meloni, P.; Carcangiu, G.; Selbmann, L.; Onofri, S.; Caneva, G.; Zucconi, L. Biological colonization on stone monuments: A new low impact cleaning method. *J. Cult. Herit.* **2018**, *30*, 100–109. [CrossRef]
82. Romano, I.; Granata, G.; Poli, A.; Finore, I.; Napoli, E.; Geraci, C. Inhibition of bacterial growth on marble stone of 18th century by treatment of nanoencapsulated essential oils. *Int. Biodeter. Biodegr.* **2020**, *148*, 104909. [CrossRef]

Article

Proposal of New Natural Hydraulic Lime-Based Mortars for the Conservation of Historical Buildings

Marco Destefani *, Laura Falchi and Elisabetta Zendri

Department of Environmental Sciences, Informatics and Statistics (DAIS), Scientific Campus,
Ca' Foscari University of Venice, 30170 Venezia, Italy; laura.falchi@unive.it (L.F.); elizen@unive.it (E.Z.)
* Correspondence: 858827@stud.unive.it; Tel.: +39-3808965178

Abstract: NHL mortars are known to be compatible materials for the conservation of architectural heritage. To improve their properties with regard to salt resistance and lower their carbon footprint, NHL-based mortars with salt inhibitor agents were studied and different formulations were produced: NHL-based mortars (MSs), composed of natural hydraulic lime; and sand and cocciopesto mortars (MSCs), in which NHL, sand and brick powder were admixed with two different products, diethylenetriaminapenta and chitosan, in different concentrations. The mortar performance was tested against freeze–thaw and salt crystallization through immersion–drying cycles in a 14% sodium sulfate solution. The results highlighted that the addition of cocciopesto was effective in increasing the salt resistance, but increased the water intake during the freeze–thaw tests. The use of DTPMP produced less thixotropic mortars and decreased the water uptake, but worsened the salt resistance of hardened mortars. Chitosan allowed a good workability of fresh mortar; its water uptake was similar to the reference mortar and slightly increased the salt resistance. In the cocciopesto samples, both additives reduced the weight variation during freeze–thaw tests; meanwhile, for the lime samples, the additives increased the weight variation during the final cycles.

Keywords: NHL-based mortars; green materials; restoration; durability; building conservation

check for
updates

Citation: Destefani, M.; Falchi, L.; Zendri, E. Proposal of New Natural Hydraulic Lime-Based Mortars for the Conservation of Historical Buildings. *Coatings* **2023**, *13*, 1418. https://doi.org/10.3390/coatings13081418

Academic Editor: Peng Liu

Received: 31 May 2023
Revised: 24 July 2023
Accepted: 7 August 2023
Published: 12 August 2023

1. Introduction

Concrete production is responsible for a negative impact on the environment, contributing 8%–10% of greenhouse gas emissions [1–3], 15% of global electrical energy consumption and the exhaustion of non-renewable resources [3]. It is, therefore, important to direct the construction sector towards a policy where conventional construction materials are replaced with by-products that will significantly reduce the environmental impact. Construction and demolition waste, glass waste, plastic waste, sludge from wastewater treatments and supplementary cementitious materials represent a starting point from which new construction materials can be designed [4–20]. Increasing the service life of a render product is another strategy to diminish the overall impact. Specific formulations need to be developed in relation to severe environmental conditions, e.g., salt resistance properties in coastal environments or freeze–thaw resistance in cold climates. A contribution to this can also come from the field of archaeological conservation, where the proposal of sustainable materials must be combined with the need for durable and compatible interventions [19–24].

In recent decades, a great amount of attention has been paid to proposing traditional mortars for the preservation of historical buildings and archaeological sites based on a reverse-engineering approach [25,26]. Among the traditional mortars, a mixture of limes with low-temperature fired-brick powder ("cocciopesto" mortar) has been demonstrated to be a possible durable and sustainable solution for obtaining hydraulic mortars [27–31]. Brick powder is a pozzolanic material capable of producing more flexible and permeable mortars [7–30]. Ca(OH)2 reacts with the pozzolanic material [32,33] to form stable compounds, increasing durability and freeze–thaw resistance [34]. The freeze–thaw phenomenon can have dramatic consequences on infrastructures; it especially occurs in cold areas where ice forms. As the water freezes, the ice expands inside the cracks and repulsive forces push apart the stone, causing erosion and internal stress.

Less studied is the effect of brick powder addition to natural hydraulic limes (NHLs), even though this kind of mortar has been demonstrated to have adequate characteristics for new renders and plasters [35,36]. NHL-based mortars are more compatible with traditional masonries than cement-based mortars and are more durable than lime-based mortars. These characteristics make this kind of binder very interesting when preserving traditional buildings. The possibility of enhancing the positive mechanical and physical characteristics of NHL-based mortars with and without brick powder by the addition of water repellents or salt inhibitors is still an open and little-investigated issue.

In a coastal environment, the salt resistance of mortars, in addition to compatibility and sustainability characteristics, is currently a challenging goal [37–43]. Soluble salts present in salt water are able to penetrate porous building materials and crystallize within the pores, causing stress and internal damage. Moreover, marine aerosols deposit conspicuous amounts of salts on architectural surfaces that may retain humidity due to salt hygroscopicity and can penetrate within the material in the presence of free water. Usually, salt protection is achieved by preventing or reducing salt solution intake in porous materials or by applying water-repellent layers (e.g., silanes, polymeric layers, etc.) [44].

The addition of salt inhibitors has recently been proposed to enhance the durability of building materials (stone and bricks) to salt crystallization. Salt inhibitors are capable of modifying the crystalline morphology of the salt (e.g., ferrocyanide) or reducing the pressure between the growing crystal and the surface of the pore. The efficacy of inhibitors applied to building materials has been generally tested by mixing them in a soluble salt solution and evaluating the effects on the materials after the absorption of the solution and salt crystallization [45–48].

In recent years, the mixing of modifiers in a mortar during its production has been proposed to mitigate the salt crystallization effects as soon as the solution enters the mortar. The experimentation has generally been performed on lime-based and cement-based mortars [49–52].

Among others, two salt-inhibitor compounds are of increasing interest in enhancing the resistance of mortars: chitosan and diethylenetriamine penta(methylene phosphonic acid) (DTPMP). Chitosan is a linear polysaccharide, which acts by reducing the crystallization pressure between the growing crystal and the surface of the pore by forming a thin polymeric layer [48,50–54]. DTPMP is a stable phosphonic acid, which acts as a chelating agent for cations present in soluble salts; it is widely used in desalination plants to prevent scale formation and corrosion [55,56].

DTPMP and chitosan have been tested as possible inhibitors of salt crystallization in cement-based mortars [50,51,53–56], but, to the best of our knowledge, there are no studies on the performances of natural hydraulic lime (NHL) mortars with these compounds added. The aim of the present work was to evaluate the performance of compatible and sustainable mortars based on a natural hydraulic binder and natural hydraulic binder with the addition of brick powder to emulate the traditional brick-crushed ("cocciopesto") mortar. To enhance their resistance against salt crystallization, DTPMP and chitosan at were used different concentrations. The effects of these anti-salt agents as bulk admixtures were investigated. Special attention was paid to the evaluation of the rheology of the fresh mortars and the effects of the admixtures on the workability. Moreover, the behavior with respect to water was considered using capillary absorption and permeability determinations and a freeze–thaw resistance evaluation. To evaluate the performance of the anti-salt agents, salt-weathering tests were conducted by subjecting the specimens to absorption/drying cycles in a sodium sulfate solution [57,58]. All test results are represented through graphs created with the help of Origin 8.5 software in order to ensure easier interpretation of the results obtained.

2. Materials and Methods

Natural hydraulic lime mortar (MS series) and brick-crushed ("cocciopesto") mortar (MSC series) (average particles diameter 0.1 μm) mockups with the addition of chitosan and DTPMP were produced. The natural hydraulic lime (NHL), white lime by "Lafarge", obtained through a calcination process at 1000/1100 °C of limestone containing about 10% of diffused silica [59], and the Ticino sand by "VAGA-Mapei Group", a limestone silica sand with a size fraction of 0.1 ÷ 0.9 mm, were selected. The cocciopesto used was obtained by crushing "San Marco Terreal S.p.A" red full brick with a size fraction of 0.1 μm. Chitosan (medium molecular weight, technical grade) and DTPMP (diethylenetriaminapenta solution 50% (T) technical grade) were provided by "Sigma-Aldrich". The lime mortars were prepared by mixing NHL and sand with a mass ratio of 1:2, the cocciopesto mortars were prepared at a mass ratio of 1:1.7:0.3 NHL/sand/cocciopesto. The addition of the two admixtures was performed by replacing the mixing water with a 0.5%–0.25% aqueous solution of chitosan or DTPMP. Furthermore, mixtures with chitosan in powder form at 0.5 and 0.25% admixture/binder were prepared. A total of 350 mockups were produced, with 25 replicas for each formulation. Table 1 reports the names and composition of the different formulations.

Table 1. Sample composition of mockups. The lime: sand and brick powder ratio (in weight) is given in brackets.

Sample	Composition
MS	Natural hydraulic lime NHL + Sand (1:2)
MSCL05	Natural hydraulic lime NHL + Sand (1:2) + Chitosan (liquid) 0.5%
MSCL025	Natural hydraulic lime NHL + Sand (1:2) + Chitosan (liquid) 0.25%
MSD05	Natural hydraulic lime NHL + Sand (1:2) + DTPMP 0.5%
MSD025	Natural hydraulic lime NHL + Sand (1:2) + DTPMP 0.25%
MSCS05	Natural hydraulic lime NHL + Sand (1:2) + Chitosan (powder) 0.5%
MSCS025	Natural hydraulic lime NHL + Sand (1: 2) + Chitosan (powder) 0.25%
MSC	Natural hydraulic lime NHL + Sand + Brick powder (1:1.7: 0.3)
MSCCL05	Natural hydraulic lime NHL + Sand + Brick powder (1:1.7:0.3) + Chitosan (liquid) 0.5%
MSCCL025	Natural hydraulic lime NHL + Sand + Brick powder (1:1.7:0.3) + Chitosan (liquid) 0.25%
MSCD05	Natural hydraulic lime NHL + Sand + Brick powder (1:1.7:0.3) + DTPMP 0.5%
MSCD025	Natural hydraulic lime NHL+ Sand + Brick powder (1:1.7:0.3) + DTPMP 0.25%
MSCCS05	Natural hydraulic lime NHL + Sand + Brick powder (1:1.7:0.3) + Chitosan (powder) 0.5%
MSCCS025	Natural hydraulic lime NHL + Sand + Brick powder (1:1.7:0.3) + Chitosan (powder) 0.25%

The materials were mixed with tap water (w/b = 0.8) with a Mortar Mixer Paddle, 120 mm, triple helix-whip, threaded shM14 at a drilling speed of 100 rpm for a total of 3 min to allow a homogeneous distribution of different components. On the fresh mixtures, a flow test was performed following EN 1015-3 [60], using a Tecnotest manual flow table to determine the consistency of the mortar. Consistency was assessed by measuring the slump diameter obtained from the regulated amount of mortar on a shaking table. A total of three repetitions were carried out at t = 0, t = 30 and t = 60 min, respectively, to follow the variation in consistency over time. In addition, the viscosity of the fresh mixes was measured continuously from the right end of mixing using an IKA ROTAVISC hi-vi II viscometer, with an SP-10 rotating arm set at 60 rpm, for a total duration of 15 minutes, in order to evaluate the dynamic viscosity of the mixes over time.

Physical Behavior of the Hardened Samples

The mixes were poured into $5 \times 5 \times 2$ cm silicone molds without demolding agents in order not to affect the surface properties. These mockups were stored at 70% RH for 48 h to promote the setting phase. The curing phase was then carried out in a controlled environment at 19 °C and 50% RH, for 28 days. To evaluate the physical behavior of the cured specimens, capillary absorption, permeability, compressive strength, freeze–thaw resistance and salt resistance were assessed. All tests were carried out with three replicates for each formulation, and average values for each point were reported in the tables. Water vapor permeability was evaluated following UNI-EN-1015/19 [61]; in particular, the WDD = density of moisture flow rate; μ = moisture resistance factor; s_d = water vapor diffusion-equivalent air thickness were determined. Capillary water absorption of the samples was tested based on standard UNI-EN 1015/18 and UNI 10859 [62,63] at 20 ± 1 °C, the absorbed water was graphed against the square of time, then the capillary absorption coefficient CA was calculated as the slope of the initial part of the absorption graphs. The capillary index IC was evaluated by considering the ratio among the total water absorbed per surface unit over time and the total water*total time: IC = $\int f(Q_i)^* dt / (Q_{tf}^* t_f$ (where Q_i = quantity of water absorbed per surface unit at time i; t = time, t_f = final time. At the end of the absorption test, with fully impregnated mock-ups, it was possible to determine the total open porosity TOP as correspondent to the volume of water intruded by capillarity.

Compressive strength was tested on hardened mortars at 28 days according to EN 1015-11 [64].

In order to assess the frost resistance of the mortars, samples were subjected to freeze–thaw cycles according to EN 12371 [65]. The freeze–thaw cycles included a wetting period of three hours in which mockups were immersed in distilled water at room temperature, followed by a freezing step at −20 °C for 21 h. The salt resistance was tested according to EN 12370:2001 [66] with the following variations: Specimens were dried to constant weight at 20 °C prior to testing, then they were immersed in a 14% sodium sulphate decahydrate solution at ambient temperature (20 ± 5 °C) for 2 h and dried at room temperature (20 ± 5 °C) for 22 h instead of in an oven at 40 °C in order to simulate more realistic crystallization conditions (it is expected that a slower drying under standard environmental conditions would result in the formation of larger salt crystals in the mortar matrix, causing greater damage). Visual inspection was carried out after each cycle, while weight was monitored after each drying period to assess the salt solution absorption of salt-aged specimens. After reaching 15 cycles, the samples were desalinated by immersion in water at room temperature, and the salt extraction was monitored by measuring the electrical conductivity of the water. The total water volume was periodically replaced until constant electrical conductivity was achieved. After desalination, the samples were weighed, and the total weight variation was recorded.

After freeze–thaw cycles and the salt resistance tests, three replicates for each formulation underwent compressive strength test according to EN1015-11.

3. Results and Discussion

3.1. Results of Tests on Fresh Samples

Figure 1a,b shows the dynamic viscosity of NHL mortars (MS) (Figure 1a), NHL with brick powder (MSC) (Figure 1b) and with the different additives. In general, the viscosity of the samples decreased significantly in the 15 minutes of mixing, from about 3000 mPas/s to about 1000 mPas/s. The final values were quite similar for all samples. The NHL mortars show a more heterogeneous trend in viscosity values. The addition of DTPMP and chitosan increased the viscosity in the first minute of mixing, and then the viscosity decreased more than in the sample without additives. The presence of brick powder seems to determine a more homogeneous behavior, and the influence of additives in the mortar formulation did not produce significant variations in the dynamic viscosity values (Figure 1b).

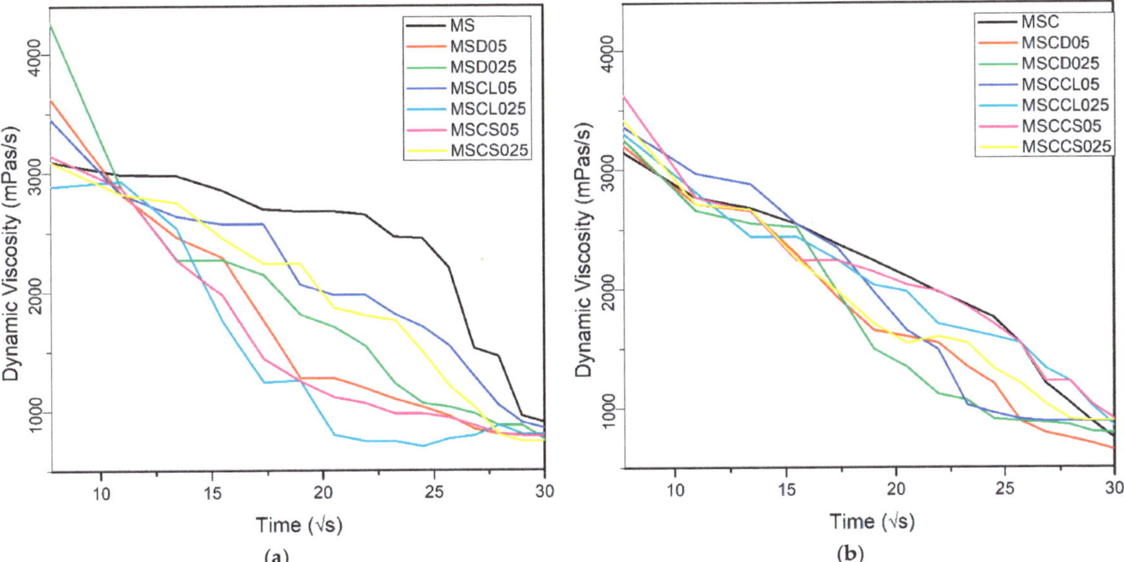

Figure 1. (**a**) Dynamic viscosity for NHL mortars (MS); (**b**) Dynamic viscosity for NHL with brick powder mortars (MSC).

The consistency of the different samples, expressed in terms of slump diameter (in mm), is shown in Table 2. Mortars based on NHL (MS) and NHL + brick powder (MSC) showed a similar consistency. The presence of DTPMP and chitosan in samples with NHL (MS series) had a different effect. DTPMP reduced the shrinkage to similar values independently of presence and amount of brick powder. The consistency of the mortars with added chitosan differs significantly: the samples produced with NHL show an increase in the slump diameter (239 and 226 mm), while samples with NHL and brick powder show values similar to those without chitosan (177 and 157 mm). The addition of chitosan to the powder has the same effect on the two series of samples (MS and MSC), with a slight reduction in consistency. The addition of DTPMP produces a light, not very thixotropic and soft mortar. This effect is more pronounced at the higher percentage (0.50%), especially in the MS sample series. Therefore, the MSD05 and MSD025 formulations do not seem to be suitable for vertical or inclined surfaces. On the other hand, chitosan produces mortars with a soft, homogeneous and thixotropic consistency, suitable for application on vertical surfaces.

Table 2. Flow test table measurements at 0, 30 and 60 min (Δ = average value in mm).

Sample	Slump Diameter (mm)			Δ(mm)
	t = 0′	t = 30′	t = 60′	
MS	180	171	156	169
MSD05	260	244	212	239
MSD025	243	225	209	226
MSCL05	160	142	126	143
MSCL025	164	153	122	146
MSCS05	175	158	145	159
MSCS025	162	152	144	153
MSC	179	169	154	167

Table 2. *Cont.*

Sample	Slump Diameter (mm)			Δ(mm)
	t = 0′	t = 30′	t = 60′	
MSCD05	184	175	171	177
MSCD025	164	157	149	157
MSCCL05	150	147	143	147
MSCCL025	160	153	145	153
MSCCS05	166	156	145	156
MSCCS025	165	152	141	153

3.2. Hardened Mortars Characterization

3.2.1. Hydric Behavior

Table 3 shows the apparent density, permeability characteristics and water absorption behavior of the mortars. Moreover, from the total absorbed water it was possible to determine the porosity of the specimens as TOP. Apparent density and total open porosity showed an inverse relationship, as expected. However, the DTPMP did not affect the porosity of MS samples, but reduced the porosity of MSC samples, while liquid chitosan slightly increased the porosity of both MS and MSC samples. Total open porosity and the permeability to water vapor are not perfectly correlated for each specimen; the differences are probably caused by a specific pore size distribution. It is possible that this difference also depends on an overall lower water absorption in the presence of certain additives. In particular, the use of DTPMP on MSC specimens caused a decrease in water absorption, whereas this behavior is not observed for MS specimens with similar ICs. Figure 2 highlights the fast water uptake of all mixtures with a saturation reached within 15 min since the beginning of the test. For MS specimens, the capillary absorption is more homogeneous between the different admixtures, while for MSC, an increased absorption is observed when chitosan is added (MSCCL05, MSCCS05, MSCCS025) Meanwhile, a decrease is observed when DTPMP is added (MSCD05 and slightly lower for MSCD025).

Table 3. Hydric behavior, data regarding water vapor permeability and capillary water absorption are listed: WDD = density of moisture flow rate; μ = moisture resistance factor; s_d = water vapor diffusion-equivalent air thickness; IC = capillary index; CA = coefficient of water absorption by capillarity; TOP = total open porosity as determined by the volume of water absorbed by capillary absorption.

Sample			Water Vapor Permeability				Capillary Water Absorption	
		Apparent density	TOP *	WDD	μ	sd	CA	IC
		g/cm^3	%	g/(m^2·s)		m	mg/cm^2*s1/2	
MS	mean	1.71	16.6	58.16	14.23	0.30	2.33	1.09
	σ	0.03	0.6	3.24	0.87	0.01	0.60	0.23
MSD05	mean	1.64	17.7	59.64	13.81	0.30	2.50	1.19
	σ	0.05	0.2	0.55	0.14	0.003	0.11	0.03
MSD025	mean	1.68	15.6	57.16	14.48	0.31	2.33	1.32
	σ	0.06	0.3	1.86	0.53	0.01	0.16	0.13
MSCL05	mean	1.61	19.0	53.13	15.69	0.34	3.44	1.17
	σ	0.03	0.4	1.87	0.61	0.01	0.35	0.07
MSCL025	mean	1.74	14.7	58.51	14.11	0.30	2.28	1.23
	σ	0.04	0	0.37	0.1	0.002	0.12	0.11
MSCS05	mean	1.69	16.5	54.10	15.42	0.33	2.17	1.31
	σ	0.01	0.3	3.42	1.03	0.02	0.09	0.14
MSCS025	mean	1.72	17.4	74.43	10.81	0.23	2.44	1.21
	σ	0.03	0.3	3.45	0.55	0.01	0.43	0.05
MSC	mean	1.62	21.8	52.29	15.96	0.34	2.20	0.71
	σ	0.18	0.9	0.70	0.23	0.004	0.45	0.01
MSCD05	mean	1.62	17.4	59.08	15.55	0.33	2.43	0.90
	σ	0.13	0.7	25.55	7.33	0.16	0.19	0.17

Table 3. *Cont.*

Sample		Apparent density	Water Vapor Permeability				Capillary Water Absorption	
			TOP *	WDD	μ	sd	CA	IC
		g/cm³	%	g/(m²·s)		m	mg/cm²*s1/2	
MSCD025	mean	1.65	16.9	72.60	11.18	0.24	2.44	1.15
	σ	0.07	2.7	8.18	1.42	0.03	0.19	0.24
MSCCL05	mean	1.55	21.1	61.35	13.95	0.30	2.17	0.95
	σ	0.05	0.6	16.55	4.14	0.09	0.09	0.10
MSCCL025	mean	1.51	23.3	77.48	10.49	0.23	2.74	0.95
	σ	0.07	0.8	13.43	2.06	0.04	0.66	0.11
MSCCS05	mean	1.54	23.3	48.34	17.37	0.38	3.43	0.96
	σ	0.11	4.4	0.65	0.25	0.01	0.36	0.15
MSCCS025	mean	1.63	19.9	40.58	21.05	0.45	2.28	1.01
	σ	0.02	0.5	3.50	1.94	0.041	0.12	0.04

* TOP was estimated by evaluating the volume of water absorbed at the end of the capillary test.

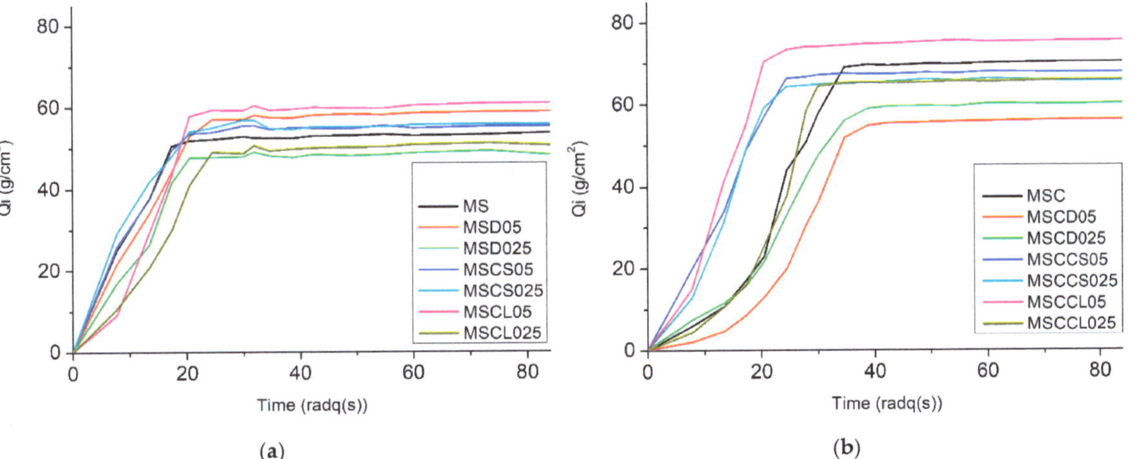

Figure 2. (**a**) Capillary absorption for MS sample series; (**b**) Capillary absorption for MSC sample series. Qi expresses the quantity of water absorbed per surface unit.

3.2.2. Determination of Frost Resistance

The weight variation in the MS samples during the cycles was characterized by a slight weight reduction in the first cycles followed by an almost constant weight (Figure 3). During the cycles, different trends were observed for admixed and non-admixed samples, especially in the presence of cocciopesto. MSC showed an increase in weight variation, possibly due to an increase in open porosity and a higher water retention rate. The lower weight variation observed with the added admixtures suggests that there is less variation in porosity and that the admixtures are effective at protecting the specimen from freeze–thaw, in particular MSCCL05, MSCD05 and MSCCS025 when the higher amount of additive is present in the mixture (0.5%). All MS mixtures showed negligible weight variation, but again MS showed a higher weight variation compared to admixed specimens. The presence of additives can mitigate, albeit partially, frost damage in the case of brick powder samples. The MS samples, on the other hand, showed a more uniform weight variation in the added mortars. At the end of the freeze–thaw cycles, most of the samples appear to be slightly crushed, without major damage. In the case of MSD05 and MSCD05 samples, the upper layer came off, as shown in Figure 4.

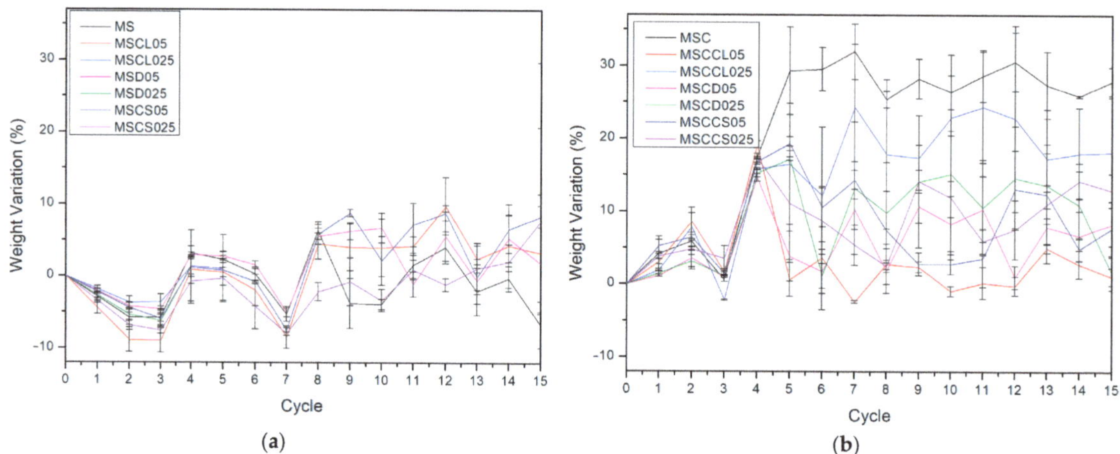

Figure 3. (**a**) Weight variation after each freeze–thaw cycle for MS sample series; (**b**) Weight variation after each freeze–thaw cycle for MSC sample series.

Figure 4. (**a**) MSD05 sample before (top left-hand corner) and after undergoing freeze–thaw cycles; (**b**) MSCD05 sample before (top left-hand corner) and after undergoing freeze–thaw cycles.

3.2.3. Determination of Salt Resistance

Table 4 reports the overall weight difference after sample desalination, and Figure 5 reports the weight variations during the salt crystallization test.

During the first cycles, the MS mortar samples show greater resistance to salt cycles. The weight variation for the first cycles proves to be almost zero, in contrast to the samples with brick powder component, which seems to increase the absorption and retention of water and salts (Figure 5a,b). Not all samples made it to the end of the 14 + 1 cycles; MSD025 and MSCL025 were severely damaged during cycles 6° and 8°, respectively, preventing further analysis (Figure 5a). At the end of the cycles and after desalination, the total weight variation of the MS samples was higher than that of the MSC samples. In addition, samples with added DTPMP and chitosan (liquid) at 0.50% showed less weight loss.

Table 4. Overall weight variation at the end of the test for resistance to salt crystallization.

Sample	Weight Variation (%)
MS	−18.48%
MSCL05	−16.73%
MSCL025	/
MSD05	−9.76%
MSD025	/
MSCS05	−10.34%
MSCS025	−14.03%
MSC	−7.92%
MSCCL05	−0.10%
MSCCL025	−4.02%
MSCD05	−1.10%
MSCD025	−4.99%
MSCCS05	−6.08%
MSCCS025	−4.94%

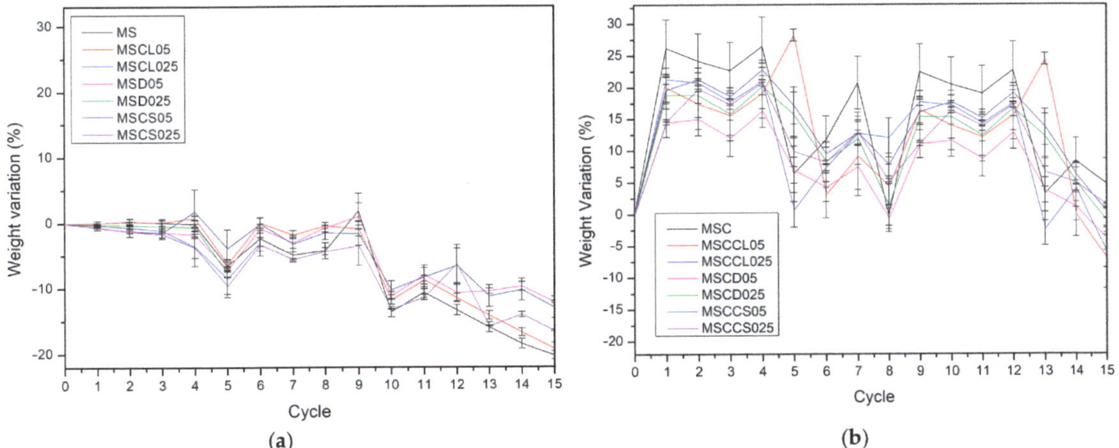

Figure 5. (**a**) Weight variation during salt crystallization test cycles of MS samples; (**b**) Weight variation during salt crystallization test cycles of MSC samples.

Salt-induced decay, based on a variation of the standardized method, determined the growth of efflorescences around each sample. The appearance of the specimens did not show any difference between the pure and added ones, but the slightest touch causes the effloresces to fall together with some mortar powder (Figure 6). Although they show a tendency to lose mass due to salt crystallization, their general appearance is less damaged than that of the samples with added brick powder (Figures S1–S4). Likely due to their higher water absorption, these samples are better able to retain the salt solution, limiting the weight loss of the damaged matrix.

(a)

(b)

Figure 6. (**a**) MSCL025 sample after 6° saline cycle; (**b**) MSD025 sample after 8° saline cycle.

3.2.4. Determination of Resistance to Compression of Hardened and Weathered Mortars

In order to evaluate the performances under compression of the mockups, tests were carried out by crushing mockups before and after salt crystallization and freeze–thaw cycles.

As Figure 7 shows, when comparing the compressive strength after 28 days and the water vapor permeability, an inverse relationship was observed. Thus, a higher porosity leads to lower compressive strength for both MS and MSC specimens independently of the admixture used. A slight positive effect of chitosan on the compressive strength of MS specimens was observed, possibly due to a higher moisture retention promoting a better NHL hydration (chitosan is a hygroscopic molecule able to retain water). A similar effect was observed when liquid chitosan was used in MSC. In MSC mixtures, the presence of DTPMP increases the density of the specimen and thus the compressive strength.

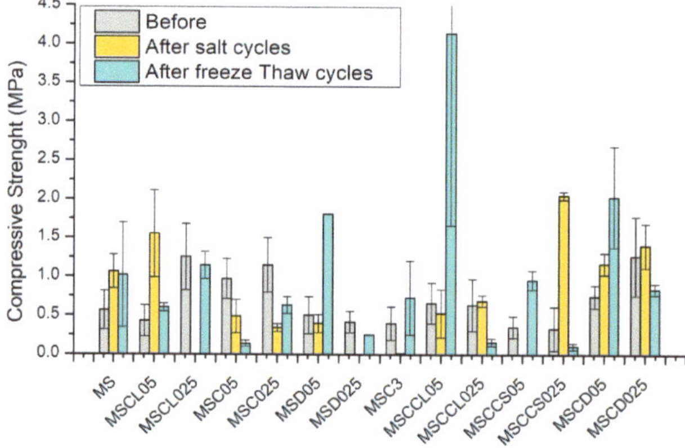

Figure 7. Compressive strength compared between undamaged, post-salt-crystallization and post-freeze–thaw-cycle samples.

The increase in compressive strength observed for MS and MSC samples after freeze–thaw cycles is attributable to a further hydration of the mortar and a pore structure able to withstand freezing cycles. The use of chitosan seems to slightly increase the resistance, with the exception of MSCS05. DTPMP guarantees compressive strength through MS and MSC samples.

In many cases, the specimens were unable to withstand the salt cycles and a general reduction in compressive strength was observed, with the notable exception of DTPMP in MSC specimens. Further hydration leading to higher compressive strength does not counterbalance the damaging effect of salt crystallization.

4. Conclusions

This study is the first to investigate the possibility of improving the performance and durability of NHL-based mortars in the presence of soluble salts and against freeze–thaw events by adding DTPMP and chitosan at different concentrations.

Chitosan has recently been re-evaluated and studied as a carbonation catalyst for lime mortars, while DTPMP, although generally used as a desalination treatment, has undergone various studies on its effect as a retarder in cement mortars. Currently, few sources have been found on the incorporation of these additives into natural hydraulic lime mortars, hence the purpose of our study. Special attention was paid to the evaluation of the rheology of the fresh mortars and the effects of the admixtures on the workability. Moreover, the behavior towards water was studied by capillary absorption and permeability determination.

Two sets of samples were prepared, based on NHL binder and NHL binder with brick powder. The structures based on cocciopesto generally exhibit improved durability and performance compared to pure lime mortars, approaching the most recent cement mortars' physical and chemical properties [31,34,35].

To evaluate the performance of the anti-salt agents, DTPMP and chitosan, salt weathering tests were carried out by subjecting the specimens to absorption/drying cycles in a sodium sulfate solution. Freeze–thaw resistance was also evaluated.

From a rheological point of view, DTPMP generally reduces the consistency of NHL-based mortars, producing soft and low thixotropic mixtures that are not always suitable for vertical or inclined surfaces. Chitosan produces mortars with a soft, homogeneous, and thixotropic consistency, suitable for application on vertical surfaces. Chitosan behaves differently depending on the composition of the samples by increasing consistency of the mortar in the presence of brick powder and decreasing it in samples without brick powder.

DTPMP is able to impart a partial hydrophobic behavior to the mortar, reducing water absorption by capillarity in both the series of samples. The resistance to salt crystallization is increased by the presence of brick powder, and, for these samples, the presence of DTPMP seems to make a positive contribution. Exceptions were the NHL samples with 0.5% of DTPMP and with 0.5% chitosan added, which proved to be not durable under the salt crystallization.

The freeze–thaw resistance does not seem to be increased by the presence of DTPMP and chitosan.

It is still difficult to predict the durability of the mortars against salt attack or thawing of mortars on the basis of their compositional, physical and chemical characteristics. Nevertheless, we have been able to provide, albeit loosely, an overview of the performance of different types of innovative mortars mixed with low environmental impactful products, of which the presence of pozzolanic mortars was the most prominent. Future investigations on these materials would clarify the dynamics of mortar hydration and the kinetics of the evolution of the chemical and physical properties of the formulations in presence of additives.

Supplementary Materials: The following supporting information can be downloaded at: https://www.mdpi.com/article/10.3390/coatings13081418/s1, Figure S1: MS series samples (part 1) before (left) and after (right) 14 saline cycles; Figure S2: MS series (part 2) samples before (left) and after (right) 14 saline cycles; Figure S3: MSC series samples (part 1) before (left) and after (right) 14 saline cycles; Figure S4: MSC series (part 2) samples before (left) and after (right) 14 saline cycles.

Author Contributions: Conceptualization, E.Z. and M.D.; Methodology, E.Z., L.F. and M.D.; Investigation, M.D. and L.F.; Resources, E.Z.; Data Curation, E.Z., L.F. and M.D.; Writing—Original Draft Preparation, M.D.; Writing—Review and Editing, E.Z., L.F. and M.D.; Supervision, E.Z. and L.F.; Funding Acquisition, E.Z. All authors have read and agreed to the published version of the manuscript.

Funding: This research received no external funding.

Institutional Review Board Statement: Not applicable.

Informed Consent Statement: Not applicable.

Data Availability Statement: Data is unavailable due to privacy restrictions.

Conflicts of Interest: The authors declare no conflict of interest.

Abbreviations

NHL	natural hydraulic lime mortars as defined by UNI EN 459-1:2015Building lime—Part 1: Definitions, specifications and conformity criteria [67]
MS	NHL Mortar Samples
MSC	Cocciopesto (minced bricks) -NHL Mortar Samples
C	Chitosan, specifically CL= chitosan in liquid solution at 0.5% or 0.25%, and CS= chitosan as solid powder
D	DTPMP, diethylenetriamine penta(methylene phosphonic acid)
05	0.5%. by binder weight
025	0.25%. by binder weight
RH	relative humidity
WDD	density of moisture flow rate according to UNI-EN-1015/19 [61]
μ	moisture resistance factor according to UNI-EN-1015/19 [61]
s_d	water vapor diffusion-equivalent air thickness according to UNI-EN-1015/19 [61]
CA	capillary absorption coefficient according to UNI-EN 1015/18 and UNI 10859 [62,63]
IC	capillary index IC according to UNI-EN 1015/18 and UNI 10859 [62,63], IC = $\int$ f(Q_i)*dt/(Q_{tf}*t_f) where Q_i= quantity of water absorbed per surface unit at time i; t = time, t_f = final time.
TOP	total open porosity determined by the water volume intruded according to UNI-EN 1015/18 [62]

References

1. Juimo Tchamdjou, W.H.; Cherradi, T.; Abidi, M.L.; Pereira de Oliveira, L.A. Influence of different amounts of natural pozzolan from volcanic scoria on the rheological properties of portland cement pastes. *Energy Procedia* **2017**, *139*, 696–702. [CrossRef]
2. Lea, F.M. *The Chemistry of Cement and Concrete*, 3rd ed.; Edward Arnold: London, UK, 1974.
3. Cabeza, L.F.; Rincón, L.; Vilariño, V.; Pérez, G.; Castell, A. Life cycle assessment (LCA) and life cycle energy analysis (LCEA) of buildings and the building sector: A review. *Renew. Sustain. Energy Rev.* **2014**, *29*, 394–416. [CrossRef]
4. Cemnet.com [Internet]. *The Global Cement Report*, 13th ed.; International Cement Review: Dorking, UK, 2019; Available online: https://www.cemnet.com/Publications/Item/182291/the-global-cement-report-13th-edition.html (accessed on 24 July 2021).
5. Gursel, A.P.; Masanet, E.; Horvath, A.; Stadel, A. Life-cycle inventory analysis of concrete production: A critical review. *Cem. Concr. Compos.* **2014**, *51*, 38–48. [CrossRef]
6. Robalo, K.; Costa, H.; do Carmo, R.; Julio, E. Experimental development of low cement content and recycled construction and demolition waste aggregates concrete. *Constr. Build. Mater.* **2021**, *273*, 121680. [CrossRef]
7. Medina, C.; Banfill, P.F.G.; Sánchez de Rojas, M.I.; Frías, M. Rheological and calorimetric behaviour of cements blended with containing ceramic sanitary ware and construction/demolition waste. *Constr. Build. Mater.* **2013**, *40*, 822–831. [CrossRef]
8. Medina, C.; Saez del Bosque, I.F.; Asensio, E.; Frías, M.; Sánchez de Rojas, M.I. Mineralogy and microstructure of hydrated phases during the pozzolanic reaction in the sanitary ware waste/Ca(OH)₂ system. *J. Am. Ceram. Soc.* **2016**, *99*, 340–348. [CrossRef]
9. Asensio, E.; Medina, C.; Frías, M.; Sánchez de Rojas, M.I. Fired clay-based construction and demolition waste as pozzolanic addition in cements. *J. Clean. Prod.* **2020**, *265*, 121610. [CrossRef]
10. Amiri, M.; Hatami, F.; Golafshani, E.M. Evaluating the synergic effect of waste rubber powder and recycled concrete aggregate on mechanical properties and durability of concrete. *Case Stud. Constr. Mater.* **2021**, *15*, e00639. [CrossRef]
11. Siddique, R.; Cachim, P. *Waste and Supplementary Cementitious Materials in Concrete: Characterisation, Properties and Applications*; Woodhead Publishing Limited: Cambridge, UK, 2018.
12. Rani, G.Y.; Krishna, T.J.; Murali, K. Strength studies on effect of glass waste in concrete. *Mater. Today Proc.* **2021**, *46*, 8817–8821. [CrossRef]

13. Ibrahim, K.I.M. Recycled waste glass powder as a partial replacement of cement in concrete containing silica fume and fly ash. *Case Stud. Constr. Mater.* **2021**, *15*, e00630. [CrossRef]

14. Abu-Saleem, M.; Zhuge, Y.; Hassanli, R.; Ellis, M.; Rahman, M.; Levett, P. Evaluation of concrete performance with different types of recycled plastica waste for kerb application. *Constr. Build. Mater.* **2021**, *293*, 123477. [CrossRef]

15. Jain, A.; Siddique, S.; Gupta, T.; Jain, S.; Sharma, R.K.; Chaudhary, S. Evaluation of concrete containing waste plastic shredded fibers: Ductility properties. *Struct. Concr.* **2021**, *22*, 566–575. [CrossRef]

16. Ojeda, J.P. A meta-analysis on the use of plastic waste as fibers and aggregates in concrete composites. *Constr. Build. Mater.* **2021**, *295*, 123420. [CrossRef]

17. Kadir, A.A.; Salim, N.S.A.; Sarani, N.A.; Rahmat, N.A.I.; Abdullah, M.M.A.B. Properties of fired clay brick incoporating with sewage sludge waste. *AIP Conf. Proc.* **2017**, *1885*, 020150.

18. Mathye, R.P.; Ikotun, B.D.; Fanourakis, G.C. The effect of dry wastewater sludge as sand replacement on concrete strengths. *Mater. Today Proc.* **2021**, *38*, 975–981. [CrossRef]

19. Kumar, M.; Shreelaxmi, P.; Kamath, M. Review on characteristics of sewage sludge ash and its partial replacement as binder material in concrete. In *Recent Trends in Civil Engineering*; Das, B.B., Nanukuttan, S.V., Patnaik, A.K., Panandikar, N.S., Eds.; Springer: Singapore, 2021; pp. 65–78.

20. Juenger, M.C.G.; Snellings, R.; Bernal, S.A. Supplementary cementitious materials: New sources, characterization, and performance insights. *Cem. Concr. Res.* **2019**, *122*, 257–273. [CrossRef]

21. Muslim, F.; Wong, H.S.; Choo, T.H.; Buenfeld, N.R. Influences of supplementary cementitious materials on microstructure and transport properties of spacer-concrete interface. *Cem. Concr. Res.* **2021**, *149*, 106561. [CrossRef]

22. Delgado Rodrigues, J.; Grossi, A. Indicators and ratings for the compatibility assessment of conservation actions. *J. Cult. Herit.* **2007**, *8*, 32–43. [CrossRef]

23. *EUROPEAN QUALITY PRINCIPLES for EU-Funded Interventions with Potential Impact upon Cultural Heritage, 2020*; ICOMOS International Secretariat: Charenton-le-Pont, France, 2020; ISBN 978-2-918086-36-9.

24. Bertolin, C.; Loli, A. Sustainable interventions in historic buildings: A developing decision making tool. *J. Cult. Herit.* **2018**, *34*, 291–302. [CrossRef]

25. Buda, A.; de Place Hansen, E.J.; Rieser, A.; Giancola, E.; Pracchi, V.N.; Mauri, S.; Marincioni, V.; Gori, V.; Fouseki, K.; Polo López, C.S.; et al. Conservation-Compatible Retrofit Solutions in Historic Buildings: An Integrated Approach. *Sustainability* **2021**, *13*, 2927. [CrossRef]

26. Apostolopoulou, M.; Aggelakopoulou, E.; Bakolas, A.; Moropoulou, A. Compatible Mortars for the Sustainable Conservation of Stone in Masonries. In *Advanced Materials for the Conservation of Stone*; Hosseini, M., Karapanagiotis, I., Eds.; Springer: Cham, Seitzerland, 2018. [CrossRef]

27. Do Rosário Veiga, M.; Fragata, A.; Ana Luisa Velosa, A.L.; Magalhães, A.C.; Margalha, G. Lime-Based Mortars: Viability for Use as Substitution Renders in Historical Buildings. *Int. J. Archit. Herit.* **2010**, *4*, 177–195. [CrossRef]

28. Moropoulou, A.; Bakolas, A.; Bisbikou, K. Investigation of the technology of historic mortars. *J. Cult. Herit.* **2000**, *1*, 45–58. [CrossRef]

29. Moropoulou, A.; Biscontin, G.; Theoulakis, P.; Bisbikou, K.; Theodoraki, A.; Chondros, N.; Zendri, E.; Bakolas, A. Study of mortars in the Medieval City of Rhodes. In *Conservation of Stone and Other Materials: Proceedings of the International RILEM/UNESCO Congress Held at the UNESCO Headquarters, Paris, France, 29 June–1 July 1993*; Unesco: Paris, France, 1993; pp. 394–401.

30. Matias, G.; Faria, P.; Torres, I. Lime mortars with heat treated clays and ceramic waste: A review. *Constr. Build. Mater.* **2014**, *73*, 125–136. [CrossRef]

31. Torres, I.; Matias, G.; Faria, P. Natural hydraulic lime mortars-The effect of ceramic residues on physical and mechanical behaviour. *J. Build. Eng.* **2020**, *32*, 101747. [CrossRef]

32. Klisińska-Kopacz, A.; Tišlova, R. The Effect of Composition of Roman Cement Repair Mortars on Their Salt Crystallization Resistance and Adhesion. *Procedia Eng.* **2013**, *57*, 565–571. [CrossRef]

33. Zendri, E.; Lucchini, V.; Biscontin, G.; Morabito, Z.M. Interaction between clay and lime in "cocciopesto" mortars: A study by 29Si MAS spectroscopy. *Appl. Clay Sci.* **2004**, *25*, 1–7. [CrossRef]

34. Sánchez de Rojas Gómez, M.I.; Frías Rojas, M. Natural pozzolans in eco-efficient concrete. In *Eco-Efficient Concrete*; Woodhead Publishing: Cambridge, UK, 2013; pp. 83–104.

35. Apostolopoulou, M.; Asteris, P.G.; Armaghani, D.J.; Douvika, M.G.; Lourenço, P.B.; Cavaleri, L.; Bakolas, A.; Moropoulou, A. Mapping and holistic design of natural hydraulic lime mortars. *Cem. Concr. Res.* **2020**, *136*, 106167. [CrossRef]

36. Silva, B.A.; Pinto, A.F.; Gomes, A. Natural hydraulic lime versus cement for blended lime mortars for restoration works. *Constr. Build. Mater.* **2015**, *94*, 346–360. [CrossRef]

37. Ashall, G.; Butlin, R.N.; Teutonico, J.M.; Martin, W. Development of Lime Mortar For-mulations for use in Historic Buildings. In *Durability of Building Materials & Components 7*; Routledge: London, UK, 2004; Volume 1.

38. Lubelli, B.; van Hees, R.P.J.; Groot, C.J.W.P. The role of sea salts in the occurrence of different damage mechanisms and decay patterns on brick masonry. *Constr. Build. Mater.* **2004**, *18*, 119–124. [CrossRef]

39. Espinosa-Marzal, R.M.; Scherer, G.W. Advances in Understanding Damage by Salt Crystallization. *Acc. Chem. Res.* **2010**, *43*, 897–905. [CrossRef]

40. Falchi, L.; Corradini, M.; Balliana, E.; Zendri, E. Urban Scale Monitoring Approach for the Assessment of Rising Damp Effects. *Venice Sustain.* **2023**, *15*, 6274. [CrossRef]

41. Henriques, F.M.A. Challenges and perspectives of replacement mortars in architectural conservation. In *International Workshop. Repair Mortars for Historic Masonry, Rilem Technical Committee*; TU Delft: Delft, The Netherland, 2009; pp. 26–28.

42. Aggelakopoulou, E.; Ksinopoulou, E.; Eleftheriou, V. Evaluation of mortar mix designs for the conservation of the Acropolis monuments. *J. Cult. Herit.* **2022**, *55*, 300–308. [CrossRef]

43. Falchi, L.; Zendri, E.; Capovilla, E.; Romagnoni, P.; De Bei, M. The behaviour of water-repellent mortars with regards to salt crystallization: From mortar specimens to masonry/render systems. *Mater. Struct.* **2017**, *50*, 66. [CrossRef]

44. Granneman, S.J.C.; Lubelli, B.; van Hees, R.P.J. Mitigating salt damage in building materials by the use of crystallization modifiers—A review and outlook. *J. Cult. Herit.* **2019**, *40*, 183–194. [CrossRef]

45. Gulotta, D.; Goidanich, S.; Tedeschi, C.; Toniolo, L. Commercial NHL-containing mortars for the preservation of historical architecture. Part 2: Durability to salt decay. *Constr. Build. Mater.* **2015**, *96*, 198–208. [CrossRef]

46. Feijoo, J.; Ergenç, D.; Fort, R.; de Buergo, M. Addition of ferrocyanide-based compounds to repairing joint lime mortars as a protective method for porous building materials against sodium chloride damage. *Mater. Struct.* **2019**, *54*, 14. [CrossRef]

47. Houck, J.; Scherer, G.W. Controlling stress from salt crystallization. In *Fracture and Failure of Natural Building Stones*; Springer: Berlin/Heidelberg, Germany, 2006; pp. 299–312.

48. Bracciale, M.P.; Sammut, S.; Cassar, J.; Santarelli, M.L.; Marrocchi, A. Molecular Crystallization Inhibitors for Salt Damage Control in Porous Materials: An Overview. *Molecules* **2020**, *25*, 1873. [CrossRef]

49. Granneman, S.J.C.; Lubelli, B.; van Hees, R.P.J. Effect of mixed in crystallization modifiers on the resistance of lime mortar against NaCl and Na2SO4 crystallization. *Constr. Build. Mater.* **2019**, *194*, 62–70. [CrossRef]

50. Ustinova, Y.V.; Nikiforova, T.P. Cement Compositions with the Chitosan Additive. *Procedia Eng.* **2016**, *153*, 810–815. [CrossRef]

51. Lasheras-Zubiate, M.; Navarro-Blasco, I.; Fernández, J.M.; Alvarez, J.I. Studies on chitosan as an admixture for cement-based materials: Assessment of its viscosity enhancing effect and complexing ability for heavy metals. *J. Appl. Polym. Sci.* **2011**, *120*, 242–252. [CrossRef]

52. Lubelli, B.; Nijland, T.G.; van Hees, R.P.J.; Hacquebord, A. Effect of mixed in crystallization inhibitor on resistance of lime–cement mortar against NaCl crystallization. *Constr. Build. Mater.* **2010**, *24*, 2466–2472. [CrossRef]

53. Muxika, A.; Etxabide, A.; Uranga, J.; Guerrero, P.; De La Caba, K. Chitosan as a bioactive polymer: Processing, properties and applications. *Int. J. Biol. Macromol.* **2017**, *105*, 1358–1368. [CrossRef]

54. Franzoni, E.; Sassoni, E.; Marrone, C. Development of hydroxyapatitechitosan-based treatments for the mitigation of salt damage in globigerina limestone. In Proceedings of the SWBSS 2021–Fifth International Conference on Salt Weathering of Buildings and Stone Sculptures Edited by Barbara Lubelli Ameya Kamat Wido Quist, Delft, The Netherlands, 22–24 September 2021; pp. 233–240.

55. Mohamed, M.F.; Bayat, P.; Kelland, M.A. Environmentally Friendly Phosphonated Polyetheramine Scale Inhibitors—Excellent Calcium Compatibility for Oilfield Applications. *Inpowderrial Eng. Chem. Res.* **2020**, *59*, 9808–9818.

56. Ochoa, N.; Baril, G.; Moran, F.; Pébère, N. Study of the properties of a multi-component inhibitor used for water treatment in cooling circuits. *J. Appl. Electrochem.* **2002**, *32*, 497–504. [CrossRef]

57. Ruiz-Agudo, E.; Lubelli, B.; Sawdy, A.; van Hees, R.; Price, C.; Rodriguez-Navarro, C. An integrated methodology for salt damage assessment and remediation: The case of San Jerónimo Monastery (Granada, Spain). *Environ. Earth Sci.* **2011**, *63*, 1475–1486. [CrossRef]

58. Kamat, A.; Palin, D.; Lubelli, B.; Schlangen, E. Tunable chitosan-alginate capsules for a controlled release of crystallisation inhibitors in mortars. *MATEC Web Conf.* **2023**, *378*, 02011. [CrossRef]

59. Lafarge White Lime Technical Data Sheet. Available online: https://www.ctseurope.com/img/cms/documentazione/23/INGLESE/LAFARGE%20WHITE%20LIME_TDS.pdf (accessed on 12 January 2022).

60. UNI-EN 1015-3:2007; Ente Italiano di Normazione, Metodi di Prova per Malte per Opere Murarie-Parte 3: Determinazione della Consistenza della Malta Fresca (Mediante Tavola a Scosse). Ente Nazionale Italiano di Unificazione (UNI): Rome, Italy, 2007.

61. UNI-EN 1015/19:2008; Ente Italiano di Normazione, Metodi di Prova per Malte per Opere Murarie-Parte 19: Determinazione della Permeabilità al Vapore D'acqua delle Malte da Intonaco Indurite. Ente Nazionale Italiano di Unificazione (UNI): Rome, Italy, 2008.

62. UNI-EN-1015/18:2004; Ente Italiano di Normazione, Metodi di Prova per Malte per Opere Murarie–Determinazione del Coefficiente di Assorbimento D'acqua per Capillarità della Malta Indurita. Ente Nazionale Italiano di Unificazione (UNI): Rome, Italy, 2004.

63. EN 10859:2000; Beni Culturali-Materiali Lapidei Naturali ed Artificiali-Determinazione Dell'assorbimento D'acqua per Capillarità. Ente Nazionale Italiano di Unificazione (UNI): Rome, Italy, 2000.

64. EN 1015-11:2019; Metodi di Prova per Malte per Opere Murarie–Parte 11: Determinazione della Resistenza a Flessione e a Compressione della Malta Indurita. Ente Nazionale Italiano di Unificazione (UNI): Rome, Italy, 2019.

65. UNI EN 12371:2003; Ente Italiano di Normazione, Metodi di Prova per Pietre Naturali–Determinazione della Resistenza al Gelo. Ente Nazionale Italiano di Unificazione (UNI): Rome, Italy, 2003.

66. *UNI EN 12370:2001*; Ente Italiano di Normazione, Metodi di Prova per Pietre Naturali–Determinazione della Resistenza alla Cristallizzazione dei Sali. Ente Nazionale Italiano di Unificazione (UNI): Rome, Italy, 2001.
67. *UNI EN 459-1:2015*; Ente Italiano di Normazione, Calci da costruzione - Parte 1: Definizioni, specifiche e criteri di conformità. Ente Nazionale Italiano di Unificazione (UNI): Rome, Italy, 2015.

Article

Photonic Applications for Restoration and Conservation of 19th Century Polychrome Religious Wooden Artworks

Victoria Atanassova [1] [ID], **Monica Dinu** [1,*] [ID], **Sultana-Ruxandra Polizu** [2,3] and **Roxana Radvan** [1]

1   National Institute of Research and Development for Optoelectronics—INOE 2000, 077125 Măgurele, Romania; victoria.atanassova@inoe.ro (V.A.)
2   Department of Systematic Theology, Practice and Sacred Art, Faculty of Orthodox Theology "Justinian the Patriarch", University of Bucharest, 040155 Bucharest, Romania
3   IORUX Restorations, 040155 Bucharest, Romania
*   Correspondence: monica.dinu@inoe.ro

Abstract: The present paper reports the multi-analytical approach for the removal of thick layers of metallic overpaints from a Brancovan iconostasis of the "Holy Trinity" church in Măgureni, România, which was built in 1694. After a restoration procedure at the beginning of the 20th century, the polychrome sculpture of the frame, which was initially gilded with a thin silver foil, was covered with a thick metallic overpaint layer imitating silver and gold. Currently, the conservation project of the church is focused on restoring the original aspect; thus, the overpainting that presented strong oxidation and soiling was removed. The adopted conservation methodology involved physicochemical characterization of the pictorial layers via optical microscopy, laser-induced breakdown spectroscopy, and Fourier-transform infrared spectroscopy, followed by the removal of the overpaints. The cleaning tests were performed by evaluating several methods in order to find the proper regime that would help preserve as much of the underlying polychrome layers as possible. Based on the tests, it was decided that the best solution was to use laser cleaning for the rough removal of the metallic paint overlayers and finalizing with chemical cleaning.

Keywords: LIBS; FTIR; laser cleaning; polychrome artworks

Citation: Atanassova, V.; Dinu, M.; Polizu, S.-R.; Radvan, R. Photonic Applications for Restoration and Conservation of 19th Century Polychrome Religious Wooden Artworks. *Coatings* **2023**, *13*, 1235. https://doi.org/10.3390/coatings13071235

Academic Editor: Marko Petric

Received: 17 May 2023
Revised: 16 June 2023
Accepted: 7 July 2023
Published: 11 July 2023

1. Introduction

The "Holy Trinity" church (also called the Cantacuzini church) is placed in the locality of Măgureni, which is a part of Prahova County in Romania. It was built during 1671–1674 as a chapel on the east side of the Cantacuzini estate in Măgureni. The painting of the church was executed in 1694. In 1838, a strong earthquake destroyed the vaults and the upper parts of the church, which were rebuilt again in 1839, but only in 1925, it was painted and the icons of the iconostasis were replaced [1].

The iconostasis, with a width of 549 cm and a height of 520 cm reaching the upper limit of the *molenia* icons, has four registers of icons arranged in a classical form, as can be observed in Figure 1. The iconostasis was manufactured in a manner specific to the end of the 17th century, the era of Constantin Brâncoveanu, but influenced by the Baroque style as well, considering the oval shape of the icons from the register of the prophets and the abundance of polychrome decorations with plant ornaments. To achieve more special chromatic effects, the areas in the background of the sculpted decorations were painted in vivid shades of orange and blue.

The metallic leaf is usually very thin (0.1–10 μm) and the embellished surfaces can be easily damaged, resulting in losses of the metal or cracking, which is due to environmental factors and aging of the underlying materials [2,3]. Probably due to the fact that it appeared degraded, the polychrome sculpture of the iconostasis was covered with paint imitating gilding (liquid bronze), hiding the details of the sculpture and affecting the overall aesthetic. The first coat of metallic paint was applied most probably during the conservation procedures in 1925, and the second was applied later, also using gold metallic paint. Such practice was well known and adopted not only in the Eastern Orthodox churches [4], but all around Europe, as reported in similar cases [2,5–8]. Since the last intervention, the appearance of the iconostasis has changed over time, manifested by the accumulation of adherent and non-adherent deposits on the surface, strong oxidation of the silvery paint applied to the sculpture, cracks, dislocations, and loss of sculptural elements. Considering all the facts, it was decided that the best approach, from a conservation point of view, is to remove the degraded overpainting and reveal the original polychrome pictorial layer. The methodology involved a multi-analytical approach consisting of documentation and examination of the morphology of the surfaces, stratigraphic layers, and constituent materials, followed by the overpaint removal by combining the action of a laser and chemical solvents.

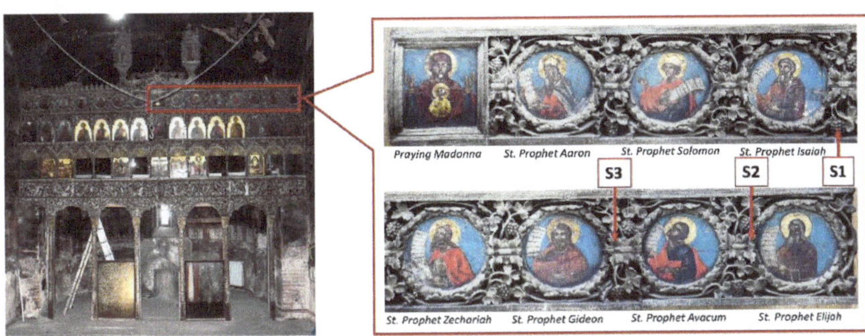

Figure 1. The iconostasis of the "Holy Trinity" church (**left**). In the red rectangle, the fragment of the 4th register cleaned with a laser is shown. Closer details and the analysed areas of the altarpiece (**right**).

Photonic technology has become an important part of the conservation and restoration of cultural heritage. Lasers have gained popularity in the field, especially as polychrome multi-layered artworks pose many challenges for conservation and restoration due to their complex composition, structure, and degradation mechanisms. Comprehensive analytical techniques, such as laser-induced breakdown spectroscopy (LIBS), complemented by Fourier transform infrared spectroscopy (FTIR), are particularly useful for studying and characterizing these materials. The combination of FTIR and LIBS can provide complementary information about the composition and structure of the artwork. FTIR can provide information on the organic components of the materials, such as binders, whereas LIBS can provide information on inorganic elements present in the materials, such as pigments and fillers. The use of these techniques can help determine the original paint composition, identify areas of restoration or overpainting, and provide information on the materials and methods used in the artwork's structure.

Laser cleaning, on the other hand, is a well-established technique for the restoration of artworks, becoming a reliable alternative to traditional cleaning methods since it enables a selective and controlled removal of unwanted materials with immediate feedback [9–11]. The most commonly used lasers are solid-state Q-switched Nd:YAG lasers, that generate pulses with a duration of several nanoseconds. They provide high efficiency in the removal of surface-adherent deposits due to the pronounced photothermal and photomechanical effects confined within a certain depth of the surface layer. There are many reports in the literature on the use of QS YAG:Nd lasers for cleaning gilded wooden artworks [2,12,13], but SFR YAG:Nd [14] and Er:YAG lasers emitting at 2940 nm wavelength have also been reported as alternative cases [8,15,16].

A combined methodology of applying solvents first and then laser cleaning was reported as well [17]. Striber et al. found that the laser irradiation weakened the bond between the overpaint and the gold leaf due to the induced photomechanical stress, and the coating could be easily removed mechanically after that [18]. On the other hand, Gaspar et al. observed that the brass-based overpaint could not be removed with a Q-switched Nd:YAG laser generating 1064 nm without damaging the gold leaf [7].

2. Materials and Methods

The altarpiece. The investigation focused on a part of the 4th register of the iconostasis representing the Mother of God in the *Oranta* position, placed in the centre and surrounded by 7 round-shaped icons on the left and right sides (Figure 1). The register has a rectangular shape with dimensions of 550 cm × 35 cm. Each medallion depicts a prophet and has a diameter of 27 cm. The register is made of two panels of linden wood, positioned horizontally. The medallions are separated from each other by a vast space decorated with carved ornamental plant motifs. Each medallion is surrounded by laurel garlands and the remaining space is decorated with wine motifs (branches, leaves, and grapes).

The preliminary characterization of the object was undertaken by employing a multi-analytical approach including several techniques, as described below.

Microscopic observations. The surface morphology and the stratigraphy of the pictorial layers of the frame were observed via optical microscopy (OM) using a Leica M205FA fluorescence stereomicroscope. This was performed on 3 samples (some of the micrographs are shown in Figure 2), which were taken from areas S1–S3 (pointed out in Figure 1), where the pictorial layer was detached from the wooden frame, so as to limit the invasiveness of the object. The photomicrographs were acquired with the 0.63 × PlanApo objective at different magnifications, ranging from 9× to 101×. Laser cleaning was evaluated using a portable Leica DMS 300 microscope with an integrated digital camera.

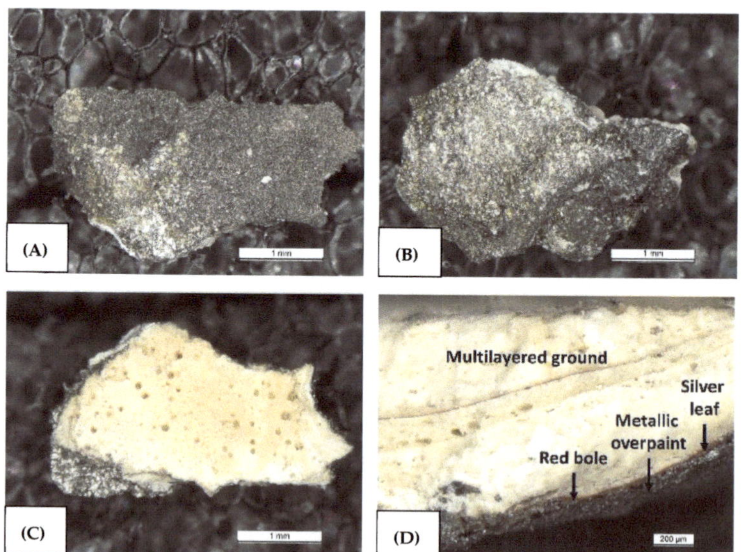

Figure 2. Microscopic images (**A**) S2, (**B**) S3, (**C**) S2—ground, and (**D**) S1—stratigraphy.

Elemental analyses. The characterization of the elemental composition of the materials used for the decoration was performed directly on the iconostasis via laser-induced breakdown spectroscopy (LIBS), providing compositional depth profiles as well. The LIBS spectra were recorded using a handheld spectrometer from SciAps operating in an Argon purge environment with 1–20 pulses. The laser used for irradiation is a Q-switched Nd:YAG emitting at 1064 nm, with an energy of 5 mJ and a laser spot of 50 µm. The system is equipped with three spectrometers that provide a spectral range of 190 nm to 950 nm.

The spectra were processed in OriginLab, and the chemical lines were identified using the SciAps software and the NIST database [19].

Molecular characterization. Fourier-transform infrared spectroscopy (FTIR) in the attenuated total reflectance (ATR) mode was applied to 3 samples (S1–3, areas shown in Figure 1) containing both the pictorial and ground layers. The equipment used was a Perkin Elmer Spectrum Two FTIR spectrometer equipped with a PIKE GladiATR accessory. The spectra were acquired in the 4000–380 cm^{-1} spectral region at a resolution of 4 cm^{-1}, averaging 32 scans. They are presented in transmission (%T), with automatic correction of the CO_2 lines and the baseline. Data processing was performed using OriginPro 2021b and interpreted based on the relevant literature.

Laser cleaning. The laser removal of the overpainting was performed on the frame of part of the register shown in the red rectangle in Figure 1. It was carried out with a Q-switched Nd:YAG laser (Palladio by Quanta System) using a wavelength of 1064 nm, pulse duration of 8 ns, pulse repetition rate of 10–20 Hz, and an optimal working energy per pulse of 350 mJ. The beam delivery system consists of an articulated arm that closely follows the surface relief. The handpiece was kept at an approximate distance of 20 cm from the surface, thus delivering a beam with a spot size of ~0.4 cm^2. Under these conditions, the working laser fluence was about 0.8 J/cm^2.

3. Results and Discussion

Following the methodology adopted from other research teams dealing with similar cases [13], the altarpiece was first examined using the outlined techniques, after which the cleaning tests were performed. The results are described below.

3.1. Preliminary Physico-Chemical Analysis

In order to decide the proper conservation approach, the altarpiece was analysed and the fundamental information concerning the morphology and chemical composition of the treated surfaces was gathered. Preliminary characterization is essential because it supports the conservation–restoration procedures and allows proper work in accordance with the original materials.

Microscopic images were acquired on some small detached fragments, and it was observed that the surface of the frame had suffered severe discoloration due to dirt deposits and the oxidation of the overpainted layers (Figure 2A,B). In addition to the yellowish color and porous structure in its base (Figure 2C), the ground seemed to be composed of several layers with a thickness ranging from 100 microns to several hundreds of microns divided by thin intermediate layers with an approximate thickness of several microns (Figure 2D). Such a multilayered structure of the ground is a common technique applied by post-Byzantine iconographers [20]. The stratigraphy showed a notable red layer after the ground, most probably a red bole, with varying thicknesses between 10 and 50 μm, which is a traditional base for the metallic leaf. The leaf itself is not so visible in the stratigraphic microscopic pictures. The thickness of the metallic overpaint was not uniform and was measured to be from several hundred microns to several milimetres. The preparation layer varied between 2 and 5 mm in thickness, covering both the elements of the relief and those in the background. The gilding of the sculptural elements was performed with silver leaf applied in the traditional manner on a layer of bole.

The FTIR analysis of the paint and the ground layers is presented in Figure 3. Since the spectra of the paint and the ground of S1 and S2 are similar, only S1 is presented in the figure. The infrared analysis indicated the presence of inorganic compounds characteristic of earth and clay minerals [21,22]. Kaolinite was identified via the O–H stretching vibrations at 3694, 3651, and 3621 cm^{-1}; the vibrations of the Si–O–Si and Si–O–Al groups at 1029 and 1008 cm^{-1}, respectively; and the Al–O–H vibration at 913 cm^{-1}. Quartz was indicated by the characteristic doublet at 797–777 cm^{-1}, and the bands at 1163, 695, and 464 cm^{-1}. The bands at 534 and 430 cm^{-1} were ascribed to iron oxides [23,24]. The presence of calcite ($CaCO_3$) was observed in all the spectra, as it was much more pronounced in the ground of S3 than in the grounds of the other two samples. This was detected through the weak combination modes at 2512 and 1795 cm^{-1} and the prominent stretching and bending modes of the carbonate group at 1412, 872, and 713 cm^{-1} [25]. Gypsum (CaSO$_4$·2H$_2$O) was found in all the spectra via the characteristic stretching (3526 and 3401 cm^{-1}) and bending (1682 and 1623 cm^{-1}) bands of the O–H bond, and the vibration modes of the sulphate group at 1106, 673, and 600 cm^{-1} [24]. The origin of the calcite and the gypsum in the paint could be from their possible use as mineral fillers or from the ground layers [26]. Given the observed multilayered structure of the ground, it is possible that the separate constituent layers could have been prepared using different recipes. It was not uncommon for the ground to consist of calcite-rich and gypsum-rich sub-layers in different ratios [20].

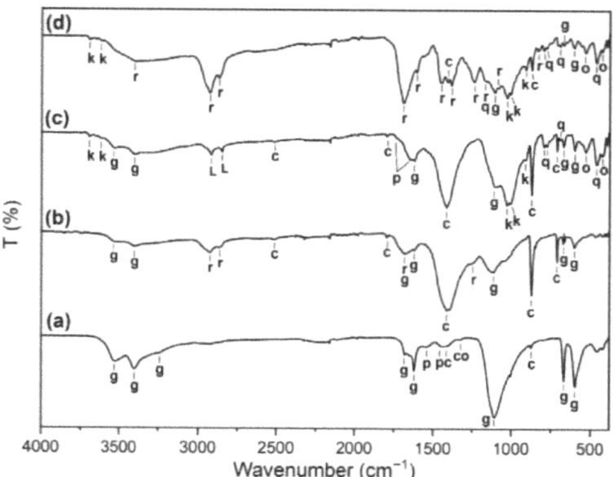

Figure 3. ATR-FTIR analysis of the paint and the ground layers of part of the samples: (**a**) S1—ground layer, (**b**) S3—ground layer, (**c**) S1—paint layer, and (**d**) S3—paint layer. Legend: k—kaolinite; g—gypsum; c—calcite; q—quartz; o—iron oxides; p—proteins; L—lipids; r—resin; co—calcium oxalate.

Some organic compounds were found as well. Proteins were present in both the paint and the grounds of S1 and S2, as determined by the emerging bands at 1646, 1544, and 1446 cm^{-1} assigned to C=O stretching (amide I), the combination of NH bending and CN stretching (amide II), and C-H bending (amide III) vibrations, respectively [27]. In the spectra of the paint in S1 and S2, only the bands associated with amide I and partly amide II appeared since amide III overlapped with the strong broad absorption of the carbonate group around 1410 cm^{-1}. On the other hand, the amide I overlapped with the bending vibration modes of the hydroxyl bond in gypsum. The vibrations emerging at 2920 and 2850 cm^{-1} in the spectra of the paint in S1 and S2, respectively, are ascribed to the stretching mode of the CH bonds. Together with the weak absorption (appearing as a shoulder) at 1736 cm^{-1}, this indicates the presence of lipids. Their occurrence is a clear indicator of an organic binder that can be associated with the saturated fatty acids of the proteinaceous medium, which in this case is most likely the egg yolk [28,29]. The identified proteinaceous species in the ground could be linked with animal glue, which is a common gluing ingredient in traditional recipes for the preparation layers in iconographic art [20]. It is highly likely that the intermediate thin layers observed in the multilayered ground are indeed animal glue.

The spectra of S3 show strong features of natural tree (pine) resin associated with the strong CH stretching modes at 2929 and 2870 cm^{-1}, the strongly broadened C=O stretching at 1693 cm^{-1}, the shoulder at 1609 cm^{-1}, the aliphatic C–H bending bands at 1483 and 1387 cm^{-1}, the C–O mode of the ester groups at 1240 cm^{-1}, the C–O stretching bands of the acid and alcohol groups at 1163 and 1081 cm^{-1}, and the band at 834 cm^{-1} [29,30]. The ground layer of S3 showed the presence of the resin, too. Since the bands were much weaker than those in the paint, it was suggested that they could come from the upper layer. The appearance of pine resin in such paintings is not a single case reported in the literature [31,32]. However, due to the lack of spectral features of the resin in the other samples, it cannot be concluded whether the resin was used as a finish layer against humidity or something else.

Calcium oxalate was also identified in the ground layers of S1 and S2 via the C=O stretching at 1326 cm^{-1}. Its occurrence could be linked to the degradation of organic layers or biodeterioration by the activity of microorganisms [33].

The LIBS results obtained from averaging the spectra from 20 pulses confirmed the presence of the following chemical elements: Ca, Al, Fe, Na, K, Si, Sr, H, and C, yet no S lines were detected probably due to the fact that sulphur is present in low concentration, below the spectrometer's limit of detection (Z300 S_{LOD} = 1%–2%). The LIBS spectra are presented in Figure 4a,b for the main elements identified. Traces of Ti, Mg, Mn, and Li were detected as well, which were linked to the earth and clay minerals in the composition of the materials. Potassium was also found as a trace element and might be associated with soiling deposits or the formation of salts on the surface [34]. These findings are in accordance with the FTIR interpretation, showing the presence of gypsum, calcite, and iron oxides.

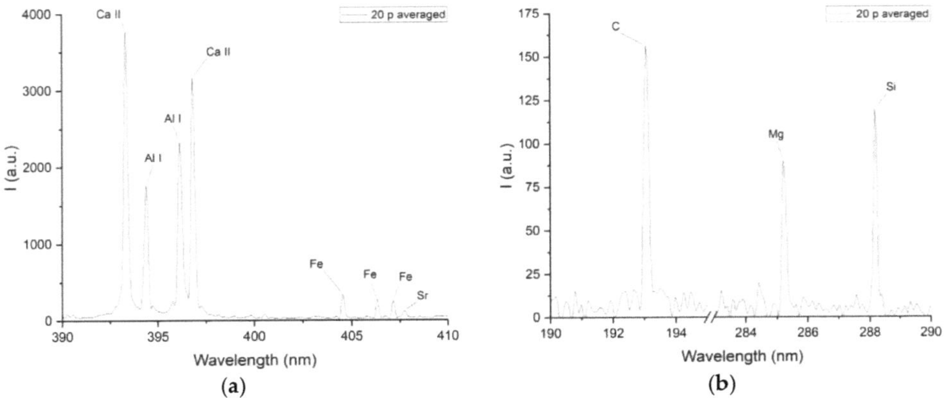

Figure 4. LIBS elemental analysis for (**a**) Ca, Al, and Fe; and (**b**) C, Mg, and Si.

LIBS stratigraphy was performed in order to obtain a glimpse of the distribution of the chemical elements by analysing each of the 20 pulses (Figure 5a–d) [35,36]. It was observed that the most analysed elements started to count only from the 5 to 6th pulse, with a peak between the 7th and the 11th pulses, corresponding to the metallic overpaint. This can be an indication that the metallic overpaint layer is covered with another layer of organic or synthetic nature, as LIBS detects only trace elements. Starting with pulse 12 till the end of the series, the distribution tends to have a quasi-constant shape for most of the elements pictured, which can be attributed to reaching the ground, as can be observed in Figure 5c,d that presents the Si and Sr distribution.

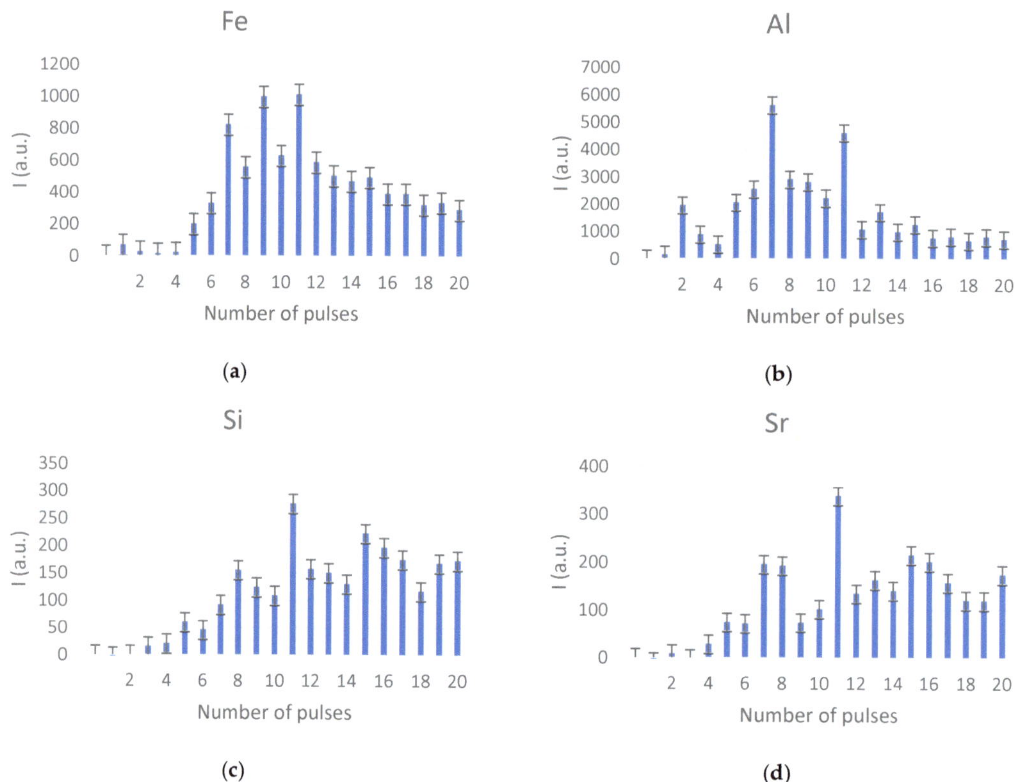

Figure 5. LIBS elemental depth distribution through 20 pulses for (**a**) Fe, (**b**) Al, (**c**) Si, and (**d**) Sr.

3.2. Laser Removal of the Metallic Overpaint

The laser cleaning regime was established empirically as a result of preliminary tests conducted on an area of approximately 4 cm^2. A low starting laser energy was set, and it was increased gradually until the ablation of the overpaint layers was observed using OM and LIBS. The ablation thresholds determined using the D^2 method are presented in Table 1. In Figure 6, microscopic images of the laser-cleaned areas are shown. The different testing areas for optimal and overcleaning are noted. It was noticed that the first overpaint layer and the adherent deposits were removed efficiently at an energy of 250 mJ (fluence of 0.62 J/cm^2), as can be seen in Figure 6C,G. As the energy was increased, at energies around 350 mJ (fluence of 0.8 J/cm^2), the thick metallic overpaint layer started ablating and the traces of the silver leaf became noticeable, as can be seen in Figure 6E. Any further increase in the energy would result in overcleaning; thus, the ablation threshold was set at 0.8 J/cm^2.

Table 1. Laser cleaning ablation thresholds for the adherent deposits and the overpainted substrates.

Substrate	Ablation Threshold
Surface adherent deposits	0.37 J/cm^2
First overpaint layer	0.62 J/cm^2
Second overpaint layer	0.8 J/cm^2

Since the overpainting layer was not uniform, nor was its thickness, the nonuniformity of the cleaning was clearly visible, and in some areas, the red bole was reached. Therefore, it was decided to use several pulses in each area (20–30 pulses) rather than using higher energy.

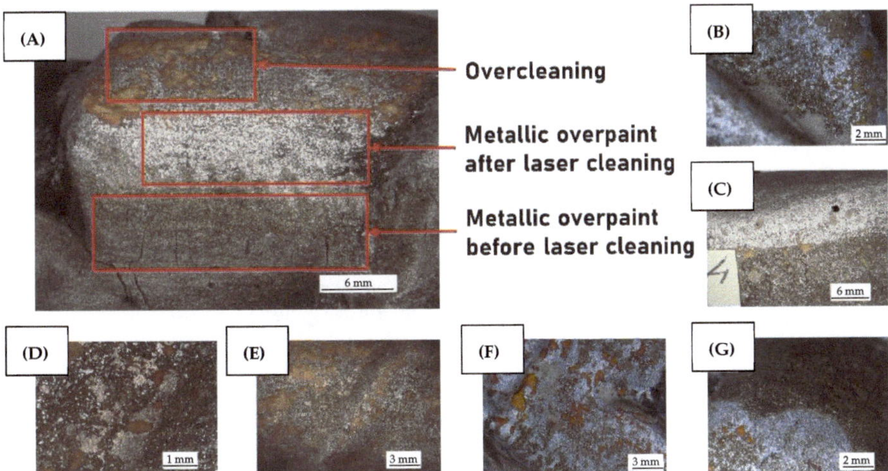

Figure 6. Micro details of the laser-cleaned surfaces. (**A**) test areas 1–3, (**B**) test area 4, (**C**) test area 5, (**D**) test area 6, (**E**) test area 7, (**F**) test area 8, (**G**) test area 9.

Laser cleaning provided good results for the rough removal of the thick layers of the metallic paints applied on top of the highly degraded pictorial substrate in a relatively short time compared to traditional cleaning techniques. Taking into consideration the thin and fragile remains of the original materials under the overpaints, and also the fact that the high fluence necessary for removal of the overpaints is over the ablation threshold of the pictorial composition reported in the literature [11,37,38], the best conservation strategy is unquestionably a hybrid approach that combines photonics with traditional methods, using the laser for the rough removal of the thick overpaint and then applying solvents to reveal the original religious motifs.

The pictorial layer was revealed by chemical cleaning using a solution composed of 85% dichloromethane, 10% methyl alcohol, and Deck 3000 with a gel-like consistency. This solution was applied using moistened pads, with the gel serving the purpose of retaining a mixture of highly volatile solvents in the treated areas for an adequate duration.

4. Conclusions

The current paper presents the interdisciplinary approach undertaken for the conservation of a 19th Century iconostasis belonging to the "Holy Trinity" church from Măgureni, România. The polychrome religious paintings were covered with what was initially thought to be a thick layer of metallic paint that was applied on top of previous restorations.

Advanced photonic methods were selected for the investigation of the chemical composition and characterization of the substrates, and laser-cleaning efficiency was tested. The stratigraphic analyses determined that the overpaint actually consisted of two layers, one of synthetic nature and the other of a metallic paint based on Fe and Al. Also, elemental analysis detected traces of many elements resulting from the terrigenous origin of the raw materials. Earth and clay minerals, such as kaolinite, calcite, gypsum, iron oxides, and quartz, were identified in a binding medium of egg yolk. The ground consisted of several sublayers of gypsum and/or calcite mixed with animal glue, which was a typical preparation layer following the conservative iconographic traditions. In one of the samples, a prominent presence of a natural tree (pine) resin was observed. The information obtained from the spectroscopic analysis was essential for establishing a conservation approach and for understanding the historical and cultural value of the polychrome wooden iconostasis.

The cleaning methodology discussed in this paper involved the use of a laser to roughly remove the overpaint layers. A Q-switched Nd:YAG laser emitting at a wavelength of 1064 nm, a pulse duration of 8 ns, and a pulse repetition rate of 10–20 Hz was used. The optimal working fluence for removing the thick metallic overpaint layers was found to be 0.8 J/cm^2, which achieved good cleaning efficiency after 5–6 passes of the beam on the surface. Taking into consideration the spectroscopic analysis and the laser cleaning tests, the best conservation strategy was decided to be a hybrid approach, combining photonic with traditional methods: using laser for the rough removal of the thick overpaint layers and solvents to reveal the original religious motifs.

Photonics plays a dynamic role in most fields and, as technology advances, so do the applications developed within the Heritage Sciences field, enabling better understanding and valorization. By providing detailed information about their composition, structure, and condition, photonic techniques can guide conservation interventions and help preserve these valuable cultural treasures for future generations.

Author Contributions: Conceptualization, V.A.; methodology, R.R. and S.-R.P.; FTIR, V.A., LIBS, M.D.; laser cleaning, S.-R.P. and M.D.; microscopy, V.A.; writing—original draft preparation, V.A.; writing—review and editing, M.D.; funding acquisition, R.R. and M.D. All authors have read and agreed to the published version of the manuscript.

Funding: This research was funded by the Romanian Ministry of Research, Innovation, and Digitalization, under Program 1—Development of the National Research-Development System, Subprogram 1.2—Institutional Performance: Projects to finance the excellence in RDI, SUPECONEX, grant no. 18PFE/30.12.2021, PNCDI 2022-2027—Core Programme 11N/03.01.2023, project no. PN 23 05 and under CNCS - UEFISCDI, project number PN-III-P4-PCE-2021-1605, within PNCDI III.

Institutional Review Board Statement: Not applicable.

Informed Consent Statement: Not applicable.

Data Availability Statement: The data that support the findings of this study are available upon request from the corresponding authors.

Conflicts of Interest: The authors declare no conflict of interest. The funders had no role in the design of the study; in the collection, analyses, or interpretation of data; in the writing of the manuscript; or in the decision to publish the results.

References

1. Draghiceanu, V.N. *Casa Cantacuzinilor din Magureni*; Ed. Ramuri: Craiova, Romania, 1924.
2. Boonrat, P.; Dickinson, M.; Cooper, M. Initial investigation into the effect of varying parameters in using an Er:YAG laser for the removal of brass-based overpaint from an oil-gilded frame. *J. Inst. Conserv.* **2020**, *43*, 94–106. [CrossRef]
3. Panzner, M.; Wiedemann, G.; Meier, M.; Conrad, W.; Kempe, A.; Hutsch, T. Laser Cleaning of Gildings. *Lasers Conserv. Artworks* **2007**, *116*, 21–28. [CrossRef]
4. Ungurean, B. Notes on the iconostas of St. Theodore's church from Iasi. Technique of execution, stylistic description and state of conservation. *Anastasis* **2018**, *5*, 86–109.
5. Vuga, M.; Semion, M.M. Typical conservation problems of polychrome wooden sculptures in Slovenia. *Conserv. Patrim.* **2015**, *22*, 17–27. [CrossRef]
6. Strzelec, M.; Marczak, J.; Koss, A.; Szambelan, R. Overpaint Removal on a Gilded Wooden Bas-Relief Using a Nd:YAG Laser at 1.064 μm. *Laser Conserv. Artworks* **2005**, *100*, 133–138. [CrossRef]
7. Gaspar, P.; Rocha, M.; Kearns, A.; Watkins, K.; Vilar, R. A study of the effect of the wavelength in the Q-switched Nd:YAG laser cleaning of gilded wood. *J. Cult. Herit.* **2000**, *1*, 133–144. [CrossRef]
8. Alabone, G.; Carvajal, M.S. The removal of bronze paint repairs from overgilded picture frames using an Erbium:YAG laser. *J. Inst. Conserv.* **2020**, *43*, 107–121. [CrossRef]
9. Pouli, P.; Selimis, A.; Georgiou, S.; Fotakis, C. Recent studies of laser science in paintings conservation and research. *Acc. Chem. Res.* **2010**, *43*, 771–781. [CrossRef]
10. Asmus, J.F. Light for Art Conservation. *Interdiscip. Sci. Rev.* **2012**, *12*, 171–179. [CrossRef]
11. Fotakis, C.; Kautek, W.; Castillejo, M. Lasers in the Preservation of Cultural Heritage. *Laser Chem.* **2006**, *2006*, 1. [CrossRef]
12. Acquaviva, S.; D'anna, E.; De Giorgi, M.; Della Patria, A.; Pezzati, L.; Pasca, D.; Vicari, L.; Bloisi, F.; Califano, V. Laser cleaning of gilded wood: A comparative study of colour variations induced by irradiation at different wavelengths. *Appl. Surf. Sci.* **2007**, *253*, 7715–7718. [CrossRef]
13. Siano, S.; Grazzi, F.; Parfenov, V.A. Laser cleaning of gilded bronze surfaces. *J. Opt. Technol.* **2008**, *75*, 419–427. [CrossRef]

14. Andreotti, A.; Bracco, P.; Colombini, M.P.; Decruz, A.; Lanterna, G.; Nakahara, K.; Penaglia, F. Novel Applications of the Er:YAG Laser Cleaning of Old Paintings. *Lasers Conserv. Artworks* **2007**, *116*, 239–247. [CrossRef]

15. Brunetto, A.; Bono, G.; Frezzato, F. Er:YAG laser cleaning of 'San Marziale in Gloria' by Jacopo Tintoretto in the Church of San Marziale, Venice. *J. Inst. Conserv.* **2020**, *43*, 44–58. [CrossRef]

16. Wiedemann, G.; Pueschner, K.; Wust, H.; Kempe, A. The Capability of the Laser Application for Selective Cleaning and the Removal of Different Layers on Wooden Artworks. *Lasers Conserv. Artworks* **2006**, *100*, 179–190. [CrossRef]

17. Striber, J.; Jovanović, V.; Jovanović, M. Easel paintings on canvas and panel: Application of Nd:YAG laser at 355 nm, 1064 nm and UV, IR and visible light for the development of new methodologies in conservation. In *Proceedings of the LACONA XI*; Elsevier: Amsterdam, The Netherlands, 2017; pp. 279–292. [CrossRef]

18. Kramida, A.; Ralchenko, Y.; Reader, J.; NIST Atomic Spectra Database Team. *Atomic Spectra Database (Version 5.3)*; National Institute of Standards and Technology: Gaithersburg, MD, USA, 2015.

19. Sawicki, M.; Bramwell–Davis, V.; Dabrowa, B. Laser cleaning from a practical perspective: Cleaning tests of varied gilded-wood surfaces using Nd:YAG Compact Phoenix laser system. *AICCM Bull.* **2011**, *32*, 44–53. [CrossRef]

20. Mastrotheodoros, G.P.; Beltsios, K.G.; Bassiakos, Y.; Papadopoulou, V. On The Grounds of Post-Byzantine Greek Icons. *Archaeometry* **2016**, *58*, 830–847. [CrossRef]

21. Cortea, I.M.; Ghervase, L.; Ratoiu, L.; Dinu, M.; Rădvan, R. Uncovering hidden jewels: An investigation of the pictorial layers of an 18th-century Taskin harpsichord. *Herit. Sci.* **2020**, *8*, 55. [CrossRef]

22. Cortea, I.M.; Ghervase, L.; Rădvan, R.; Serițan, G. Assessment of Easily Accessible Spectroscopic Techniques Coupled with Multivariate Analysis for the Qualitative Characterization and Differentiation of Earth Pigments of Various Provenance. *Minerals* **2022**, *12*, 755. [CrossRef]

23. Chukanov, N.V.; Chervonnyi, A.D. Some General Aspects of the Application of IR Spectroscopy to the Investigation of Minerals. In *Infrared Spectroscopy of Minerals and Related Compounds*; Springer: Berlin/Heidelberg, Germany, 2016; pp. 1–49. [CrossRef]

24. Genestar, C.; Pons, C. Earth pigments in painting: Characterisation and differentiation by means of FTIR spectroscopy and SEM-EDS microanalysis. *Anal. Bioanal. Chem.* **2005**, *382*, 269–274. [CrossRef]

25. Gunasekaran, S.; Anbalagan, G.; Pandi, S. Raman and infrared spectra of carbonates of calcite structure. *J. Raman Spectrosc.* **2006**, *37*, 892–899. [CrossRef]

26. Carlyle, L.; Roy, A. Artists' Pigments: A Handbook of Their History and Characteristics, Volume 2. *J. Am. Inst. Conserv.* **1996**, *37*, 892–899. [CrossRef]

27. Derrick, M.; Stulik, D.; Landry, J. *Infrared Spectroscopy in Conservation Science. The Effects of Brief Mindfulness Intervention on Acute Pain Experience: An Examination of Individual Difference*; Getty Publications: Los Angeles, CA, USA, 2015.

28. Cortea, I.M.; Ratoiu, L.; Chelmuș, A.; Mureșan, T. Unveiling the original layers and color palette of 18th century overpainted Transylvanian icons by combined X-ray radiography, hyperspectral imaging, and spectroscopic spot analysis. *X-ray Spectrom.* **2022**, *51*, 26–42. [CrossRef]

29. Lazidou, D.; Lampakis, D.; Karapanagiotis, I.; Panayiotou, C. Investigation of the Cross-Section Stratifications of Icons Using Micro-Raman and Micro-Fourier Transform Infrared (FT-IR) Spectroscopy. *Appl. Spectrosc.* **2018**, *72*, 1258–1271. [CrossRef] [PubMed]

30. Vahur, S.; Teearu, A.; Peets, P.; Joosu, L.; Leito, I. ATR-FT-IR spectral collection of conservation materials in the extended region of 4000-80 cm^{-1}. *Anal. Bioanal. Chem.* **2016**, *408*, 3373–3379. [CrossRef]

31. Bretz, S.; Baumer, U.; Stege, H.; Von Miller, J.; Von Kerssenbrock-Krosigk, D. A German house altar from the sixteenth century: Conservation and research of reverse paintings on glass. *Stud. Conserv.* **2008**, *408*, 3373–3379. [CrossRef]

32. Guttmann, M.J. Transylvanian glass icons: A GC/MS study on the binding media. *J. Cult. Herit.* **2013**, *14*, 439–447. [CrossRef]

33. Antunes, V.; Candeias, A.; Mirão, J.; Carvalho, M.L.; Dias, C.B.; Manhita, A.; Cardoso, A.; Francisco, M.J.; Lauw, A.; Manso, M. Analytical characterization of the palette and painting techniques of Jorge Afonso, the great 16th century Master of Lisbon painting workshop. *Spectrochim. Acta-Part A Mol. Biomol. Spectrosc.* **2018**, *193*, 264–275. [CrossRef]

34. Dinu, M.; Cortea, I.M.; Cortea, L.; Stancu, M.C.; Mohanu, I.; Cristea, N. Optoelectronic investigation of the mural paintings from Drăguțești wooden church, Argeș County, Romania. *J. Optoelectron. Adv. Mater.* **2020**, *22*, 303–309.

35. Pacher, U.; Dinu, M.; Nagy, T.O.; Radvan, R.; Kautek, W. Multiple wavelength stratigraphy by laser-induced breakdown spectroscopy of Ni-Co alloy coatings on steel. *Spectrochim. Acta-Part B At. Spectrosc.* **2018**, *146*, 36–40. [CrossRef]

36. Simileanu, M. Libs quantitative analyses of bronze objects for cultural heritage applications. *Rom. Reports Phys.* **2016**, *68*, 203–209.

37. Dascalu, G.R.; Stancu, M.C.; Dinu, M.; Puscas, N. Laser cleaning of polychrome artworks. Case study on graffiti. *UPB Sci. Bull. Ser. A Appl. Math. Phys.* **2020**, *82*, 307–316.

38. Kautek, W. Lasers in Cultural Heritage: The Non-Contact Intervention. *Springer Ser. Mater. Sci.* **2010**, *130*, 331–349. [CrossRef]

Article

Surface-Active Ionic-Liquid-Based Coatings as Anti-Biofilms for Stone: An Evaluation of Their Physical Properties

Marika Luci [1,2], Filomena De Leo [1], Donatella De Pascale [3], Christian Galasso [2], Mauro Francesco La Russa [4], Sandra Lo Schiavo [1], Michela Ricca [4], Silvestro Antonio Ruffolo [4,*], Nadia Ruocco [2] and Clara Urzì [1]

[1] Department of Chemical, Biological, Pharmaceutical and Environmental Sciences, University of Messina, Viale F. Stagno d'Alcontres, 31, 98166 Messina, Italy; marika.luci@szn.it or marika.luci@studenti.unime.it (M.L.); fdeleo@unime.it (F.D.L.); sloschiavo@unime.it (S.L.S.); clara.urzi@unime.it (C.U.)

[2] Department of Ecosustainable Marine Biotechnology (BLUBIO), Stazione Zoologica Anton Dohrn, Calabria Marine Centre, C.da Torre Spaccata, 87071 Amendolara, Italy; christian.galasso@szn.it (C.G.); nadia.ruocco@szn.it (N.R.)

[3] Department of Ecosustainable Marine Biotechnology (BLUBIO), Stazione Zoologica Anton Dohrn, Via Acton, 55, 80133 Naples, Italy; donatella.depascale@szn.it

[4] Department of Biology, Ecology and Earth Sciences, University of Calabria, Via P. Bucci Cubo 12 B, 87036 Arcavacata di Rende, Italy; mauro.larussa@unical.it (M.F.L.R.); michela.ricca@unical.it (M.R.)

* Correspondence: silvestro.ruffolo@unical.it

Abstract: The biodeterioration of stone surfaces can be a threat to the conservation of built heritage. Much effort has been put into finding treatments and processes to mitigate biocolonization and its effects, both in terrestrial and underwater environments. Recently, the use of surfactant ionic liquids has been shown to have biocidal and antifouling effects on stone. However, little information is currently available on the morphological and physical properties of such coatings. In this paper, we report on the physical characterization of coatings based on an ionic liquid (IL) consisting of N-(2-hydroxyethyl)-N,N-dimethyl-1-do-decanaminium cation and a combination of bromide and dodecylbenzenesulfonate (DBS) anions in a molar ratio of 3:1, respectively. Nanosilica and tetraethyl orthosilicate were used as binders to promote the adhesion of the ionic liquid to the stone surface. The coatings were applied on Carrara marble samples and analyzed using Scanning Electronic Microscopy (SEM), static contact angles, colorimetric measurements and capillary water absorption. The resistance to UV radiation and seawater was also investigated. The results show that the IL behaves differently depending on the binder. The latter influences the arrangement of the IL and its wettability, which decreases in the case of NanoEstel, whereas this parameter increases in the case of Estel. In addition, the coatings show good resistance to the degradation agents.

Keywords: ionic liquid; antibiofilm; coatings for stone; coating behavior

Citation: Luci, M.; De Leo, F.; De Pascale, D.; Galasso, C.; La Russa, M.F.; Lo Schiavo, S.; Ricca, M.; Ruffolo, S.A.; Ruocco, N.; Urzì, C. Surface-Active Ionic-Liquid-Based Coatings as Anti-Biofilms for Stone: An Evaluation of Their Physical Properties. *Coatings* **2023**, *13*, 1669. https://doi.org/10.3390/coatings13101669

Academic Editor: Peter Zilm

Received: 24 August 2023
Revised: 18 September 2023
Accepted: 20 September 2023
Published: 23 September 2023

1. Introduction

The surface of an inorganic material exposed to air or in an underwater environment suffers the action of agents that cause its deterioration [1,2]. Air pollutants, micro-organisms and continuous climatic variations [3] produce physical and chemical phenomena such as breakdown, exfoliation, efflorescence, cracks and black crusts. Similarly, exposed stone surfaces in a submerged environment undergo different forms of degradation that depend on different and mutable exposure conditions [4]. Stone materials in contact with the aquatic environment behave as attachment substrates for a variety of micro-organisms that can modify the stone's surface, resulting in alteration and deterioration processes [5]. When these processes damage our historical and artistic heritage, it becomes necessary to look for a way to prevent them [6,7].

The microbial biodeterioration of stone materials in subaerial environments is a well-documented process [8–10], but the knowledge of the impact of micro-organisms on archaeological artifacts in marine environments has large gaps. The process of bioerosion in the underwater environment is due to a process called "marine biofouling" that starts immediately after the exposure of a given surface in a water environment [11]. For this reason, the large historical and archaeological heritage submerged and/or recovered from underwater environments is sometimes threatened by degradation and unsuitable conservation strategies [12]. When an inorganic material comes into contact with the marine environment, it is quickly colonized by micro-organisms that produce microfouling [13,14]. Microbial biofilm in most cases offers a basis for the settlement of macrofoulers, with the consequent loss of the readability of the artefacts and of the material itself, or acts as a barrier against environmental stresses and further biofouler colonization [15].

Therefore, the development of methods to prevent the colonization of aerial and submerged cultural heritage must be based on strategies that, in addition to making the surface not available for microbial colonization, take into account the intrinsic properties of stone and of the environment, according to the mechanics of degradation processes and environmental factors [16,17].

Due to its characteristic of providing lasting protection, the preventive approach is the most in-demand and pursued one. Effective solutions, mainly deriving from bio- and nanotechnologies, have been proposed in this context [17–19]. These, to be considered innovative, have to be eco-sustainable. In short, they have to match "green restoration criteria", recently suggested as "all the eco-sustainable practices to be used in the conservation and restoration of Cultural Heritage assets, alternatives to traditional products and methods which are often toxic and harmful for the users and the environment" [20]. In this scenario, concepts such as Safe by Design (SbF) structural–functional relationships, non-destructive reversible procedures, bioreceptivity materials and biodegradation have to be considered in the development of innovative products/processes [21]. Such premises make the design of true sustainable protocols difficult to realize. In fact, today, surface-active quaternary ammonium (QAC)-based formulations, which adopt more sustainable protocols, are being used the most.

Indeed, ionic liquids (ILs) were proposed as a class of materials potentially capable of fulfilling the "green criteria" [20]. These, in brief, are a class of low-melting-point organic salts, exhibiting a plethora of tunable properties spanning from physico-chemical [22,23] to biological [24], which have found application in many fields. Such a versatility arises from their ionic nature, which allows synthetic control of their properties using the right combination of cation/anion coupling. Due to their widespread use, ILs' environmental impact, including toxicity and biodegradability, has been deeply investigated. This has led to the so-called "ILs of third generation" based on naturally occurring and or biodegradable ions (cholinium, morfolinium amino acids and drug anions) [25–27]. Interestingly, for stone conservation, the introduction of surfactant moieties inside the IL ion pair makes them intriguing, in terms of surface activity and, hence, bioactivity. Consistently, ILs fall into the class of Safe by Design (SbD) due to their antimicrobial features. Although few data are available in this context, the results appear promising [20,28–30].

De Leo and coworkers [28] reported on the antimicrobial/antifouling properties of a family of surfactant ILs designed by taking into account environmentally eco-sustainable criteria. These feature a series of mono- and di-cholinium cations bearing alkyl chains of different lengths and, as anions, conventional halides or surfactant dodecylbenzenesulfonate (DBS) in order to also ascertain the influence of the nature of anions on the overall bioactivity of the species. Minimum inhibitory concentration (MIC) and minimum bactericidal concentration (MBC) analyses and agar diffusion test vs. Gram (+) tests of yeasts and black fungi revealed that the biocide activity of these systems was firstly dependent on the nature of cations, in that only those bearing long alkyl chain lengths displayed antimicrobial activity. This confirms once more the lipophilicity–antimicrobial activity relationship as observed for QAC. Interestingly, from these studies, the tuning role played by the surfactant DBS anion on the bioactivity of ILs also emerged. Based on agar diffusion tests, it was established that the cholinium ionic liquid based on the N-(2-Hydroxyethyl)-N,N-dimethyl-

1-dodecanaminium cation and a combination of bromide and dodecylbenzenesulfonate (DBS) anions in the molar ratio 3:1 (IL) was the most performant (Figure 1).

As part of a project aimed at evaluating the potential of IL in the development of antifouling coatings for stones, we report here on the physico-chemical investigations performed on silica@IL layers, deposited on Carrara marble specimens. A nanosilica suspension and tetraethyl orthosilicate as silica precursors were used to promote the adhesion of the IL on the stone surface, as well as act as stone consolidating agents. The physico-chemical behavior of the treated surfaces was tested in terms of micromorphology, colorimetric variations, wettability and water absorption capability. Furthermore, the effect of UV radiation and seawater on the features of the coatings, in order to assess their durability against the weathering agents, was also reported. In particular, the results obtained are consistent with the hydrophilic coatings exhibiting a good resistance to atmospheric agents, even if long-term trials are required.

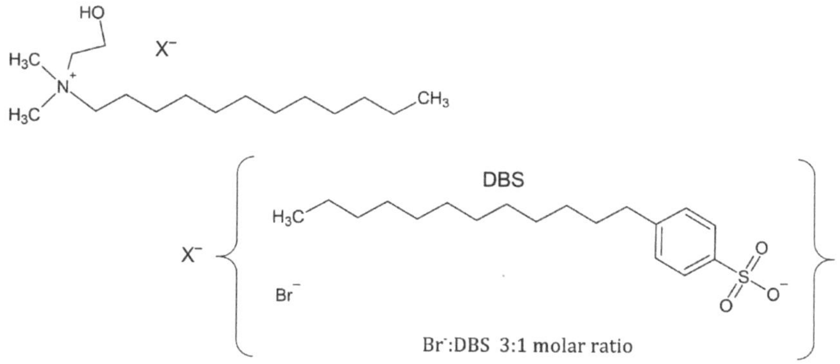

Figure 1. Chemical structure of the used ionic liquid (IL).

2. Materials and Methods

2.1. IL Synthesis and Binders

All the chemicals, unless otherwise stated, were purchased from Sigma-Aldrich and used as supplied.

The ionic liquid (IL) under study here (Figure 1), featuring N-(2-Hydroxyethyl)-N,N-dimethyl-1-dodecanaminium as the cation and bromide and dodecylbenzenesulfonate (DBS) anions in the molar ratio 3:1, was obtained by following a procedure already adopted for the synthesis of analogous cholinium ILs [27].

In brief, a methanol suspension (30 mL) of NaDBS (0.82 g, 2.34 mmol) was added to a solution (170 mL) of the N-(2-Hydroxyethyl)-N,N-dimethyl-1-dodecanaminium bromide in the same solvent (3.17 g, 9.375 mmol). The resulting mixture was left under stirring for 12 h and then left to stand at 10 °C for 3 h. The IL containing "mother liquors" was pipetted off and then diluted to 250 mL to reach a concentration of ca. 0.0375 M in cholinium-based cation. Before using, a strong sonication of the IL mixture is highly recommended to avoid significant aggregation phenomena. Indeed, the IL methanol mixture appeared transparent at the naked eye. Nevertheless, DLS measurements revealed the presence of agglomerations even at a concentration of 0.4 mM (work in progress), consistent with the presence of surfactant species.

The IL Pyrene-1-sulfonate derivative (ILPyrS) was obtained as follows: 0.074 g (0.45 mmol) (ca. 6% by mol) of sodium Pyrene-1-sulfonate was added to a methanol mixture (200 mL) of IL (concentration 0.0375 M) and the resulting mixture was left under stirring overnight. The resulting light-yellow suspension was pipetted off and used without further work-up (IL-PyrS). ILPyrS was used to check the presence of IL on the surface, its distribution and its resistance to washout.

Regarding the binders, NanoEstel (CTS srl, Altavilla Vicentina, Italy), Estel 1000 (hereinafter named Estel) (CTS srl, Altavilla Vicentina, Italy) and pure tetraethyl orthosilicate (TEOS) were used. NanoEstel is a waterborne suspension of silica nanoparticles having a mean diameter of 30 nanometers and a concentration of silica 30% wt, while Estel is based on TEOS in white spirit (concentration 75% wt).

2.2. Treatments

Carrara marble was used as the stone material in the experimentation. From a mineralogical point of view, it is constituted mainly of calcite and it has a low porosity (about 2%). It was widely used in the past, especially since the Roman epoch, and is still used nowadays. It has high homogeneity.

Several specimens were cut into two sizes: $5 \times 5 \times 2$ cm and $2.7 \times 2.7 \times 1$ cm. The choice of the total number of samples for the tests was made in relation to the number of treatments to be carried out, and to the different tests planned for the trials. All tests were performed in triplicate. Before treatment, all specimens were cleaned and washed in bidistilled water to remove any residual impurities and dried in the oven for 24 h at 80 °C. Then, each probe was properly signed and subjected to the corresponding treatment.

Three different treatments on specimens sized $5 \times 5 \times 2$ cm were performed: (a) NanoEstel treatment and application of IL after the setting of the binder (bilayer treatment, NES + I); (b) a mixture of TEOS and IL applied together on the stone (monolayer treatment, ES + I); and (c) Estel treatment and application of IL after the setting of the binder (bilayer treatment, ES + I). All applications were performed using a brush.

In more detail, NanoEstel was used diluted 1:2 with bidistilled water, and the treated specimens were left to dry at room temperature for 4 days before IL application. Treatments with Estel were carried out by applying the product on stone without any further dilution, leaving it to dry and setting it at room temperature for 4 weeks before IL application. The treatment with TEOS and IL was carried out by adding 50 mL of absolute ethanol to 73 g (77.6 mmol) of TEOS, and the mixture was then added to 250 mL of IL mixture. This mixture was then applied on the stone surface. In order to evaluate the effect of the bare binders, treatments were also carried out without IL.

Some specimens (sized $2.7 \times 2.7 \times 2$ cm) were treated following the procedures mentioned above, except, in these cases, IL-PyrS was used. This is due to the luminescence of the product, which can be checked using UV light, which represents a tag to check for the presence, distribution and persistence of the IL on the stone surface.

In Table 1, a summary of the treatments is reported.

Table 1. Summary of treatments with IL (N-(2-Hydroxyethyl)-N,N-dimethyl-1-dodecanaminium Br:DBS 3:1) and IL-PyrS (IL-Pyrene-1-sulfonate).

ID	Binder			IL/IL-PyrS	
	Product	Amount (g/m^2)		Application	Amount (mmol/m^2)
		Product	Active Matter		
Untreated	-	-	-	-	-
NES	NanoEstel	80	12	-	-
NES + I/ILPyrS	NanoEstel	80	12	Top layer	0.57
TE	TEOS	80	24	-	-
TE + I/IL-PyrS	TEOS	80	24	Mixed with binder	0.57
ES	Estel 1000	80	60	-	-
ES + I/IL-PyrS	Estel 1000	80	60	Top layer	0.57

2.3. Measurements

The micro-morphological features of the coatings were explored using scanning electron microscopy (SEM). For this reason, an ultra-high-resolution SEM (UHR-SEM) ZEISS CrossBeam 350 equipment, coupled with the spectrometer EDS-EDAX OCTANE, was used.

Colorimetric measurements of the stone samples were carried out with the ZL 310 colorimeter. The measures were carried out according to the guidelines given by 28. UNI EN ISO/CIE 11664-4:2019 [31]. The instrument provided data on L*, a* and b* coordinates, which were incorporated into the CIE system, obtaining information about the brightness and chromaticity of the materials' surfaces. The chromatic variation in the treated surfaces was evaluated by calculating $\Delta E = (\Delta L^2 + \Delta a^2 + \Delta b^2)^{1/2}$. Five surface analysis points of similar aspect were selected for each stone specimen. Colorimetric analysis was performed on all specimens before respective treatment, after treatment and after UV-daylight weathering. This allowed for the characterization of the colorimetric properties of the stone surfaces of each sample without falling into random errors due to the original chromatic variations present on one sample's surface. Next, the ΔE values of the samples treated with the same coating were averaged to obtain a single value and standard deviation.

The evaluation of the possible variation in the amount and speed of capillary water absorption on the treated surfaces was carried out according to UNI EN 15801:2010 [32]. This test was performed on specimens treated with the coatings containing the ionic liquids and on the same after UV-daylight weathering. A layer of absorbent paper about 1 cm thick was placed in the bottom of a container. Demineralized water was added to the container until the absorbent layer was saturated. The specimens were weighed and placed on the laying bed with the test surface in contact with the wet paper. The specimens were removed from the support and weighed after 10, 20, 30 and 60 min and after 4, 6, 24, 48, 72 and 96 h. After 5 days from the start of the trial, the test was stopped because the specimens showed a saturated state. The amount of water absorbed by the specimen as a function of time was expressed as $Q_i = [(m_i - m_0)/A]$, which is the difference between the mass of the specimen at time t_i and the mass of the dry specimen in relation to the surface area in contact with water. Next, the Q_i values of the equally treated samples were averaged to obtain a single value and standard deviation.

Using the contact angle test, the variations in the degree of hydrophilicity/hydrophobicity of the stone surfaces following treatment with coatings containing antifouling products were assessed. The measure of the contact angle was performed with an LSA LAUDA Scientific Surface Analyzer System. Static contact angle measurements [33] were conducted on all specimens before and after treatment and after UV-daylight aging. The results of the averages were in turn averaged to obtain a single value and standard deviation. The result was indicative of the type of interaction between water and the whole contact surface of one specimen.

2.4. UV-Daylight Weathering and Washout Tests

The samples were subjected to artificial daylight radiation using a SUNTEST XLS+ chamber to check variations in the properties of treated materials that would occur with exposure to solar radiation. The specimens were left for 507 h under the action of a daylight artificial fluorescent lamp combining visible and UV outputs (λ = 300–800 nm, irradiance = 500 W/m^2) at a temperature of about 35 °C. To check for any forms of degradation of the newly designed coatings, such as brightening, yellowing or loss of durability, the aged specimens were subsequently resubjected to the tests described above.

Marble specimens treated with pyrene sulfonate-tagged coatings were subjected to washout tests to evaluate the durability of coatings containing IL in an underwater environment. Aging tests were carried out in contact with sterile seawater to evaluate possible forms of degradation of coatings not due to the presence of micro-organisms but to the action exerted by seawater. Untreated and treated marble probes were sterilized under UV light for 30 min upside down and for 2 h facing the test surface in contact with radiation to ensure that the whole probe surface was sterile. Then, each sample was placed inside a 120 mL screw-cap sterile container and 80 mL sterile seawater was added. The water was previously prepared via vacuum filtration in a Stericup system and subsequent sterilization in an autoclave at 121 °C for 20 min. The resulting systems were incubated for 15 days in a thermostatic shaker incubator with rotating speed 100 r/min. At the end of incubation, the samples were washed with bidistilled water and dried in the oven for 2 h at 40 °C. The treated specimens were weighed before and after the test to verify the complete removal of moisture.

3. Results and Discussions

3.1. Behavior of the Coatings and Effect of UV Aging

In Figure 2, SEM images of treated and untreated surfaces are reported. A comparison between the untreated surfaces and the surface treated with NanoEstel (NES and NES + I) reveals that this binder produces a very cracked surface. This may be ascribed to a poor penetration of the nanosilica particles into the stone and to the shrinking that takes place during the condensation of the particles. On the Estel-treated specimens (ES and ES + I), a coating with sporadic cracks is visible. TEOS treatments (TE and TE + I) led to surfaces that appeared similar to the untreated ones, suggesting that the product penetrated into the stone. In all cases, it seems that the addition of IL did not affect, significantly, the morphology of the coating.

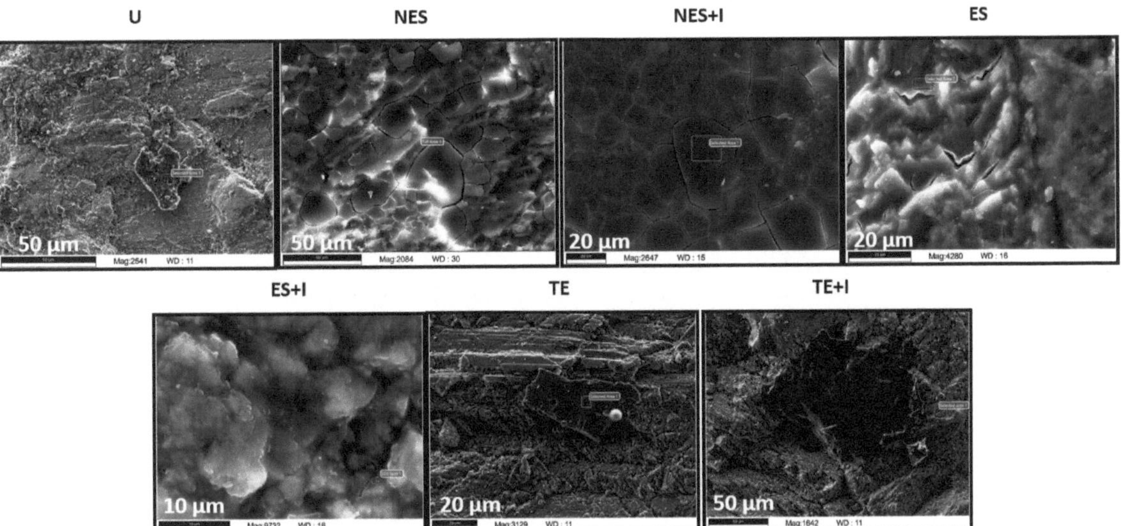

Figure 2. SEM images of untreated and treated marble surfaces.

The colorimetric variations induced by binders and IL were also investigated. This information is useful for analyzing the features of the coatings from an aesthetic point of view. In Figure 3, the effect of treatment on bare marble is shown. For all the three treatments, the colorimetric variation was below 5, which is generally considered as a threshold above which the color variation is detectable by the naked eye and, thence, unacceptable in the field of cultural heritage conservation. In the cases of NanoEstel and TEOS used as binders, the sole binders induced a color variation below 2, while the addition of IL made the variation slightly higher. A different behavior was detected for Estel treatment. In this case, the bare binder induced a stronger color variation, of a magnitude of about 4.5. A deeper analysis that takes into account the single components (L*, a* and b*) reveals that the coordinate L* seems to be the most impactful parameter on the Delta E value. In particular, for Estel treatment, the L* values dropped by about 4 units, due to a wetting effect ascribable to the solvent of the consolidant that remains trapped in the stone as the ethyl silicate cures [34]. This is a transient effect, although it can last for several months after treatment.

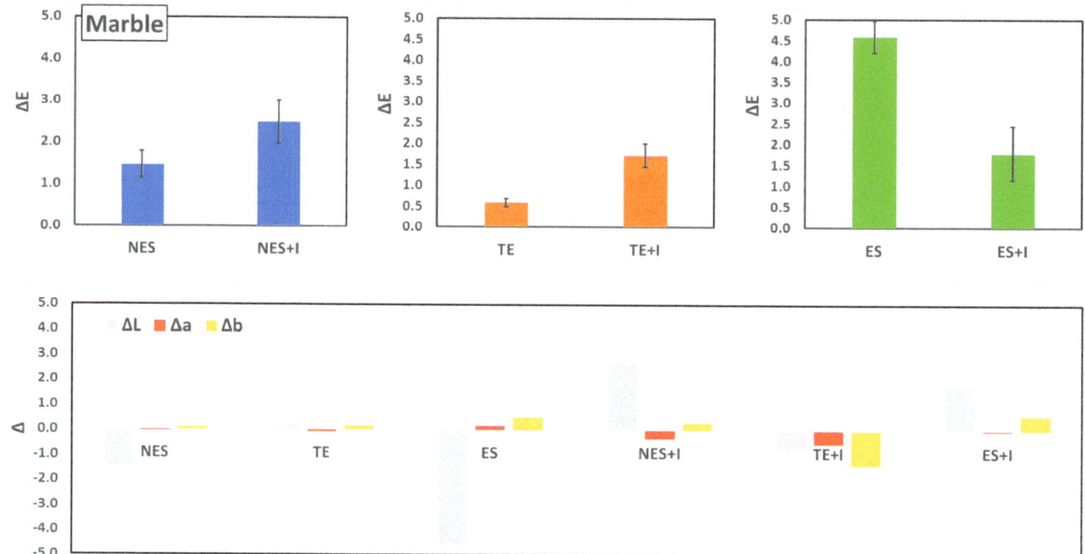

Figure 3. Colorimetric differences between untreated and treated specimens. For each treatment, the ΔE value with its components (ΔL, Δa, Δb) is reported.

The calculation of colorimetric variations between treated and aged specimens can provide information about the effects of UV-daylight radiation on the treated surface. The variations in terms of ΔE and of the single components (ΔL, Δa and Δb) are reported in Figure 4. For all treatments, a variation in ΔE below 3.5 has been detected. In all three cases, binders and IL suffered a higher variation with respect to the bare binder. The color induced by TE + I treatment suffered less from UV radiation; this is probably due to the fact that the IL was not applied on the surface in a two-step process, as it was present in bulk due to being all applied at once. Looking at the single components, even in this case the L* coordinate is the most variable. It is worth noting that for NES + I and ES + I the L* coordinate goes in the opposite direction with respect to treated–untreated variation (Figure 3), as a consequence of the coating equilibrium processes induced by UV-daylight radiation.

Capillary absorption tests were carried out on untreated, treated and aged marble specimens (Figure 5). Using this assessment, it is possible to understand the effect of binders and IL on water uptake. It should be pointed out that, although marble has a low porosity and, therefore, a low water uptake, our findings show that the effect of the treatments is not negligible. A significant variation in water uptake can be due either to a lowering of the porosity, as pores are filled with the consolidant, or to the hydrophobicity of the surface.

The behavior of the TEOS- and NanoEstel-treated specimens is quite similar to that of the untreated ones. This means that there is not a significant decrease in the porosity induced by the consolidant, nor in the hydrophobicity. On the contrary, Estel treatment induces a significant decrease in water absorption. This result may be attributable to a transient hydrophobicity, caused by the entrapment of solvent in the coating during the curing process, and is consistent with the colorimetric data detected for the same treatment. The comparison between the absorption curves of treated samples with and without IL suggests that the addition of IL exerts a slight effect on the Estel binder, leading to a faster water uptake, even if the final absorption value is lower than that recorded for the sample treated with only Estel.

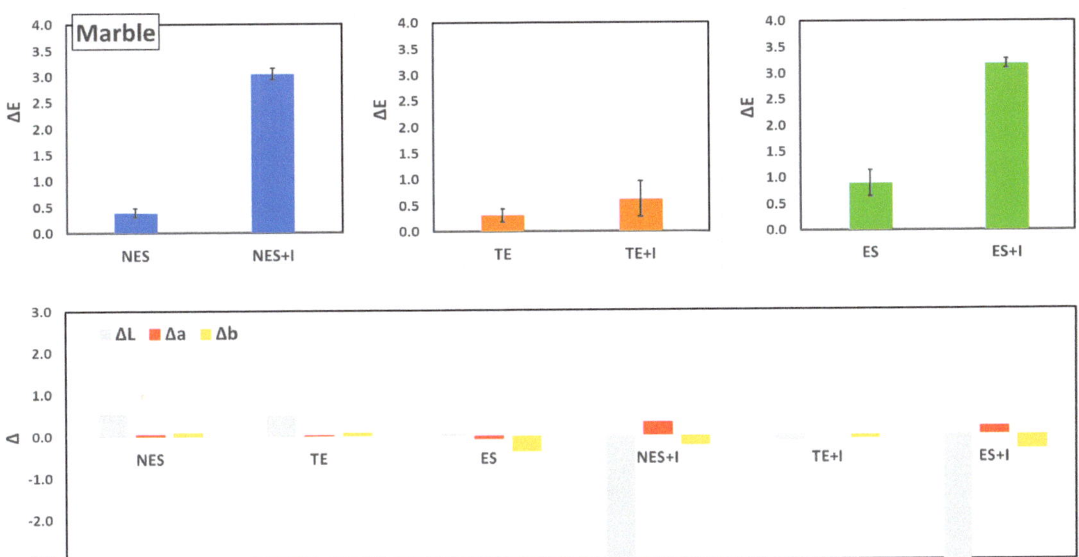

Figure 4. Colorimetric differences between aged and treated specimens. For each treatment, the ΔE value with its components (ΔL, Δa, Δb) is reported.

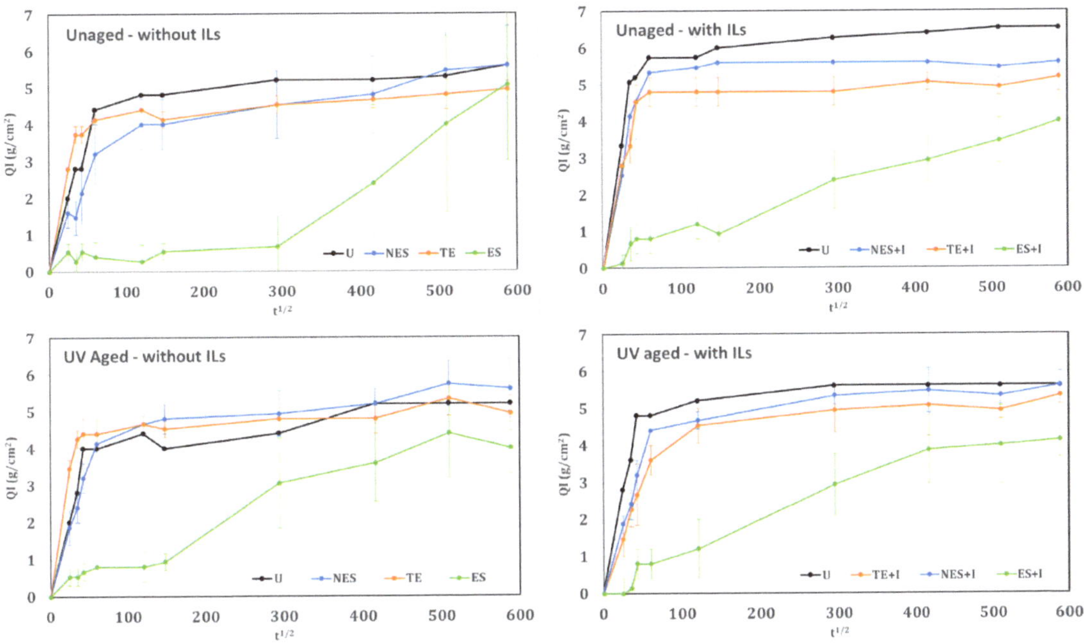

Figure 5. Capillary absorption of untreated, treated and aged samples.

The same trend of speeding up the water uptake for Estel is detected for the aged samples. This can be due to a "desiccant effect" of the UV-daylight radiation.

The contact angle measurements are reported in Figure 6. The untreated marble specimen shows a contact angle of about 40°. NES treatment leads to a hydrophilic surface since the measured contact angle is 0. This is due to nanosilica, which, once cured, leaves a negative charge on its surface due OH groups [28]. This effect can also be enhanced by the cracking on the surface observed via SEM (Figure 2). An interesting behavior is observed after IL application, as the NES + I treatment shows a similar wettability to the untreated sample.

Considering the ionic nature of IL, it is reasonable to assume that it interacts with the negatively charged stone silica surface via electrostatic bonds, with the surfactant cation playing the main role. In this light, an arrangement, in which the positive head is oriented towards the stone surface and the non-polar tail upward, is conceivable and explains the contact angle increasing [35].

TEOS treatment does not significantly change the wettability of the surface; moreover, IL addition does not affect its behavior. For Estel treatment, a great increase in the contact angle is detected. Again, this behavior is ascribable to the residue solvent in the coating. The application of the IL leads to a dropdown in the contact angle. Here, the dynamics can be the opposite with respect to those observed for NES + I. In this case, the coating is mostly nonpolar, and when IL is applied, the anion and cation are preferably oriented with tails downward and polar heads upward; this arrangement leads to a higher wettability of the surface, and then to a low/null contact angle. Aging can exert just a physical effect related to the evaporation of the solvent, which is the primary cause of the whole dynamics. This result is consistent with the colorimetric measurements, in which the color variation after aging is compatible with sample drying and capillary absorption, and occurs slightly faster for the aged sample as well as for IL application.

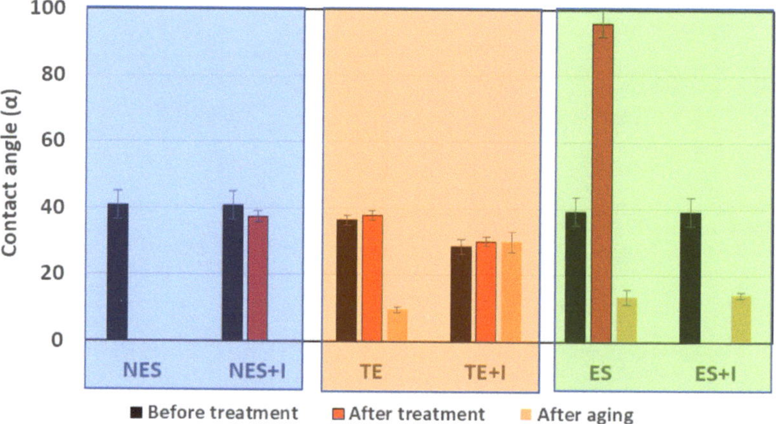

Figure 6. Contact angle measured for untreated, treated and aged samples (in blue it has been highlighted the treatments with nanosilica binder, in orange those with TEOS, in green those with ESTEL).

3.2. Washout Test

Another feature explored is the resistance of the coatings, in terms of composition and surface microstructure, toward seawater. This can simply leach the IL away, since the bond between the IL and the binder may exhibit a reversible nature. The seawater can also have an effect on the microstructure by rearranging the IL orientation.

In Figure 7, the photos taken with UV radiation are presented, highlighting the luminescence of I—PyrS. The difference between the blank sample and treated samples before the washout test is evident; the former is completely devoid of luminesce, while the latter shows a very visible luminesce. After the washout test, all samples show a similar luminescence to the unwashed ones, suggesting that the IL has a resistance toward seawater, although a quantitative assessment would need more complex experimentation.

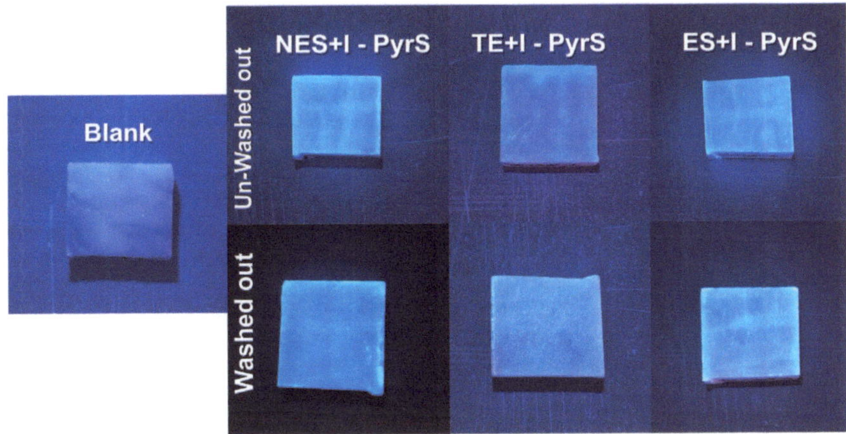

Figure 7. Images of untreated and treated/washed-out samples taken under UV light.

In addition, the measurement of the contact angle on washed-out samples (Figure 8) was repeated. Seawater seems to have no effect on surfaces treated with TEOS, while for the NES- and NES + I-treated samples a slight increase in wettability is detected, attributable to an increase in OH groups on the surface. A more complex pattern is observed for TEOS treatment. Samples treated with the bare consolidant show a decrease in the contact angle, analogously to that observed after UV aging (Figure 6), while for those treated with TEOS and IL, a slight increase is detected.

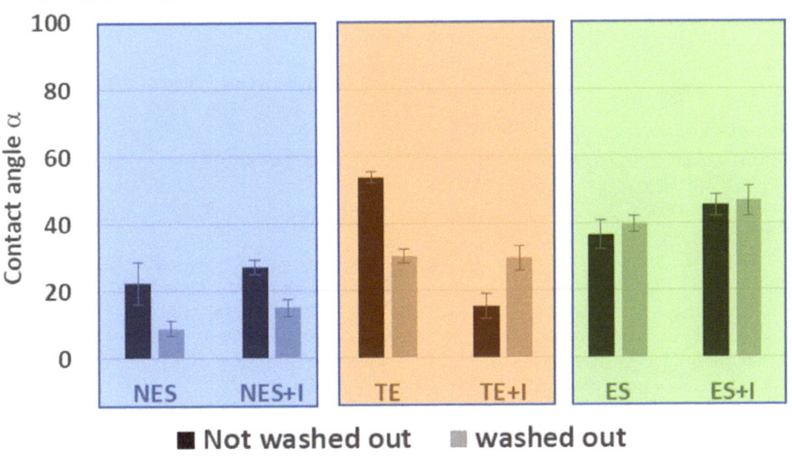

Figure 8. Contact angle measurements for treated and washed-out samples (in blue it has been highlighted the treatments with nanosilica binder, in orange those with TEOS, in green those with ESTEL).

4. Conclusions

In this paper, an ionic liquid (IL), consisting of N-(2-hydroxyethyl)-N,N-dimethyl-1-do-decanaminium cation and a mixture of bromide and dodecylbenzenesulfonate (DBS) anions in a molar ratio of 3:1, was applied on marble specimens together with the consolidants/binders NanoEstel, TEOS and Estel, used as adhesion promoters. The results suggest that the coatings are resistant to seawater and UV-daylight radiation. Moreover, IL seems to have a negligible effect on the micromorphology of the coating, but it has a significant effect on the wettability. In particular, it decreases the wettability in the case of NanoEstel, while this parameter increases for Estel with IL. This was attributed to the hydrophilicity induced by NanoEstel and hydrophobicity induced by Estel. When IL is applied, the arrangement of their ions compensates for the previous hydrophilicity/hydrophobicity of the surface. Further work has been planned in order to explore how the different IL dispositions influence the biocidal activity, as well as how longer-term exposure to weathering agents affects the coatings' features.

Obtaining hydrophilic surfaces bodes well for the antifouling activity of products in an aquatic environment. Water molecules on hydrophilic surfaces may generate a hydration layer, which, in turn, prevents the colonization process of micro-organisms on the surface [36]. All the silica@IL coatings studied herein displayed hydrophilic behavior. This result suggests that surfactant ionic liquids and, more generally, IL technology may significantly contribute to the development of novel antifouling/hydrophilic coatings for stone conservation, even in a submerged environment.

Author Contributions: Conceptualization, F.D.L., S.L.S., C.U. and S.A.R.; methodology, S.L.S., M.L., M.R. and D.D.P.; investigation, M.L., S.L.S., S.A.R., C.G. and N.R.; data curation, S.A.R., M.L. and M.R.; writing—original draft preparation, M.L., S.L.S. and S.A.R.; writing—review and editing, M.R., M.F.L.R., C.G. and N.R.; supervision, M.F.L.R., C.U. and D.D.P.; funding acquisition, N.R., D.D.P. and M.F.L.R. All authors have read and agreed to the published version of the manuscript.

Funding: This research was funded by the UNALTERABLE project, funding program CRIMAC-Centro Ricerche ed Infrastrutture Marine Avanzate in Calabria—Fondo FSC 2014–2020—Piano Stralcio «Ricerca e Innovazione 2015–2017»—Programma Nazionale Infrastrutture di Ricerca (PNIR), linea d'azione 1.

Institutional Review Board Statement: Not applicable.

Informed Consent Statement: Not applicable.

Data Availability Statement: Not applicable.

Conflicts of Interest: The authors declare no conflict of interest.

References

1. Patil, S.M.; Kasthurba, A.K.; Patil, M.V. Characterization and assessment of stone deterioration on Heritage Buildings. *Case Stud. Constr. Mater.* **2021**, *15*, e00696. [CrossRef]
2. Doehne, E.; Price, C.A. *Stone Conservation: An Overview of Current Research*; Getty Conservation Institute: Los Angeles, CA, USA, 2010.
3. Camuffo, D. *Microclimate for Cultural Heritage: Measurement, Risk Assessment, Conservation, Restoration, and Maintenance of Indoor and Outdoor Monuments*, 3rd ed.; Elsevier: Amsterdam, The Netherlands, 2019; pp. 1–552.
4. Sacco Perasso, C.; Antonelli, F.; Calcinai, B.; Casoli, E.; Gravina, M.F.; Ricci, S. The Bioerosion of Submerged Archeological Artifacts in the Mediterranean Sea: An Overview. *Front. Mar. Sci.* **2022**, *9*, 888731. [CrossRef]
5. Flemming, H.S.; Wingender, J.; Szwqzyk, U.; Steinberg, P.; Rice, S.A.; Kjelleberg, S. Biofilms: An emergent form of bacterial life. *Nat. Rev. Microbiol.* **2016**, *14*, 563–575. [CrossRef] [PubMed]
6. Bruno, F.; Muzzupappa, M.; Barbieri, L.; Gallo, A.; Ritacco, G.; Lagudi, A.; La Russa, M.F.; Ruffolo, S.A.; Crisci, G.M.; Ricca, M.; et al. The CoMAS project: New materials and tools for improving the in situ documentation, restoration, and conservation of underwater archaeological remains. *Mar. Technol. Soc. J.* **2016**, *50*, 108–118. [CrossRef]
7. La Russa, M.F.; Ruffolo, S.A.; de Buergo, M.Á.; Ricca, M.; Belfiore, C.M.; Pezzino, A.; Crisci, G.M. The behaviour of consolidated Neapolitan yellow Tuff against salt weathering. *Bull. Eng. Geol. Environ.* **2017**, *76*, 115–124. [CrossRef]
8. Gaylarde, C.C.; Morton, L.H.G. Deteriogenic biofilms on buildings and their control: A review. *Biofouling* **1999**, *14*, 59–74. [CrossRef]
9. De Leo, F.; Marchetta, A.; Urzì, C. Black Fungi on Stone-Built Heritage: Current Knowledge and Future Outlook. *Appl. Sci.* **2022**, *12*, 3969. [CrossRef]
10. Urzi, C. Microbial deterioration of rocks and marble monuments of the Mediterranean basin: A review. *Corros. Rev.* **2004**, *22*, 441–457. [CrossRef]

11. Ricca, M.; La Russa, M.F. Challenges for the protection of underwater cultural heritage (Uch), from waterlogged and weathered stone materials to conservation strategies: An overview. *Heritage* **2020**, *3*, 402–411. [CrossRef]

12. Aloise, P.; Ricca, M.; La Russa, M.F.; Ruffolo, S.A.; Belfiore, C.M.; Padeletti, G.; Crisci, G.M. Diagnostic analysis of stone materials from underwater excavations: The case study of the Roman archaeological site of Baia (Naples, Italy). *Appl. Phys. A Mater. Sci. Process.* **2014**, *114*, 655–662. [CrossRef]

13. Callow, M.E.; Callow, J.A. Marine biofouling: A sticky problem. *Biologist* **2002**, *49*, 10–14. [PubMed]

14. Bromley, R.G. A stratigraphy of marine bioerosion. *Geol. Soc. Spec. Public* **2004**, *228*, 455–479. [CrossRef]

15. Flemming, H.-C.; Wingender, J. The biofilm matrix. *Nat. Rev. Microbiol.* **2010**, *8*, 623–633. [CrossRef] [PubMed]

16. Bruno, L.; Casieri, C.; Gabriele, F.; Ranaldi, R.; Rugnini, L.; Spreti, N. In situ application of alginate hydrogels containing oxidant or natural biocides on Fortunato Depero's mosaic (Rome, Italy). *Int. Biodeterior. Biodegrad.* **2023**, *183*, 105641. [CrossRef]

17. La Russa, M.F.; Macchia, A.; Ruffolo, S.A.; De Leo, F.; Barberio, M.; Barone, P.; Crisci, G.M.; Urzi, C. Testing the antibacterial activity of doped TiO$_2$ for preventing biodeterioration of cultural heritage building materials. *Int. Biodeter. Biodegr.* **2014**, *96*, 87–96. [CrossRef]

18. Bergamonti, L.; Alfieri, I.; Lorenzi, A.; Montenero, A.; Predieri, G.; Barone, G.; Lottici, P.P. Nanocrystalline TiO$_2$ by sol–gel: Characterisation and photocatalytic activity on Modica and Comiso stones. *Appl. Surf. Sci.* **2013**, *282*, 165–173. [CrossRef]

19. Palla, F.; Bruno, M.; Mercurio, F.; Tantillo, A.; Rotolo, V. Essential Oils as Natural Biocides in Conservation of Cultural Heritage. *Molecules* **2020**, *25*, 730. [CrossRef]

20. Lo Schiavo, S.; De Leo, F.; Urzì, C. Present and Future Perspectives for Biocides and Antifouling Products for Stone-Built Cultural Heritage: Ionic Liquids as a Challenging Alternative. *Appl. Sci.* **2020**, *10*, 6568. [CrossRef]

21. Semenzin, E.; Giubilato, E.; Badetti, E.; Picone, M.; Volpi Ghirardini, A.; Hristozov, D.; Brunelli, A.; Marcomini, A. Guiding the development of sustainable nano-enabled products for the conservation of works of art: Proposal for a framework implementing the safe by design concept. *Environ. Sci. Pollut. Res.* **2019**, *26*, 26146–26158. [CrossRef] [PubMed]

22. Xin, B.; Hao, J. Imidazolium-based ionic liquids grafted on solid surfaces. *Chem. Soc. Rev.* **2014**, *43*, 7171. [CrossRef]

23. Cardiano, P.; Fazio, E.; Lazzara, G.; Manickam, S.; Milioto, S.; Neri, F.; Mineo, P.G.; Piperno, A.; Lo Schiavo, S. Highly untangled multiwalled carbon nanotube@polyhedral oligomeric silsesquioxane ionic hybrids: Synthesis, characterization and nonlinear optical properties. *Carbon* **2015**, *86*, 325–337. [CrossRef]

24. Pendleton, J.N.; Gilmore, B.F. The antimicrobial potential of ionic liquids: A source of chemical diversity for infection and biofilm control. *Int. J. Antimicrob. Agents* **2015**, *46*, 131–139. [CrossRef] [PubMed]

25. Ferraz, R.; Branco, L.C.; Prudencio, C.; Noronha, J.P.; Petrovski, Z. Ionic Liquids as active pharmaceutical ingredients. *ChemMedChem* **2011**, *6*, 975–985. [CrossRef]

26. Santos, J.I.; Gonçalves, A.M.M.; Pereira, J.L.; Figueiredo, B.F.H.T.; Silva, F.A.; Coutinho, J.A.P.; Ventura, S.P.M.; Gonçalves, F. Environmental safety of cholinium-based ionic liquids: Assessing structure–ecotoxicity relationships. *Green Chem.* **2015**, *17*, 4657–4668. [CrossRef]

27. Petkovic, M.; Seddon, K.R.; Rebelo, L.P.N.; Pereira, C.S. Ionic liquids: A pathway to environmental acceptability. *Chem. Soc. Rev.* **2011**, *40*, 1383–1403. [CrossRef] [PubMed]

28. De Leo, F.; Marchetta, A.; Capillo, G.; Germanà, A.; Primerano, P.; Lo Schiavo, S.; Urzì, C. Surface Active Ionic Liquids Based Coatings as Subaerial Anti-Biofilms for Stone Built Cultural Heritage. *Coatings* **2021**, *11*, 26. [CrossRef]

29. Eyssautier-Chuine, S.; Franco-Castillo, I.; Misra, A.; Hubert, J.; Vaillant-Gaveau, N.; Streb, C.; Mitchell, S.G. Evaluating the durability and performance of polyoxometalate-ionic liquid coatings on calcareous stones: Preventing biocolonisation in outdoor environments. *Sci. Total Environ.* **2023**, *884*, 163739. [CrossRef] [PubMed]

30. De Leo, F.; Cardiano, P.; De Carlo, G.; Lo Schiavo, S.; Urzì, C. Testing the antimicrobial properties of an upcoming "environmental-friendly" family of ionic liquids. *J. Mol. Liq.* **2017**, *248*, 81–85. [CrossRef]

31. *ISO/CIE 11664-4:2019*; Colorimetria-Parte 4: Spazio colore L* a* b* CIE 1976. Ente Nazionale Italiano di Unificazione: Milano, Italy, 2019.

32. *ISO/CIE 15801:2010*; Conservazione dei beni Culturali-Metodi di Prova-Determinazione Dell'assorbimento Dell'acqua per Capillarità. Ente Nazionale Italiano di Unificazione: Milano, Italy, 2010.

33. *ISO/CIE 15802:2010*; Conservazione dei beni Culturali-Metodi di Prova-Determinazione Dell'angolo di Contatto Statico. Ente Nazionale Italiano di Unificazione: Milano, Italy, 2010.

34. Manoudis, P.N.; Karapanagiotis, I.; Tsakalof, A.; Zuburtikudis, I.; Kolinkeová, B.; Panayiotou, C. Superhydrophobic films for the protection of outdoor cultural heritage assets. *Appl. Phys. A Mater. Sci. Proc.* **2009**, *97*, 351–360. [CrossRef]

35. Wang, Y.; Li, L. Uncovering the Underlying Mechanisms Governing the Solidlike Layering of Ionic Liquids (ILs) on Mica. *Langmuir* **2020**, *36*, 2743–2756. [CrossRef]

36. Donnelly, B.; Sammut, K.; Tang, Y. Materials Selection for Antifouling Systems in Marine Structures. *Molecules* **2022**, *27*, 3408. [CrossRef] [PubMed]

Article

UV-C Irradiation and Essential-Oils-Based Product as Tools to Reduce Biodeteriorates on the Wall Paints of the Archeological Site of Baia (Italy)

Paola Cennamo [1,*], Roberta Scielzo [1], Massimo Rippa [2], Giorgio Trojsi [1], Simona Carfagna [3] and Elena Chianese [4]

[1] Department of Humanities, University of Naples Suor Orsola Benincasa, 80132 Naples, Italy
[2] Institute of Applied Sciences and Intelligent Systems "Eduardo Caianiello", CNR, 80078 Pozzuoli, Naples, Italy
[3] Department of Biology, University of Naples Federico II, 80126 Naples, Italy
[4] Department of Science and Technology, Parthenope, University of Naples, 80143 Naples, Italy
* Correspondence: paola.cennamo@unisob.na.it

Abstract: This study is aimed to compare, through laboratory experimentations, the efficiency of UV-C irradiation and an essential-oils-based product as tools to reduce the biofilm identified in a semi-hypogeum room located in the archaeological park of Baia, Italy. During this study, the autotrophic component of the original biofilm, mostly composed of Chlorophyceae and Cyanophycean, was isolated in the laboratory, while simultaneously, the composition of the pigments used for the fresco paintings was examined in situ through X-ray fluorescence. These examinations were necessary for the creation of test samples that were similar to the original surfaces and used for subsequent experiments. The plaster testers were contaminated with artificial biofilm, exposed to UV-C at a distance of 80 cm for a fixed time interval and treated with *ESSENZIO©*, a product based on oregano and thyme essential oils, to eradicate the biological species. The treatment's effectiveness was then assessed by employing optical microscopy and spectrometric techniques applied to the areas previously occupied by the biofilm on the different test samples. To obtain an additional parameter to evaluate the treatments efficacy, the concentrations of the photosynthetic pigments were also measured by spectrophotometry. Results showed that biofilms were successfully removed by the irradiation of the surfaces and by the essential-oils-based product at a dilution of 50% in demineralized water with a time of application of 1 h and 30 min; in addition, no visible change of the pigments used on the testers were observed, demonstrating the high efficiency of the treatments against biodeteriogens. The two methods and their different mechanisms of action have provided interesting aspects that suggest a combined strategy to contrast and prevent biological growth in archaeological contexts.

Keywords: biodeterioration; essential oils; UV-C irradiation; biofilms; conservation; microbial growth control

Citation: Cennamo, P.; Scielzo, R.; Rippa, M.; Trojsi, G.; Carfagna, S.; Chianese, E. UV-C Irradiation and Essential-Oils-Based Product as Tools to Reduce Biodeteriorates on the Wall Paints of the Archeological Site of Baia (Italy). *Coatings* **2023**, *13*, 1034. https://doi.org/10.3390/coatings13061034

Academic Editors: Patricia Sanmartín and Ajay Vikram Singh

Received: 11 April 2023
Revised: 22 May 2023
Accepted: 26 May 2023
Published: 2 June 2023

1. Introduction

Among the most common conservation problems for historical and artistic monuments, biodeterioration represents one of the most important causes of decorated surface damage. Archaeological areas are highly influenced by the environmental context, which is one of the main factors that lead to the deterioration of artifacts. These peculiar environmental conditions, together with high humidity levels and slight fluctuation in temperatures, allow for the proliferation of heterogeneous biological species typical of semi-confined areas, which contribute to the biodeterioration phenomena [1–3]. The diffusion of biological species over painted surfaces leads to the chromatic variation and degradation of the artifacts [3,4]. The progress in the treatment of biological patinas has motivated the experimentation of alternative methods (e.g., essential oils, UV-C irradiation, laser cleaning) to control the proliferation of biofilms on historical and artistic objects. The most common biocides in conservative procedures, such as Biotin T, that are characterized by proven effectiveness for the treatment of biodeteriogens are very harmful to the environment and the operators. The most recent experimentations have shown the efficacy of UV-C radiation and essential oil-based products to control the biological growth of different typologies of artifacts [4,5].

UV-C irradiation has been used to treat stone monuments since the 1960s, as reported in Borderie et al. [4] and Liverani et al. [6]. Recently, this method was also tested by Cennamo et al. to create a pilot system to prevent biological growth on the painted surfaces of the tomb ES-07 in Porta Nocera necropolis, located in the Archeological Park of Pompeii [7].

Recently, essential oils were tested in different contexts to remove the biodeteriogens from different kinds of artifacts; as an example, Devreux et al. used these substances for the treatment of stone statues located in the Vatican Gardens as a pilot experiment for new methods aimed to control biological growth on artistic artifacts [8,9].

The Archeological Park of Baia, with its characteristic environmental conditions— high humidity and constant temperatures due, in particular, to the presence of underground thermal water—represents the perfect context for the formation of heterotrophic and autotrophic biofilm. For this reason, the individuation of effective methods to remove biodeteriogens and to prevent their future growth represents a fundamental topic. The area selected to collect and study biofilm was room SB-E0-R07, located in the north area of the Mercury Sector in the Archeological Park of Baia, where a thick layer of biological species was present all over the painted surfaces. A pilot system was then developed to irradiate some pigmented samples with UV-C rays, and other testers were treated with an essential-oil-based product to evaluate and to compare the efficiency of both methods in the removal of the biofilm.

2. Materials and Methods

Room SB-E0-R07 is situated in the north zone of the Mercury Sector, in the Archeological Park of Baia (Figures 1 and 2). This area is characterized by the presence of underground salty thermal water, which is one of the main factors that contributes to the high level of humidity recorded here.

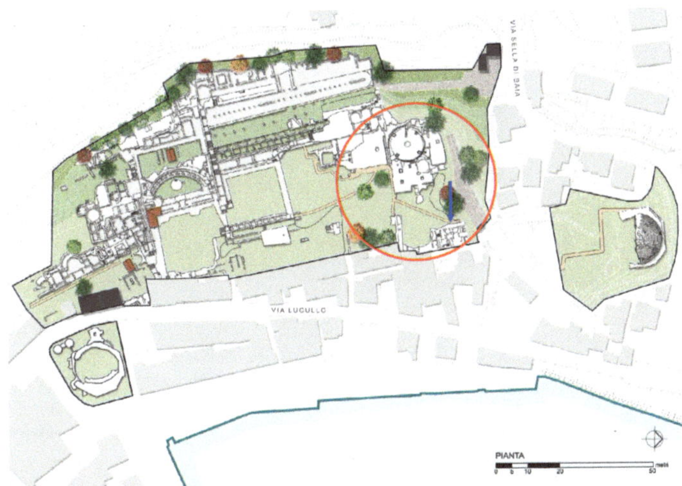

Figure 1. Planimetry of the Archeological Park of Baia. The red circle highlights the Mercury Sector, while the blue arrow indicates the SB-E0-R07 room.

Figure 2. Biofilms growing on the wall paintings of the SB-E0-R07 room.

In particular, in the chamber object of the present study, we investigated the variations in the thermohygrometric parameters for one year with the support of four dataloggers (*ORIA©*). After the monitoring period, data were processed to detect any variation in humidity and temperature; as a result, we observed a high value for the average humidity (93%) and a mild value for the average temperature (in the range of 12–14 °C) (Figures 3 and 4).

These factors are the main causes of the proliferation of the biodeteriogens on the fresco paintings in this site; the identification of microbial communities was previously carried out [7,8], showing the presence of different autotrophic and heterotrophic microorganisms, in particular, Cyanobacteria and algae.

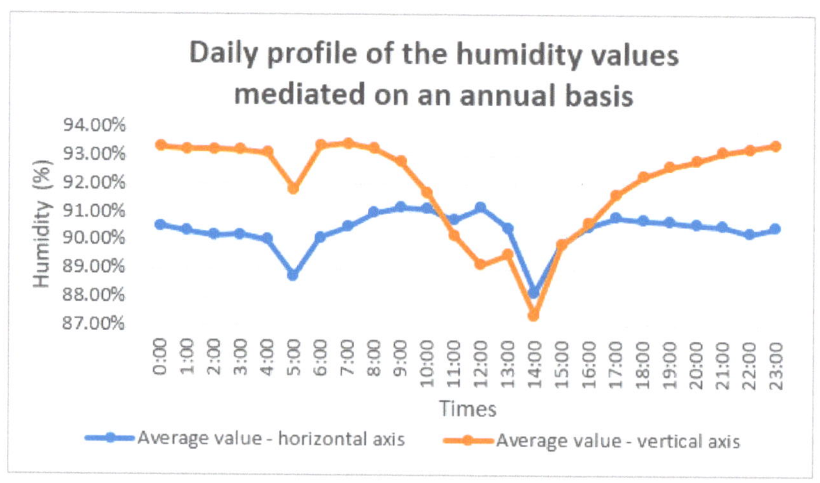

Figure 3. Variations in the daily profile of the humidity of room SB-E0-R07 on an annual basis.

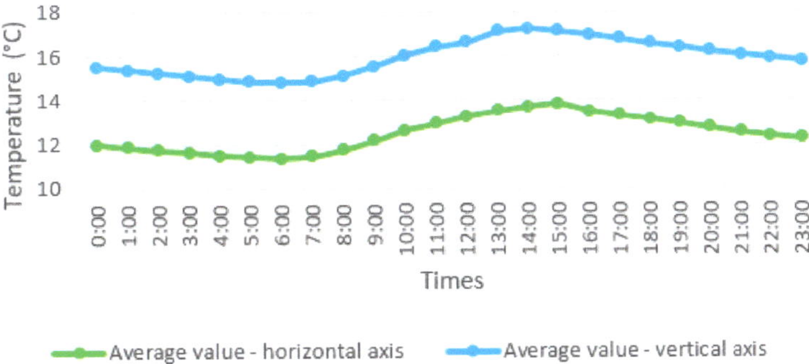

Figure 4. Variations in the daily profile of the temperature of room SB-E0-R07 on an annual basis.

2.1. Chemical-Physical Analysis (XRF)

X-ray fluorescence analysis Assing, Rome, Italy) were performed to detect the elementary composition of the original pigments of the wall paintings. Red, yellow, white and black colours, in addition to the composition investigation of the original plaster, were analysed with a portable XRF-Q Assing spectrometer (Assing, Rome, Italy), a tungsten tube, and a PiN silicon diode detector with beryllium window, at operating conditions of 30 KV and 0.5 mA (counting time: 30 s).

2.2. Cultivation of Algae

Following the detection of microalgae belonging to Chlorophyceae and Cyanophycean, two different growth media, BG11 and BBM [10,11], were used to determine growth in the laboratory. In order to obtain exponential growth, the cultures were stored in a climatic chamber at 37 °C for 12 days. Artificial biofilm preparation: A series of test samples were artificially reproduced using the same construction materials present in the nymphaeum and pigments used for the original surfaces (Figure 1). A total of two test samples, four for each of the three original pigments, were prepared and inoculated with the isolated cultures in the laboratory until they were completely covered with biofilm; in Figures 5 and 6 are shown (a) black samples; (b) yellow samples; (c) white samples and red samples (d), following the procedure of Cennamo et al. [10]. The biofilm growth on the substrate was monitored in the following days.

2.3. Essential Oils-Based Product Treatments

The samples were treated with *ESSENZIO©*, a product based on thyme and oregano essential oils and produced by *IBIX©*, applied by brush and diluted in demineralized water as suggested by the procedure of Devreux et al. [9], at different concentrations (10%, 20%, 50%, and pure), and it was removed with demineralized water after different durations of action (from 30 to 90 min) (Figure 5).

2.4. UV-C Treatments

The test samples were exposed to radiation using two UV-C lamps (Fluorescent Compact G9W Lynx, Sylvania, OA, USA, 2 × 9 W each = 18 W total, λ max = 254 nm). A total of 5 mL of algal culture was extracted from the test samples to be used as a control. The irradiations were conducted at a distance of 80 cm for a duration of 8 h a day every day, for a total duration of exposure of 24 h (Figure 5).

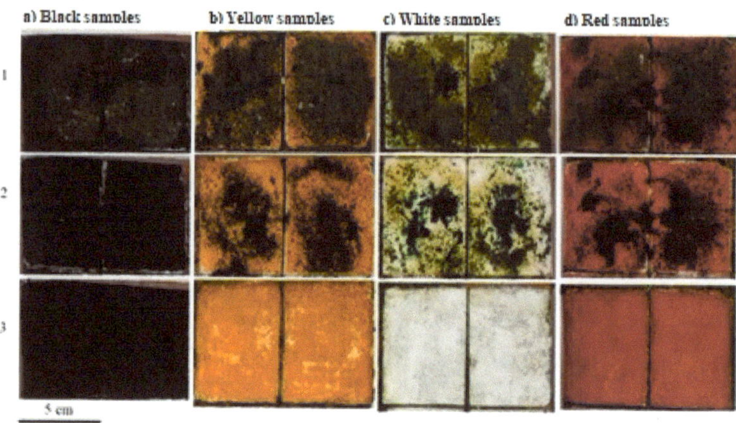

Figure 5. Samples were used for inoculation with a mixture of cyanobacteria and green algae for the UV-C treatments. (**a**) Black samples; (**b**) yellow samples; (**c**) white samples and (**d**) red samples. 1—before the treatments; 2—after the treatments; 3—24 h after the treatment.

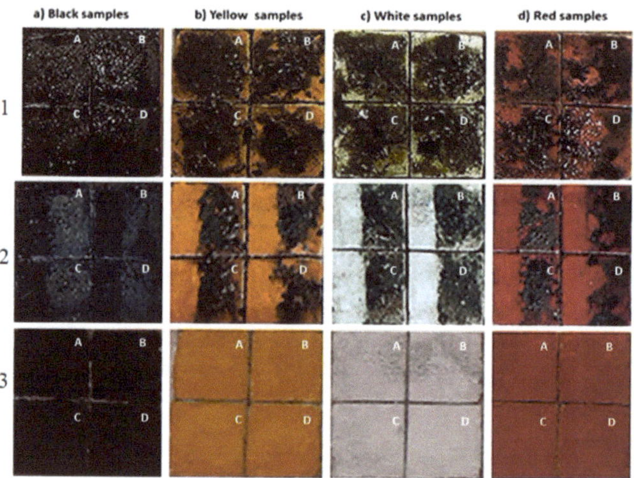

Figure 6. Samples treated with *ESSENZIO©* at different concentrations in demineralized water (10% (**A**), 20% (**B**), 50% (**C**), and pure (**D**)) and time of applications from 0 min (1) 30 min (2) to 90 min (3). (**a**) Black samples; (**b**) yellow samples; (**c**) white samples; and (**d**) red samples.

2.5. Absorption of Main Photosynthetic

Pigments following centrifugation (1′, 5000× *g*) and cell pigments were extracted using 2 mL of N,N-dimethylformamide (Spectrophotometer 7315, Jenway, Hong Kong, China), following the methodology illustrated by Carfagna et al. [12]. Chlorophyll content of cells was estimated spectrophotometrically (Spectrophotometer 7315, Jenway, Hong Kong, China). The absorbance at wavelengths 647 and 664 nm was used to determine the concentration of total chlorophyll as well as that of chlorophyll a and b, using the following equations:

$$\text{Chl Tot} = (A647 \times 17.9) + (A664 \times 8.08); \quad \text{Chla} = (A664 \times 12.7) - (A647 \times 2.79);$$
$$\text{Chl b} = (A647 \times 20.78) - (A664 \times 4.88).$$

2.6. Colorimetric Analyses

Colorimetric investigations were conducted by light absorption in diffuse reflection using a MAYA 2000 pro (Ocean Insight, Oxford, UK) spectrophotometer in a Pump-Probe configuration. The light source used was a halogen lamp (CIE standard illuminant D65, Ocean Insight, Oxford, UK) with an emission spectrum in the VIS-NIR range 400–1000 nm. Before measurements, the device was calibrated using a white ceramic disk and a black trap portion. The commercial software OceanView 2.0 (Ocean Insight, Oxford, UK), with which the device was supplied, was used for the acquisition of the reflected spectra, the colorimetric data and for basic operations. Colorimetric parameters were calculated in the CIE L* a* b* 1976 colour space [13–15]. In this system, L* coordinate is lightness (L* = 0: black and L* = 100: white), a* coordinate represents the green/red component (negative values represent the green component and positive values represent the red component, through grey close to zero), and the b* coordinate refers to the blue/yellow component (negative values representing the blue component and positive values the yellow component, through grey close to zero). Samples with 4 different coloured areas—black, yellow, red and white—were investigated (Figures 5 and 6). The parameters L*, a* and b* were measured for the four colours in four different experimental conditions (M): before biofilm growth (M1), before biofilm growth but after the treatments considered (M2), after biofilm growth (M3), and after the treatments considered on biofilm (M4). For each colour, ten measurements in different points of the interested area were performed, and then the values were mediated to obtain the final parameters. Subsequently, for all treatments considered, the variations ΔL*, Δa* and Δb* were calculated considering the difference in measurements M4–M1, M4–M3 and M2–M1.

3. Results

3.1. XRF Analyses of the Pigments

The analysis of the red and yellow pigments shows the presence of calcium and iron, ascribable to a red ochre (probably hematite) and a yellow ochre (probably goethite) painted on a calcium carbonate plaster. In addition, the black pigment (as well as the white colour) revealed a peak only for the calcium, a sign of the organic nature of this colour, probably carbon black.

3.2. Effect of Essential Oil and UV-C on Artificial Biofilm and Control Test Samples

The samples completely covered by artificial biofilms consisting of several layers of intense green autotrophic microorganisms were subjected to the action of the essential oils of *ESSENZIO©* and the UV-C rays. The presence of an intense green biofilm indicates a high photosynthetic capacity.

After the treatments, the effects were clearly observable both on the samples treated with the oils and on those treated with UV-C; the biofilms grown on the exposed test samples were visibly bleached after the treatments (Figure 5). The progressive bleaching of the biofilms was related to the degradation of chlorophyll, as evidenced by spectrophotometric determinations.

The exposed test samples reported in Figures 5 and 6 show that all microorganisms were eradicated by both the essential oils and the UV-C treatment used in this study. A selection of test samples free from artificial biofilms, representative of each pigmentation, were then exposed to UV-C radiation. The colourimetric analysis performed on these samples did not show any chromatic alteration, proving that the treatment does not interfere with the support.

3.3. Measurement of Chlorophyll Concentration

The autotrophic microorganisms exposed to the different treatments showed a significant decrease in chlorophyll concentration. The decrease was more evident in the treatment with the essential-oils-based product (50% concentration in demineralized water, with a duration of application of 1 h and 30 min), compared to UV-C (irradiation conducted at

80 cm for a duration of 8 h a day every other day, for a total duration of twenty-four hours of exposure) (Figure 7). Coherently with the data for cell viability, these fluctuations were expressed as percentages.

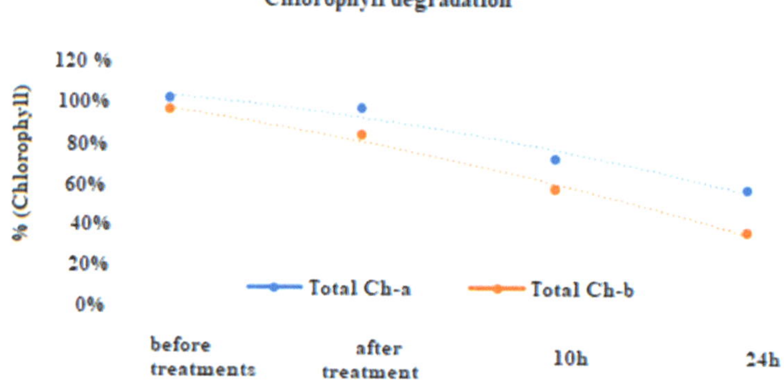

Figure 7. Total chlorophyll content in samples subjected to UV-C (ChlA) and *ESSENZIO©* (ChlB) treatment at different exposure times (before treatment, after treatment, 10 h, 24 h).

3.4. Colorimetric Analyses

Below, the colourimetric data for the black, yellow, red, and white colours of the samples treated with twelve different protocols based on the use of essential-oils-based products and with one protocol based on the use of UV-C radiation are reported. In particular, Tables 1 and 2 show the variations in the colourimetric parameters relating to the four colours of the samples achieved between measurements realized after the *ESSENZIO©* treatments on biofilm (M4) and, respectively, before the growth of the biofilm (M1) and after the growth of the biofilm (M3). In Tables 3 and 4, the same quantity referring to UV-C treatment are reported.

Table 1. Variations (M4–M1) in the colorimetric data of the black, yellow, red and white colours achieved between measurements realized on the samples after the twelve *ESSENZIO©* treatments considered on biofilm (M4) and before the growth of the biofilm (M1).

| Treatment: *ESSENZIO* | | Colour | | | | | | | | | | | |
| | | Black | | | Yellow | | | Red | | | White | | |
Percentage in Water (%)	Time of Action (min)	ΔL^*	Δa^*	Δb^*	ΔL^*	Δa^*	Δb^*	ΔL^*	Δa^*	Δb^*	ΔL^*	Δa^*	Δb^*
10	30	2.28	0.30	8.93	7.84	6.01	4.76	10.17	−0.45	0.88	−4.60	−0.68	9.56
10	60	3.41	−0.09	9.42	16.10	8.87	13.13	1.97	−6.28	73.2	−8.17	−0.57	9.58
10	90	2.46	−0.27	−6.13	14.66	10.44	11.66	2.93	−2.08	−4.63	−11.76	−1.16	3.56
20	30	2.29	−0.24	−6.24	17.85	9.57	12.34	6.75	1.09	−0.18	−13.85	0.32	21.69
20	60	1.92	−0.22	−6.18	14.74	9.09	10.96	11.05	6.13	4.75	−22.59	−5.04	14.51
20	90	1.16	−0.23	−6.63	10.96	5.26	4.02	11.79	3.43	5.20	−7.64	−0.87	6.39
50	30	3.27	−0.37	−5.99	8.25	2.63	1.95	4.36	−5.83	−6.64	−13.64	−0.46	6.16
50	60	4.24	−0.1	−5.51	9.21	2.65	2.75	6.65	−1.53	−2.61	−13.95	−0.57	5.76
50	90	2.00	−0.29	−6.79	16.41	5.46	11.36	4.35	−2.24	−3.01	−2.58	−0.97	6.06
100 (pure)	30	1.40	−0.42	−6.57	13.90	5.72	8.36	4.33	−4.03	−6.40	−10.08	−0.54	6.45
100 (pure)	60	1.69	−0.27	−6.94	23.48	12.58	19.86	2.95	−3.11	−5.08	−3.12	−1.47	3.42
100 (pure)	90	2.87	−0.39	−7.32	18.75	7.26	11.72	7.78	0.98	−1.51	−12.10	−0.69	6.99

Table 2. Variations (M4–M3) in the colorimetric data of the black, yellow, red and white colours achieved between measurements realized on the samples after the twelve *ESSENZIO©* treatments considered on biofilm (M4) and after the growth of the biofilm (M3).

Treatment: *ESSENZIO*		Colour											
		Black			Yellow			Red			White		
Percentage in Water (%)	Time of Action (min)	ΔL^*	Δa^*	Δb^*	ΔL^*	Δa^*	Δb^*	ΔL^*	Δa^*	Δb^*	ΔL^*	Δa^*	Δb^*
10	30	−16.31	3.34	−9.28	11.49	23.55	19.66	7.24	33.23	7.73	47.95	2.55	−15.64
10	60	−15.18	2.95	−8.79	19.75	26.41	28.03	1.02	27.40	80.05	44.38	2.66	−15.62
10	90	−16.13	2.77	−24.34	18.31	27.98	26.56	0	31.60	2.22	40.79	2.07	−21.64
20	30	−16.30	2.80	−24.45	21.5	27.11	27.24	3.82	34.77	6.67	38.70	3.55	−3.51
20	60	−16.67	2.82	−24.39	18.39	26.63	25.86	8.12	39.81	11.60	29.96	−1.81	−10.69
20	90	−17.43	2.81	−24.84	14.61	22.80	18.92	8.86	37.11	12.05	44.91	2.36	−18.81
50	30	−15.32	2.67	−24.20	11.90	20.17	16.85	1.43	27.85	0.21	38.91	2.77	−19.04
50	60	−14.35	2.94	−23.72	12.86	20.19	17.65	3.7	32.15	4.24	38.60	2.66	−19.44
50	90	−16.59	2.75	−25.00	20.06	23.00	26.26	1.42	31.44	3.84	49.97	2.26	−19.14
100 (pure)	30	−17.19	2.62	−24.78	17.55	23.26	23.26	1.4	29.65	0.45	42.47	2.69	−18.75
100 (pure)	60	−19.90	2.77	−25.15	27.13	30.12	34.76	0.02	30.57	1.77	49.43	1.76	−21.78
100 (pure)	90	−15.72	2.65	−25.53	22.40	24.80	26.62	4.85	34.66	5.34	40.45	2.54	−18.21

Table 3. Variations (M4–M1) in the colorimetric data of the black, yellow, red and white colours achieved between measurements realized on the samples after the UV-C radiation treatments on biofilm (M4) and before the growth of the biofilm (M1).

Treatment: UV-C Radiation			
Colour	ΔL^*	Δa^*	Δb^*
Black	−0.72	0.12	0.86
Yellow	3.12	3.32	1.81
Red	−1.53	−7.28	−6.04
White	−2.42	2.72	10.34

Table 4. Variations (M4–M3) in the colorimetric data of the black, yellow, red and white colours achieved between measurements realized on the samples after the UV-C radiation treatments on biofilm (M4) and after the growth of the biofilm (M3).

Treatment: UV-C Radiation			
Colour	ΔL^*	Δa^*	Δb^*
Black	−19.31	−3.34	−14.91
Yellow	6.77	20.29	16.71
Red	−4.46	26.40	0.81
White	50.15	5.95	−14.86

In Figure 8, for each colour, the reflectance spectrum measured before the biological growth (red lines) and after the contamination with the subsequent biocidal treatments (black lines) of the samples treated with UV-C radiation are reported.

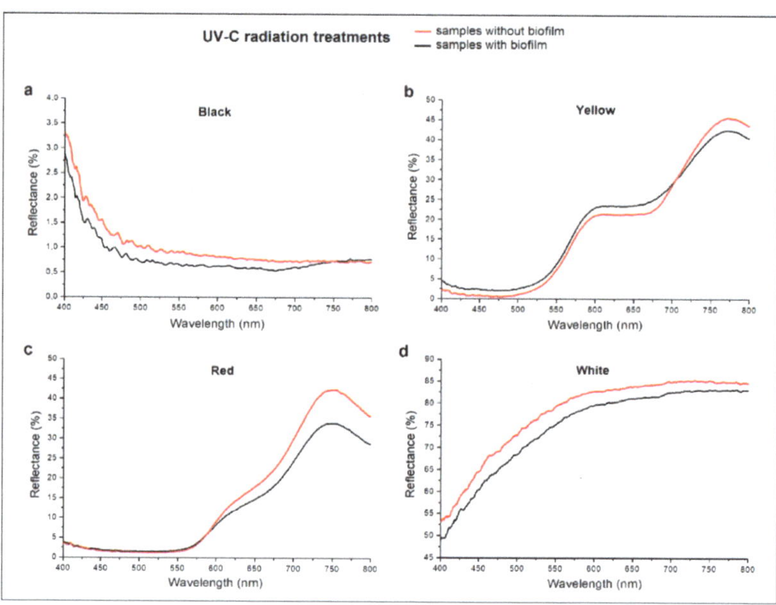

Figure 8. Reflectance spectra measured for samples exposed to UV-C radiation before biofilm growth (red lines) and after the contamination with the subsequent biocidal treatments (black lines) referring to the colours (**a**) black, (**b**) yellow, (**c**) red and (**d**) white.

4. Discussion

The extreme microclimatic conditions of the archaeological sites led to the proliferation of different biological species stimulated by the formation of a polysaccharide matrix that contributes to the genesis of the biofilm [1,16–22]. In our study, we observed the development of different organisms on the wall paintings present in room SB-E0-R07 of the Archeological Park of Baia. In this study, we propose two alternative methods to reduce the presence of biodeteriogens on the painted surfaces: essential-oils-based products and UV-C. Our experiments show that both these methods are effective in the reduction of biodeteriogens under laboratory conditions. In detail, tests were performed using the thyme and oregano essential-oils-based product *ESSENZIO©*; in particular, the product was applied by brush at different percentages of dilution (10%, 20%, 50% in demineralized water and pure) and left on for different durations of action (30 min, 1 h and 1 h and 30 min). Both of these essential oils contain phenols (respectively, thymol and carvacrol), chemical molecules known for their antimicrobial action due to the presence of oxidrilic groups. These chemical groups could cause an alteration in the enzymatic reactions of the biological species, inhibiting their proliferation [23–26]. Due to their antimicrobial proprieties, Origanum vulgare and Thymus vulgaris essential oils were recently used to control biological growth on artistic manufacts, highlighting their effectiveness in conservation strategies [23,27–30], as shown in the study carried out in the Vatican Gardens by Devreux et al. [9], on stone walls in Persepolis and Salluzzo by Favero Longo et al. [31], and by the experiments on some microbial strains isolated from different cultural objects executed by Rotolo et al. [29].

Based on the results obtained through the experiments carried out, the procedures have shown the greater effectiveness of *ESSENZIO©*, particularly at a dilution of 50% in demineralized water with a duration of application of 1 h and 30 min.

The colorimetric analysis confirmed the success of the treatments performed on the samples, with different results for the pigments investigated, as shown in Tables 1 and 2.

The black samples treated with *ESSENZIO©* show, in all the procedures of application, a slight whitening of the testers due to a thinning of the painted layer and incomplete carbonation at the moment of the experimentation and during the removal of the product, in agreement with modest increases in L* values between 1.16 and 4.34 (Table 1). The negative rates of the L* achieved comparing the values in the presence of the biofilm and after the treatments are indicative, on the other hand, of a good reestablishment of the luminosity of the tester after the treatments, confirming the success of the procedures, ulteriorly validated by the increase in the Δa* and decrease in the Δb* parameters (Table 2).

The yellow samples treated with *ESSENZIO©* show, as in the case of the black testers, an increase in the L* parameters between 7.84 and 23.48 (Table 1) caused, once again, by a thinning of the painted layer after the rinsing operations of the product. The increase in the Δa* and Δb* parameters (Tables 1 and 2), instead, indicate the effective restoration of the original conditions of the pigment, with a slight colour change caused by the proliferation of the biological species that was not visible to the eye.

The red testers indicated an increase in the L* values between 1.97 and 11.79 (Table 1), in agreement with what was said earlier. The positive tendency of this parameter confirms the success of the treatments in the removal of the biofilm. This aspect is confirmed also by the increase in the Δa* and Δb* parameters (Table 2) which show, however, slight variations in the original values (Table 1) due to a modest alteration of the original colour caused by the biodeteriogens.

The white samples, instead, indicate a decrease in the L* parameters, a sign of the darkening of the surface visible to the naked eye, particularly in the testers treated with *ESSENZIO©* diluted at 10% and 20% in demineralized water (Table 1). However, the positive values of this parameter in the range 29.96 and 49.97 compared to the measurements taken on the contaminated samples show the general success of the treatments (Table 2).

The presence of visible green spots in the frame treated with *ESSENZIO©* diluted at 10% and 20% on the samples is indicated also by the decrease in the Δa* and increase in the Δb* values compared to the original parameters, signs of a colour change in the green and the yellow regions (Table 1). However, the increase in the Δa* and increase Δb* after the treatments indicate a good outcome of the procedures applied (Table 2).

Concerning the UV-C, the samples were exposed to irradiation at a distance of 80 cm for a duration of 8 h a day every day, for a total duration of 24 h of exposure. Prolonged exposure to radiation causes damage to the biological organisms due to the high absorbance of their proteins [32,33]. Over the last decades, the absence of any adverse effect on operators, environment and cultural heritage, combined with the effectiveness of this method, has led to the application of UV-C rays to the conservation procedures [16,22,34]; this was demonstrated, in particular, in the Cennamo et al. study conducted on Necropoli of Porta Nocera (Pompei) wall paintings [7] and by the application of this procedure in Moidons Caves (France) carried out by Borderie et al. [32,35,36], showing encouraging results.

The black samples treated with UV-C at a distance of 80 cm showed a huge decrease in L* (Table 4), with a slight increase compared to the initial value (Table 3) due to the colour change caused by the biodeteriogens. The Δa* and Δb* values (Table 3), on the other hand, indicate a small increment, confirming the release of pigments by the biological species. However, the decrease in these parameters in relation to the contaminated surfaces indicate the success of the biocidal treatment (Table 4). The reflectance spectrum (Figure 8a) shows a weak reflectance reduction in the sample after the UV-C treatment, in agreement with the mild modification of the colour caused by the biodeteriogens.

The yellow samples show an increase in the ΔL* compared to the initial value, referable, once again, to the mechanical removal of the biological species after the treatment. This aspect is confirmed also by the slight increment of the reflectance of the sample after the UV-C treatment, as visible in Figure 8b. The positive value of the ΔL*, however, indicate the success of the biocidal method. This observation is further confirmed by the increase in the Δa* and Δb* parameters after the application of the biocidal protocol (Tables 3 and 4).

The red sample indicates a slight decrease in ΔL^* equal to -1.53 due to a darkening of the original pigment visible to the naked eye caused by the proliferation of the artificial biofilm. This observation is also confirmed by the decrease in the Δa^* and Δb^* parameters measured with respect to the untreated sample (Table 3), which indicates a colour change of the pigment towards the green and yellow regions. This aspect is confirmed also by the decrease in the reflectance of the sample after the UV-C treatment, as visible in the reflectance spectrum in Figure 8c. However, the increase in these two parameters shown in Table 4 (especially for Δa^*) indicate the effectiveness of the biocidal treatment, with a good restoration of the original proprieties of the painted layer.

Finally, the white sample exposed to UV-C radiation indicated a decrease in the ΔL^* equal to -2.42 due to a darkening of the surface caused by the disfiguring action of the biodeteriogens, a consequence also of the increase in the Δa^* and Δb^* (Table 3). The reflectance spectrum in Figure 8d confirms this data, with a decrease in the curve after the biocidal treatment. However, comparing the ΔL^* values shown in Tables 3 and 4, it is possible to notice a huge increase when it is calculated in the presence of the biofilm, demonstrating the effectiveness of the treatment. This observation is further confirmed by the increase and decrease in Δa^* and Δb^*, respectively, compared to the values recorded before the biocidal treatment (Table 4).

In reference to the graphs in Figure 8, it should be noted that in the case of the black (Figure 8a), red (Figure 8c) and white (Figure 8b) colours, the reflectance measured after treatment (black line) is on average slightly lower on the whole spectrum than that measured before biological growth (red line). These data indicate a slight increase in the opacity of these colours due to the combined action of the biological film and the consequent UV treatment carried out. Differently, in the case of the yellow colour (Figure 8b), the reflectance measured after treatment (black line) is slightly higher than that measured before (red line) up to about a wavelength of 700 nm, while after this value, it is lower, which is in agreement with what was found with the other colours. This last achievement highlights how the effects on this colour of both the growth of the biofilm and the treatment performed affect the final state of the surface and are considered lower.

Colorimetric analysis was also used to investigate the effects of the treatments with *ESSENZIO* and UV-C radiation on samples without biofilm. Parameters ΔL^*, Δa^* and Δb^* were calculated by considering the difference of the measurements performed before biofilm growth (M1) and before biofilm growth but after the treatments considered (M2). The results obtained show small variations in parameters L^*, a^* and b^* for both approaches, which in each case were less than 0.03; therefore, they show the absence of substantial chromatic variations in the pigments caused by the tested protocols.

5. Conclusions

The results of the experimentations carried out during the present study successfully demonstrated the effectiveness of UV-C irradiation and *ESSENZIO©* as alternative methods to traditional biocides for the reduction of biological growth on artistic artifacts. In particular, the following conclusions have been drawn:

1. The protocol conducted with *ESSENZIO©* at a dilution of 50% in demineralized water with a duration of application of 1 hour and 30 min showed the best results for the removal of the artificial biofilm on the painted samples, despite not demonstrating any biocidal power—only a cleaning action. The Origanum vulgare and Thymus vulgaris essential oils, the main ingredients of *ESSENZIO©*, contain, respectively, thymol and carvacrol, phenolic compounds able to inhibit the enzymatic reactions of the biological species due to the acidic nature of hydroxyls within in the aromatic groups [26].

2. UV-C treatments, conducted at a distance of 80 cm and with a wavelength of λ max = 254 nm and a duration of exposure of 8 h, proved to be very useful in countering the proliferation of biological species. As demonstrated by previous studies [7], UV-C radiation plays an important role in the degradation of the proteins that make up the two photosystems and some of the enzymes involved in photosynthesis [18]. The alterations

caused by UV-C on photosynthetic microorganisms are described as "whitening" due to the evident reduction in the chromatic intensity of the pigments [19]; a similar result was also obtained with the use of *ESSENZIO©*.

3. The colorimetric analysis of the samples treated with essential-oils-based products indicated an increase in the L* parameter after the biocidal protocol compared to the values registered on the painted surfaces contaminated with artificial biofilm. These data confirm the effectiveness of the treatments. The slight variations in the Δa^* and Δb^* parameters, on the other hand, indicate a mini-mum colour change caused mainly by the proliferation of the biological species, with a peak in the values registered on the testers treated with *ESSENZIO©* diluted at 10% and 20% in demineralized water, a sign of incomplete removal of biodeteriogens.

4. The colorimetric investigations on the samples treated with UV-C showed an increase in the L* parameter after the biocidal protocol, confirming the success of the procedures and the restoration of the optical proprieties of the testers before the contamination with artificial biofilm. Instead, the variation in the Δa^* and Δb^* parameters indicates a slight variation in the original colour of the surfaces due to the proliferation of the artificial biofilm. This aspect is further confirmed by the increase in the reflectance spectra after the application of the biocidal protocol (Figure 8).

5. Spectrophotometric readings indicate a major decrease in the chlorophyll concentration in the samples treated with *ESSENZIO©* (50% concentration in demineralized water with a time of application of 1 h and 30 min), compared to UV-C (irradiation conducted at 80 cm for a duration of 8 h a day every other day, for a total duration of 24 h of exposure). These data demonstrate the higher effectiveness of the essential-oils-based product compared to UV-C irradiation for the treatment of the artificial biofilm.

6. The comparison of the two methods and the results of the subsequent diagnostic investigations have led, in consequence, to suggest a combined use of *ESSENZIO©*, efficient in the first removal of the biofilms, and UV-C rays, ideal for preventing the biological proliferation of the biodeteriogens. This combined protocol could be a valid alternative to chemical biocides, which are highly toxic for operators and the environment, guaranteeing, at the same time, the control of biodeteriogen diffusion on artistic crafts without any side effects for the restorers and the ambiance.

Author Contributions: Conceptualization, P.C. and G.T.; methodology, R.S., E.C. and M.R.; software, M.R. and S.C.; validation, P.C., R.S. and G.T.; formal analysis, R.S., G.T. and M.R.; investigation, R.S.; resources, P.C. and G.T.; data curation, P.C. and G.T.; writing—original draft preparation, R.S., P.C. and G.T.; writing—review and editing, P.C., R.S. and G.T.; visualization P.C., R.S. and G.T.; supervision, P.C.; project administration, P.C. and G.T.; funding acquisition, P.C. All authors have read and agreed to the published version of the manuscript.

Funding: This research was funded by PNRR_PE CHANGES-Cultural Heritage Active Innovation for Sustainabel Society. SPOKE n 6 –"History, conservation and restoration of cultural heritage".

Institutional Review Board Statement: Not applicable.

Informed Consent Statement: Not applicable.

Data Availability Statement: Not applicable.

Conflicts of Interest: The authors declare no conflict of interest.

References

1. Caneva, G.; Nugari, M.P.; Salvadori, O. *La Biologia Vegetale per i Beni Culturali*, 2nd ed.; Nardini Editore: Firenze, Italy, 2007; Volume I.
2. Mora, P. *Causes of Deterioration of Mural Paintings*; International Centre for the Study of the Preservation and the Restoration of Cultural Property: Roma, Italy, 1964.
3. Cennamo, P.; Montuori, N.; Trojsi, G.; Fatigati, G.; Moretti, A. Biofilms in churches built in grottoes. *Sci. Total Environ.* **2016**, *543*, 727–738. [CrossRef] [PubMed]

4. Borderie, F.; T'ete, N.; Cailhol, D.; Alaoui-Sehmer, L.; Bousta, F.; Rieffel, D.; Aleya, L.; Alaoui-Sossé, B. Factors driving epilithic algal colonization in show caves and new insights into combating biofilm development with UV-C treatment. *Sci. Total Environ.* **2014**, *484*, 43–52. [CrossRef] [PubMed]

5. Pfendler, S.; Borderie, F.; Bousta, F.; Alaoui-Sosse, L.; Alaoui-Sosse, B.; Aleya, L. Comparison of biocides, allelopathic substances and UV-C a streatments for biofilm proliferation on heritage monuments. *J. Cult. Herit.* **2018**, *33*, 117–124. [CrossRef]

6. Liverani, P. Le necropoli sotto la Basilica di San Pietro. Conservazione e Restauro. In *Le Necropoli Vaticane, La città dei Morti, Milano*; Liverani, P., Spinola, G., Zander, P., Eds.; Jacabook: Milan, Italy, 2010; pp. 287–310.

7. Cennamo, P.; Ebbreo, M.; Quarta, G.; Trojsi, G.; De Rosa, A.; Carfagna, S.; Caputo, P.; Martelli Castaldi, M. UV-C Irradiation as a Tool to Reduce Biofilm Growth on Pompeii Wall Paintings. *Int. J. Environ. Res. Public Health* **2020**, *17*, 8392. [CrossRef] [PubMed]

8. Cappitelli, F.; Cattò, C.; Villa, F. The Controlof Cultural Heritage Microbial Deterioration. *Microorganisms* **2020**, *8*, 1542. [CrossRef]

9. Devreux, G.; Santamaria, U.; Morresi, F.; Rodolfo, A.; Barbabietola, N.; Fratini, F.; Reale, R. Fitoconservazione, trattamenti alternativi sulle opere in materiale lapideo nei giardini vaticani. In Proceedings of the Atti del Congresso Nazionale IGIIC Lo Stato Dell'arte XIII Venaria Reale, Torino, Italy, 22–24 October 2015; pp. 199–206.

10. Bischoff, H.W.; Bold, H.C. Some soil algae from Enchanted Rock and related algal species. In *Phycological Studies IV*; No. 6318; University of Texas: Austin, TX, USA, 1963; pp. 1–95.

11. Castenholz, R.W. Culturing methods for Cyanobacteria. *Methods Enzymol.* **1988**, *167*, 68–93.

12. Carfagna, S.; Lanza, N.; Salbitani, G.; Basile, A.; Sorbo, S.; Vona, V. Physiological and morphological responses of Lead or Cadmium exposed Chlorella sorokiniana 211–8K (Chlorophyceae). *SpringerpPlus* **2013**, *2*, 147–154. [CrossRef]

13. Johnston-Feller, R. Color science in the examination of museum objects. In *Nondestructive Procedures*; The Getty Conservation Institute: Los Angeles, CA, USA, 2001.

14. Wyszecki, G.; Stiles, W.S. *Color Science: Concepts and Methods, Quantitative Data and Formulae*, 2nd ed.; John Wiley & Sons: Hoboken, NJ, USA, 2000.

15. Nassau, K. *Color for Science, Art and Technology*; Elsevier Science: Amsterdam, The Netherlands, 1999.

16. Tinzl, C.; Oldenbourg, C.; Petersen, K.; Fricke-Tinzl, H.; Hilge, C.; Katkov, M. UV-C-irradiation for removal and control of lichen growths from wall paintings and stone sculptures: An alternative to using biocides? In *Restauratorenblätter (Bd. 16)*; Koller, M., Prandtstetten, R., Eds.; Mayer & Comp.: Wien, Austria, 1995; pp. 127–138.

17. Warscheid, T.; Braams, J. Biodeterioration of Stone: A review. *Int. Biodeterior. Biodegrad.* **2000**, *46*, 343–368. [CrossRef]

18. Hermann, H.; Häder, D.P.; Köfferlein, M.; Seidlitz, H.K.; Ghetti, F. Effects of UV radiation on photosynthesis of phytoplankton exposed to solar simulator light. *J. Photochem. Photobiol.* **1996**, *34*, 21–28. [CrossRef]

19. Zvezdanović, J.; Cvetić, T.; Veljović-Jovanović, S.; Marković, D. Chlorophyll bleaching by UV-irradiation in vitro and in situ: Absorption and fluorescence studies. *Radiat. Phys. Chem.* **2009**, *78*, 25–32. [CrossRef]

20. Sarghein, S.H.; Carapetian, J.; Khara, J. Effects of UV-radiation on photosynthetic pigments and UV absorbing compounds in *Capsicum longum* (L.). *Int. J. Bot.* **2008**, *4*, 486–490. [CrossRef]

21. Stewart, J.; More, A.; Simpson, P. The use of ultraviolet irradiation to control microbiological growth on mosaic pavements: A preliminary assessment at Newport Roman Villa. 2023. Available online: https://www.iwhistory.org.uk/archive/NewportConservation.pdf (accessed on 10 April 2023). Unpublished.

22. Palla, F.; Bruno, M.; Mercurio, F.; Tantillo, A.; Rotolo, V. Essential Oils as Natural Biocides in Conservation of Cultural Heritage. *Molecules* **2020**, *25*, 730. [CrossRef] [PubMed]

23. Kalemba, D.; Kunicka, A. Antibacterial and antifungal properties of essential oils. *Curr. Med. Chem.* **2003**, *10*, 813–829. [CrossRef] [PubMed]

24. Reichling, J.; Schnitzler, P.; Suschke, U.; Saller, R. Essential oils of aromatic plants with antibacterial, antifungal, antiviral, and cytotoxic properties—An overview. *Forsch. Komplement.* **2009**, *16*, 79–90. [CrossRef] [PubMed]

25. Saad, N.; Muller, C.D.; Lobstein, A. Major bioactivities and mechanism of action of essential oil and their components. *Flavour Fragr. J.* **2013**, *28*, 269–279. [CrossRef]

26. Salem, M.Z.M.; Zidan, Y.E.; Mansour, M.M.A.; El Hadidi, N.M.N.; Abo Elgat, W.A.A. Antifungal activities of two essential oils used in the treatment of three commercial woods deteriorated by five common mold fungi. *Intern. Biodeter. Biodegr.* **2016**, *106*, 88–96. [CrossRef]

27. Casiglia, S.; Bruno, M.; Scandolera, E.; Senatore, F. Influence of harvesting time on composition of the essential oil of *Thymus capitatus* (L.) Hoffmanns. & Link. growing wild in northern Sicily and its activity on microorganisms affecting historical art crafts. *Arab. J. Chem.* **2019**, *12*, 2704–2712.

28. Rotolo, V.; De Caro, M.L.; Di Carlo, E.; Giordano, A.; Palla, F. Plant extracts as green potential strategies to control the biodeterioration of cultural heritage. *Intern. J. Conservat. Sci.* **2016**, *7*, 839–846.

29. Lavin, P.; Gómez de Saravia, S.; Guiamet, P. *Scopulariopsis* sp. and *Fusarium* sp. in the Documentary Heritage: Evaluation of their biodeterioration ability and antifungal effect of two essential oils. *Microb. Ecol.* **2016**, *71*, 628–633. [CrossRef]

30. Favero-Longo, S.E.; Laurenzi Tabasso, M.; Brigadeci, F.; Capua, M.C.; Morelli, A.; Pastorello, P.; Sohrabi, M.; Askari Chaverdi, A.; Callieri, P. A first assessment of the biocidal efficacy of plant essential oils against lichens on stone cultural heritage, and the importance of evaluating suitable application protocols. *J. Cult. Herit.* **2022**, *55*, 68–77. [CrossRef]

31. Borderie, F.; Alaoui-Sossé, B.; Aleya, L. Heritage materials and biofouling mitigation through UV-C irradiation in show caves: State-of-the-art practices and future challenges. *Environ. Sci. Pollut. Res.* **2015**, *22*, 4144–4172. [CrossRef] [PubMed]

32. Stapleton, A.E. Ultraviolet radiation and plants: Burning question. *Plant Cell* **1992**, *4*, 1353–1358. [CrossRef] [PubMed]
33. Rakotonirainy, M.S.; Lavédrine, B. Screening for antifungal activity of essential oils and related compounds to control the biocontamination in libraries and archives storage areas. *Intern. Biodeter. Biodegrad.* **2005**, *55*, 141–147. [CrossRef]
34. Afifi, H.A.M. Comparative efficacy of some plant extracts against fungal deterioration of stucco ornaments in the Mihrab of Mostafa Pasha, Ribat, Cairo, Egypt. *Am. J. Biochem. Mol. Biol.* **2012**, *2*, 40–47. [CrossRef]
35. Borderie, F.; Laurence, A.-S.; Naoufal, R.; Bousta, F.; Orial, G.; Rieffel, D.; Badr, A.-S. UV-C irradiation as a tool to eradicate algae in caves. *Int. Biodeter. Biodegr.* **2011**, *65*, 579–584. [CrossRef]
36. Pozo-Antonio, J.S.; Sanmartín, P. Exposure to artificial daylight or UV irradiation (A, B or C) prior to chemical cleaning: An effective combination for removing phototrophs from granite. *Biofouling* **2018**, *34*, 851–869. [CrossRef]

Article

A New Green Coating for the Protection of Frescoes: From the Synthesis to the Performances Evaluation

Raffaella Lamuraglia [1,2], Andrea Campostrini [1], Elena Ghedini [1], Alessandra De Lorenzi Pezzolo [3], Alessandro Di Michele [4], Giulia Franceschin [2], Federica Menegazzo [1,2,*], Michela Signoretto [1] and Arianna Traviglia [2]

1   CATMAT Lab, Dipartimento di Scienze Molecolari e Nanosistemi, Università Ca' Foscari Veneziaand, INSTM RUVe, 30170 Venice, Italy
2   Center for Cultural Heritage Technology, Istituto Italiano di Tecnologia, 30170 Venice, Italy
3   Dipartimento di Scienze Molecolari e Nanosistemi, Università Ca' Foscari Venezia, 30170 Venice, Italy
4   Dipartimento di Fisica e Geologia, Università di Perugia, 06123 Perugia, Italy
*   Correspondence: fmenegaz@unive.it

Abstract: This work presents the formulation and characterization of a new product for the protection of outdoor frescoes from aggressive environmental agents. The formulation is designed as an innovative green coating, prepared through a zero-waste one-pot-synthetic method to form silver nanoparticles (AgNPs) directly in a chitosan-based medium. The AgNPs are seeded and grown in a mixed hydrogel of chitosan, azelaic, and lactic acid, by the reduction of silver nitrate, and using calcium hydroxide as precipitating agent. The rheological properties of this coating base are optimized by the addition of a solvent mixture of glycerol and ethanol with a 1:1 volume ratio. The new formulation and two commercial products (Paraloid® B72 and Proconsol®) are then applied by brush to ad hoc mock-ups to be evaluated for chemical stability, color and gloss variations, morphological variation, hydrophobicity, and water vapor permeability via Fourier-transform infrared spectroscopy (FT-IR) in attenuated total reflection (ATR) mode, spectrophotometer analysis, stereomicroscope observations, UNI EN 15802, and UNI EN 15803, respectively. The results show that the application of the hybrid chitosan-AgNPs coating is promising for the protection of outdoor frescoes and that it can underpin the development of new products that address the lack of conservation strategies specifically designed for wall painting.

Keywords: green coating; chitosan; AgNPs; frescoes; zero-waste one-pot synthesis

Citation: Lamuraglia, R.; Campostrini, A.; Ghedini, E.; De Lorenzi Pezzolo, A.; Di Michele, A.; Franceschin, G.; Menegazzo, F.; Signoretto, M.; Traviglia, A. A New Green Coating for the Protection of Frescoes: From the Synthesis to the Performances Evaluation. *Coatings* **2023**, *13*, 277. https://doi.org/10.3390/coatings13020277

Academic Editor: Patricia Sanmartín

Received: 20 December 2022
Revised: 20 January 2023
Accepted: 24 January 2023
Published: 26 January 2023

1. Introduction

Historical buildings, open-air mural paintings, and monuments located outdoors are constantly exposed to many different degradation factors i.e., humidity, ultraviolet (UV) light, pollution, microbiological colonization, whose effects may be exacerbated by the influence of climate changes [1–4]. The degradation processes can impact both the bulk and the surface of materials [5–7]. The decorated surface of mural paintings represents the layer in which the exchange of pollutants and water from the environment occurs predominantly, and, thus, is the part that is most susceptible to alteration of its material properties [8]. Water and other water-related decay mechanisms—such as crystallization of insoluble salts, erosion, and pollutant deposition—are described in the literature as the most impactful degradation phenomena for mural paintings. Acting together, these mechanisms are responsible for triggering other severe decay forms, such as biological colonization, detachments, or even more complex characteristic phenomena such as *flos tectori* [9].

Most of the conservation practices currently adopted to preserve wall-painting exposed to not-controlled environments, however, are limited to physical barriers meant to decrease the impact of damaging environmental agents such as rain (e.g., external structures such as roofing and canopies, as seen in Pompeii, Herculaneum, or Malta), resulting in limited protection and in the persistence of adverse micro-climate conditions [10–12].

Despite several documented forms of degradation, specific products for the in situ protection of frescoes are not commercially available, and even research on this particular type of material is still scarce. Restorers often address this lack of product by using materials that were originally designed for other applications (for instance, products for consolidation and generic fixing of common stone materials) and often ignore the long-term effects of these compounds on delicate systems such as frescoes. In most cases, they are applied with the main purpose of restoring the bulk of the wall rather than the surface layer. Three classes of treatments for fresco restoration can currently be identified on the market according to their properties as either consolidant, filler, or adhesive. Some of the most employed products are consolidants, made of organic or inorganic polymers, nanomaterials, and bio-consolidants, with generic adhesive, cohesive, and film-forming properties [13]. Their own chemical nature is key to their poly-functionality. A common example is Proconsol® (Casoria Arpino, Italy), which is a commercial product sold as a protective and consolidant solution of calcium hydroxide ($Ca(OH)_2$) specifically for wall-painting and frescoes by the Cimmino Calce company [14]. Paraloid® B72 (C.T.S. s.r.l., Altavilla Vicentina, Italy), an ethyl-methacrylate copolymer, is another example and one of the most popular products used by restorers for different kinds of applications, including wall and stone material protection and consolidation [15].

New products designed for Cultural Heritage applications must meet the requirements imposed by the Restoration Charters, which prescribe them to be as compatible as possible with the material on which they are applied, so that they do not modify the external appearance of the treated surface [16]. Protective coatings can be designed to completely embrace this concept and, for this reason, in the last decades, they have gained increasing interest within the scientific community. In fact, while complying with the widely accepted 'gold standards' of restoration, they also meet additional, relevant requirements: transparency, reversibility or re-treatability, compatibility with the surface, long-term lifetime, ease of synthesis, low-cost maintenance, and non-toxicity [17,18]. In addition, particular attention should be placed on developing products that comply with green principles and are risk-free for conservators, restorers, and the environment [18]. The discussion around the feasibility and opportunity of some of these requirements, however, is still ongoing: Andreotti et al. (2018), for example, have argued that achieving the full reversibility of treatments is often just an idealistic goal and that a more viable method could be that of retreating the surfaces [19]. Biopolymers and waste-derived materials are considered as valid alternatives to synthetic ones. The increasing interest in bio-polymers is due to the fact that they naturally comply to the principles of reversibility and/or re-treatability: the polymer is expected to disappear from the surface once its water-repellent properties disappear without affecting subsequent conservative treatments [19]. In recent studies [19–21], biopolymers, such as zein, pectin, chitosan, polyhydroxybutyrate (PHB), and poly-l-lactic acid (PLA), were employed to replace synthetic polymers in many applications, with appreciable results [22]. Among biopolymers, chitosan was estimated to be particularly promising due to its biodegradability, biocompatibility, and antimicrobial activity; therefore, it has recently been used for the formulation of sustainable coatings as an anti-corrosion coating for indoor bronze metal substrates [23,24]. Chitosan, which derives from the partial deacetylation of chitin [25], is considered as a non-toxic biopolymer for its excellent antimicrobial and antifungal activities against a wide range of microorganisms when compared to other polymers and biopolymers [26,27]. Xin Wang et al. successfully designed and developed a controlled release-system against microorganisms based on thymol-loaded chitosan nanoparticles, demonstrating that it is possible to obtain natural biocides as a viable alternative to the toxic products currently in use for the same purpose [28]. Another example of the application of chitosan based on its physical and antimicrobial characteristics is the treatment of ethnographic objects (made of cotton and linen) by L. Indrie et al. [29]. Notwithstanding this potential for threating several types of cultural heritage, however, chitosan has not been tested yet on frescoes to the best of our knowledge.

Nanoparticles are another important element in the formulation of protective coatings, because of their ability to add specific physical and chemical properties to the resulting film. Among others, even silver nanoparticles (AgNPs) have attracted more and more attention in different research fields due to their potential as antimicrobial agents [30]. The antimicrobial behavior of silver is enhanced when the material is organized in the form of NPs, thanks to the electronic effect given by the changes in the local electronic order on the surfaces of objects with an average size <100 nm. The antimicrobial property of AgNPs could be enhanced in hybrid formulation with polysaccharides due to plasmonic resonance effects, eventually inducing photocatalytic phenomena [31–33].

Based on the above, the authors have selected chitosan as the most appropriate element to underpin the formulation of an innovative organic–inorganic hybrid system to be applied to frescoes and a porous matrix of chitosan hosting Ag active nanoparticles which has been created through a novel one-pot synthesis method. As inorganically charged, AgNPs are grown in-situ in the synthesis medium using a one-pot process, starting from silver nitrate as a precursor. The novelty of this work is represented by the synthesis method, which consists of growing AgNPs within the chitosan-based hybrid matrix prepared using azelaic and lactic acid instead of acetic acid, while the hybrid matrix acts as a reducing, capping, and stabilizing agent simultaneously. Both acetic and azelaic acid are green compounds; however, the latter presents intrinsic antimicrobial features which are able to prevent biological attacks on a treated surface. Furthermore, azelaic acid has a long chain structure that enhances its film-forming ability. Azelaic acid combined with lactic acid can effectively hydrolyze chitosan to exert its reducing and capping action. The resulting hybrid material is the core of the product, and it will be reported in this work as coating base. To optimize the rheological properties of the final product and obtain a ready-to-use formulation which easily applicable by brush and in situ, the coating base is mixed with suitable solvents (i.e., glycerol and ethanol), selected for their own affinity with the coating base and their own rheological properties.

To validate the protective features of the proposed coating, the final properties are compared with those of two commercial products, i.e., Paraloid® B72 (C.T.S. s.r.l., Altavilla Vicentina, Italy) and Proconsol® (Casoria Arpino, Italy). The analyses are carried out on both the coatings and the treated surfaces of fresco mock-ups.

2. Materials and Methods

2.1. Preparation of Chitosan and AgNPs Coating Base

Chitosan (CAS 9012-76-4), azelaic acid (CAS 123-99-9), lactic acid (CAS 50-21-5), and sodium hydroxide (CAS 1310-73-2) were purchased from Sigma-Aldrich® (Merck Life Science S.r.l., Milan, Italy), and silver nitrate (CAS 7761-38-8) and calcium hydroxide (CAS 1305-62-0) were purchased from Acros Organics, (Fisher Scientific, Thermo Fisher Scientific, Waltham, MA, USA). Both are used for the synthesis.

A chitosan-based medium (Medium Molecular weight) is prepared by adding the appropriate amount of reagents, as reported in Table 1, to 15 mL of deionized water.

Table 1. Amount of reagents used for the coating base's synthesis.

Reagent	Quantity
Chitosan	30 mg
Azelaic Acid	9.5 mg
Lactic Acid	300 µL (5% v/v)

Silver NPs are synthesized in-situ in a one-pot synthesis together with chitosan, as suggested by Goy, R.C., et al. (2009) [26]. Different concentrations (0.5 M, 0.1 M, and 0.01 M) of silver nitrate in water are investigated as AgNPs precursor. The effects of AgNPs precipitation are tested in a solution of NaOH 0.3 M, as normally done in the literature [27], and in a 0.015 M solution of $Ca(OH)_2$. The beaker containing the gel is aged at 95 °C for 4 h with constant stirring at 300–400 rpm. The hydro-gel color changes after the first hour, turning to intense yellow/orange/brown after the fourth hour, indicating AgNPs formation.

2.2. Addition of Solvents

Glycerol (CAS 56-81-5) and ethanol (CAS 64-17-5) purchased from Sigma-Aldrich® (Merck Life Science S.r.l., Milan, Italy) are used as solvents without further purification. The two solvents are mixed in a proportion of 1:1. The final formulation is prepared by blending in, at a proportion of 1:1, the coating-base with the solvents mixture.

Once the new coating is ready to be applied, the fresco mock-ups are treated with it and with the commercial products Proconsol® (Casoria Arpino, Italy) and Paraloid B72® (C.T.S. s.r.l., Altavilla Vicentina, Italy) using a brush to compare and validate the protective features of the newly formulated coating (0.6 L m^{-2} applied by brush).

2.3. Mock-Ups Preparation

Sixteen bricks of 5 cm × 5 cm × 2.5 cm are prepared to replicate the characteristics of original fresco samples, following the recipe reported by Vitruvius in *De Architectura*. After 12 h immersion in water, three/four parallel grooves are dug with a 10 mm-diameter drill on each mock-up, to simulate the wall inhomogeneity and to enhance the plaster grip on the material.

To obtain a wider perspective, the *arriccio* layer is prepared using slaked lime (C.T.S. s.r.l., Altavilla Vicentina, Vicenza, Italy) and three different inert materials: *cocciopesto*, *pozzolana*, and white river sand (all purchased from C.T.S. s.r.l., Altavilla Vicentina, Vicenza, Italy). Three plaster doughs are prepared using a proportion of 3:2 between inert material and slaked lime. The first inert material (sand-based) is prepared mixing at a proportion of 1:2 coarse and thin-grained white river sand. The second (*cocciopesto*-based) and the third (*pozzolana*-based) inert materials are prepared by mixing, at a proportion of 1:1:1, coarse-grained sand, thin-grained white river sand, and *cocciopesto* or *pozzolana* powder, respectively.

The bricks with the *arriccio* layer are then dried for approximately 16 h in air. Afterwards, the surface is moistened with water, by means of a sponge float, in order to apply the thinner *intonachino* layer. The *intonachino* is prepared with slaked lime and thin-grained white river sand (in proportion 1:1), on which the pigments (yellow ochre #40301 and red ochre #48651 purchased from Kremer Pigmente GmbH & Co. KG, DE 88317, Aichstetten) are dispersed in water and applied by brush.

2.4. Benchmark Products Selected and Tested

Proconsol® and ParaloidB72® are tested as reference benchmark products. Proconsol® (water dispersion of $Ca(OH)_2$) was kindly provided by Cimmino Calce s.r.l., Casoria Arpino, Naples, Italy. To date, it is the only commercially-available product sold as protective coating for frescoes. The product is applied without further dilution. ParaloidB72® (C.T.S. s.r.l., Altavilla Vicentina, Vicenza, Italy), an ethyl-methacrylate copolymer, is diluted at 5% (w/v) in ethyl acetate (CAS 141-78-6, purchased from Sigma-Aldrich®, Merck Life Science S.r.l., Milan, Italy). It is one of the most common benchmark products used by restorers for wall and stone material protection and consolidation.

2.5. Characterization Techniques for Coating

To identify the optimal formulation parameters, UV spectroscopy, Scanning Electron Microscopy, and Transmission Electron Microscopy analyses are carried out. Rheology analyses are run to measure the viscosity of the different coatings.

A UV-Vis spectrophotometer Cary300 by Agilent technologies is used for the analyses. The spectra are collected in the 800–200 nm wavelength range. The analyses are run through the Cary WinUV Scan software 4.20 and processed with the software Origin 2021 by OriginLab, Northampton, MA, USA.

The SEM microscopy analyses on the different NP samples are carried out by an FE-SEM (Field Emission Scanning Electron Microscopy) FEG LEO 1525 ZEISS, with detector in-lens for secondary electrons. The acceleration potential voltage is maintained at 15 keV and measurements are carried out using an AsB detector (Angle selective Backscattered detector). Samples are deposited on a silicon wafer, dried in air, and observed without metallization.

TEM images are obtained using a Philips 208 Transmission Electron Microscope. The samples are prepared by putting one drop of AgNPs solution on a copper grid pre-coated with a Formvar film and dried in air.

The rheology analyses are run to measure the viscosity of the different coatings. For each sample, around 5 mL is poured into the rheometer cylindrical cup (approx. volume of 19 mL and radius of 14.452 mm). The instrument used for these analyses is a Modular Compact Rheometer (MCR 102) from Anton Paar GmbH, with a No. 3912 coaxial cylinder measuring system. A measuring Bob with radius of 13.331 mm, Ratio of Radii of 1.084, and Gap Length of 39.998 mm is employed. The cone angle is 120° and the measuring gap is 1.121 mm. For the time settings, 21 data points are collected with a point duration of 30 s, an interval duration of 630 s, and an averaging over minimum of 1 raw value. The measuring profile is set with a logarithmic shear rate of $d\gamma/dt$ from 10 to 1000 s^{-1}. The data are collected with the RheoCompass software and processed with the software Origin 2021 of OriginLab, Northampton, MA, USA.

2.6. Characterization Techniques for Applied Coatings

Before and after the ageing process, the mock-ups are characterized to evaluate the physical and chemical properties using ATR spectroscopy, measures of the color variance (ΔE), and the surface gloss, by stereomicroscopic investigation of the morphology and by evaluating the surface hydrophobicity through the contact angle measure.

The measurements are carried out using a VERTEX 70 v FT-IR Spectrometer (Bruker, Billerica, MA, USA) equipped with a PIKE MIRacle™ universal ATR sampling accessory with a diamond/ZnSe internal reflection element. The spectra are collected in the 4000–650 cm^{-1} spectral range with a 2 cm^{-1} resolution. Two hundred scans are averaged in each run and subjected to Norton Beer (medium) apodisation before Fourier transformation. The obtained spectra are min-max normalized and automatic-baseline corrected with OMNIC Atlus Software 6.0.

The morphological analyses are performed by taking pictures of each mock-up at 1× and 2× magnification by means of a NYKON stereomicroscope model SMZ 745T (European Headquarters Nikon Europe BV, Tripolis 100, Burgerweeshuispad 101, 1076 ER Amsterdam, Netherlands).

Color variations are measured with a Konica Minolta CM 700d spectrophotometer before and after ageing, as well as before and after coating application. The instrument has an 8-degree viewing angle geometry, with a diffusion light Xenon lamp and a high-resolution monolithic polychromator. The analyzed area is circular, with a 3-mm diameter, and the results are obtained by averaging six measurements carried out on different points over the mock-up.

Measurements are performed in the CIELAB1976 space. Results are elaborated by Spectra Magic NX 6. The total color variation, expressed as ΔE, is calculated by Equation (1):

$$\Delta E = \sqrt{[(\Delta L^*)^2 + (\Delta a^*)^2 + (\Delta b^*)^2]}, \tag{1}$$

where ΔL^*, Δa^*, and Δb^* are the differences between the values obtained on the same sample before and after each step. In the Heritage Science field, it is assumed that, when the ΔE values are higher than 3, the chromatic difference is visibly detectable to the naked eye [34–37].

Hydrophobicity tests are done following the UNI EN 15802 document [38], by measuring the contact angle which a water drop makes on the surface of the fresco mock-ups. The procedure is filmed with a Canon camera, mounted with a Canon macro lens of 100 mm.

The water vapor permeability is measured on fresco mock-ups following the steps described in the UNI EN 15803 document [39].

2.7. Ageing Parameters

The fresco mock-ups are artificially aged, before and after the application of the coatings, in a Q-Sun Xe-1S climatic chamber (Q-LAB Corporation, Westlake, OH 44145-1419 SUA) for a total of 72 h to study the effect of temperature, light, and humidity fluctuations. The ageing parameters adopted are the following: 12 h UV-light step with 65 W m^{-2} radiance, T = 50 °C, and 5 s water spray every 4 h ($p = 1.4$ L min^{-1}), alternated with 12-h dark step with T = 25 °C and 5 s water spray every 4 h ($p = 1.4$ L min^{-1}). Twenty-four-hour runs composed by 30 min cycles are repeated three times. Between each repetition, the positions of the samples are swapped to ensure that the artificial ageing process homogeneously impacts all of the mock-ups.

3. Results and Discussion

3.1. Coating Base Formulation and Characterization

The comparison of the UV spectra of the different NPs formulations is shown in Figure 1, where all are prepared in azelaic and lactic acid with NaOH, only varying the AgNO$_3$ concentration.

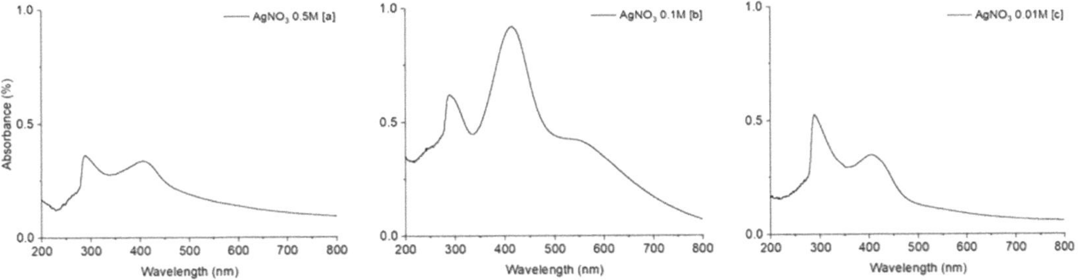

Figure 1. UV-Vis spectra of AgNPS in azelaic and lactic acid solution, with NaOH as precipitating agent. (**a**) AgNO$_3$ 0.5 M, (**b**) 0.1 M, and (**c**) 0.01 M.

UV spectroscopy is useful to study the plasmonic resonance bands of the components present in different coating base samples. The resonance bands at 410 nm are related to the presence of small silver nanoparticles [40]. In the proposed synthesis process, chitosan constitutes the capping and reducing agent: the availability of functional groups such as carboxyl, hydroxyl, and amine determine the number of available nucleation centers and, consequently, the amount of AgNPs, regardless of the precursor concentration. Moreover, when using AgNO$_3$ 0.1 M, another resonance band around 600 nm appears, indicating metal particles with larger dimensions and suggesting a double distribution of the NPs. The resonance bands at 300 nm are due to the presence of unreacted chitosan. Since the solution with AgNPs synthesized by AgNO$_3$ 0.5 M sutures the analysis, the sample is diluted 3:1 with water. As reagent for the precipitation of the NPs, a solution of Sodium Hydroxide (NaOH), as normally used the in literature, and one of Calcium Hydroxide Ca(OH)$_2$, which may increase the compatibility in the application on calcium carbonate-based materials, are tested. The hydroxides are involved in the reduction process by the formation of an Ag(OH)x intermediate with subsequent reduction in silver in an aqueous environment [41]. UV-Vis analyses show that AgNPs are successfully developed when also employing Ca(OH)$_2$. Since the frescoes' bases are principally made of CaCO$_3$, produced by the carbonatation of Ca(OH)$_2$, the addition of Ca(OH)$_2$ would not lead to any unwanted chemical reaction. For this reason, $Ca(OH)_2$ is chosen as the reagent for the precipitation of the silver nanoparticles. Figure 2 presents the UV spectra of two samples, both prepared with azelaic acid and $AgNO_3$ 0.1 M, employing the two different precipitating agents. In the sample with NaOH, the % absorbance of the band at around 410 nm is higher, which can be correlated with a higher concentration of nanoparticles in the final product. In addition, the presence of $Ca(OH)_2$ seems to enhance the phenomenon of the double distribution of the nanoparticles' sizes, as shown by the increased intensity of the 600 nm band.

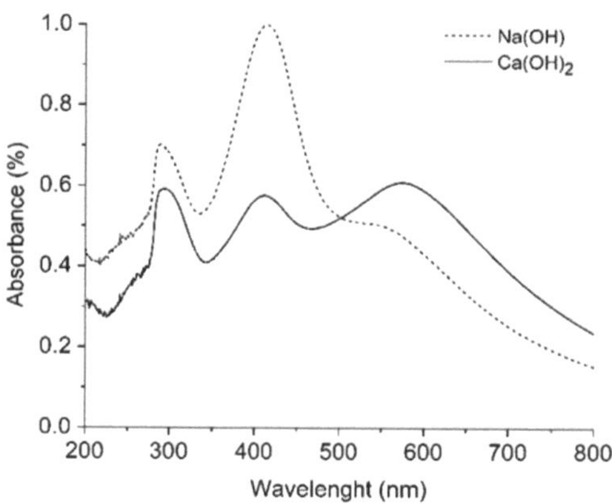

Figure 2. UV-Vis spectra of AgNPs in azelaic and lactic acid solution, with $AgNO_3$ 0.1 M, and NaOH and $Ca(OH)_2$, as precipitating agents.

Thanks to SEM and TEM analyses, it is possible to investigate the silver NPs distributions, shapes, and sizes (Figure 3). SEM images show that all of the samples contain silver NPs homogeneously distributed in the chitosan gel. Thanks to TEM images, it is possible to observe that the NPs are not aggregated, spherical in shape, or nano-dimensional. This allows for confirmation that the sample in which 0.1 M silver nitrate and $Ca(OH)_2$ are used has a double distribution of the NPs sizes.

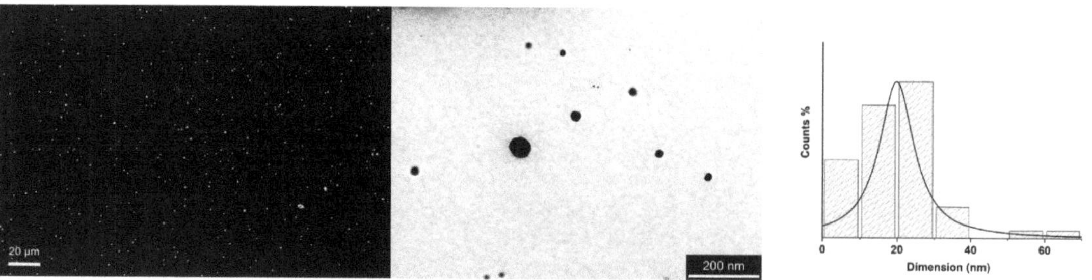

Figure 3. (From left to right) SEM (EHT = 15.00 kV, WD = 7.6 mm, Mag = 1.00 K X, Signal A = AsB), TEM, and distribution curve of AgNPs in lactic and azelaic acid solution, with $AgNO_3$ 0.1 M and $Ca(OH)_2$ as precipitating agent. ImageJ software (LOCI, University of Wisconsin, Madison, WI, USA) is used to estimate the particle size distribution of NPs by TEM images.

To obtain suitable rheological properties, an additional solvent is added to the formulation. The first tests are performed using the same solvents employed in the commercial products for cultural heritage restorations: ethyl acetate ($CH_3COOC_2H_5$) (as in Paraloid®) and water (as in Proconsol®). Other solvents are also taken into account (i.e., glycerol, ethanol, isopropanol), and the product behavior at variable ratios of 1:10 or 1:5 base to solvent is investigated. Ethyl acetate is a moderately-polar solvent that has the advantages of being volatile, relatively non-toxic, and non-hygroscopic. Nevertheless, it is quite unstable in the presence of strong aqueous bases and acids. Water is the best solvent in terms of toxicity, availability, and cost; however, it has low volatility. Glycerol, as well, has all the advantages of water, being non-toxic, easily available, and economic. Ethanol has very high volatility but is still a good solvent for its low toxicity and easy synthesis. Finally, isopropanol has high volatility as well but is also quite toxic. The dilution with ethyl acetate, unfortunately, damages the coating, altering the chitosan matrix stability and producing a white precipitate; therefore, it is excluded. On the contrary, with water, glycerol, ethanol, and isopropanol, no visible damage is encountered, even after two months. The crucial point is to study the nanoparticles' stability in different environments and to achieve a product with medium volatility and an adequate evaporation rate. A solvent with low volatility (as glycerol) delays the evaporation, favoring the coating application, but increasing the risk of leakage. On the other hand, a solvent with excessively-high volatility (such as isopropanol) impedes a homogeneous coating application on the surface. Finally, even if it could be an ideal solvent for applications in Cultural Heritage, its low volatility is a limiting factor that leads to discarding water.

The final coating is prepared by blending the above-described gel in a proportion of 1:1 with a mixture of glycerol and ethanol that combines the properties of the two solvents and reaches a good compromise between volatility and applicability. As expected, the viscosity values obtained for the new formulation and for the commercial products (i.e., Paraloid® and Proconsol®) are different, in line with their different chemical compositions. All the samples present a rheological behavior appropriate to an application by brush: Proconsol® and Paraloid® have an average viscosity value, respectively, of 2 and 3 mPa·s (Table 2), whereas the Chitosan gel with AgNPS showed a higher viscosity (13 mPa·s). This is a promising aspect, which would allow for keeping the chitosan coating on the outer layers without nullifying the properties of the application [42].

Table 2. Viscosity average values for each coating.

Product	Average Viscosity [mPa·s]
Proconsol®	2
Paraloid B72®	3
Chitosan AgNPs product	13

3.2. Coating Application

The ATR spectroscopy analyses, performed before and after ageing, show that the three coatings have good chemical stability. The spectra of the samples coated with Proconsol® (Figure 4a) and with the new chitosan-AgNPs (Figure 4c) show distinctive features of the first two strata (pigment and *intonachino*) underlying the coating films. The absorptions at about 1430–1395, 872 and 712 cm^{-1} (fundamental bands), and 2873, 2512, and 1795 cm^{-1} (overtone and combination bands) are ascribable to the calcite present both in the *intonachino* layer and, as a filler, in the yellow ochre pigment employed in the sample set investigated by ATR-FTIR. The goethite contained in the pigment can be related to a very weak band present in the spectra at 677 cm^{-1}, and possibly to another one, superimposed to one due to quartz, falling at about 800 cm^{-1}. Other characteristic absorptions of quartz, of which the thin-grained aggregate fraction of the *intonachino* layer is made, are found at about 1170, 1085, and 780 cm^{-1}. Signals from amorphous calcite due to the carbonation of the slacked lime are found superimposed to the calcite broad v_3 band with contributions at about 1395 cm^{-1} and a high-wavenumber shoulder [43], while a very weak peak at 3640 cm^{-1} due to residual $Ca(OH)_2$ can be recognized, indicating an almost complete carbonation. All the spectra show a sharp peak of medium intensity at about 3690 cm^{-1} due to $Mg(OH)_2$, which proves, on the one hand, the presence of a dolomitic component in the commercial slacked lime, and, on the other, the different time scale of the carbonation of magnesium hydroxide process as compared to that of calcium hydroxide [44]. On the contrary, the two spectra of the Paraloid®-treated samples (Figure 4b) present a quite-different profile where, among the features due to the first two strata below the coating layer of the mock-ups, only the calcium carbonate features at about 875 and 712 cm^{-1} are visible as weak absorptions together with the well-defined sharp peak at about 3690 cm^{-1}, corresponding to magnesium hydroxide. This is indicative of the thickness of the protective layer created by the product.

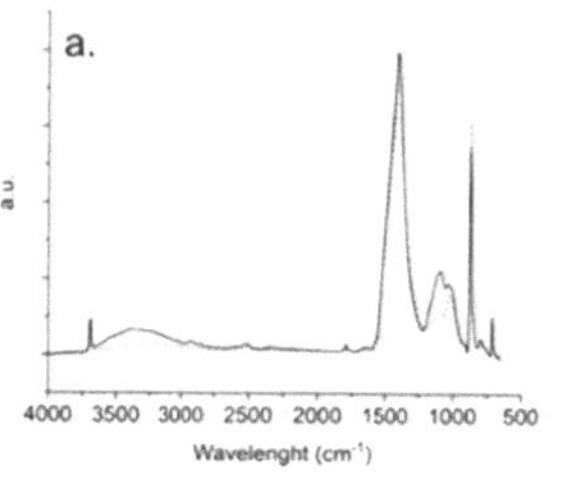
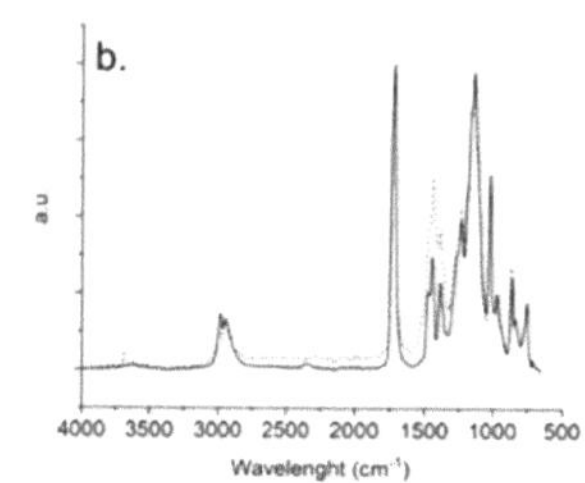
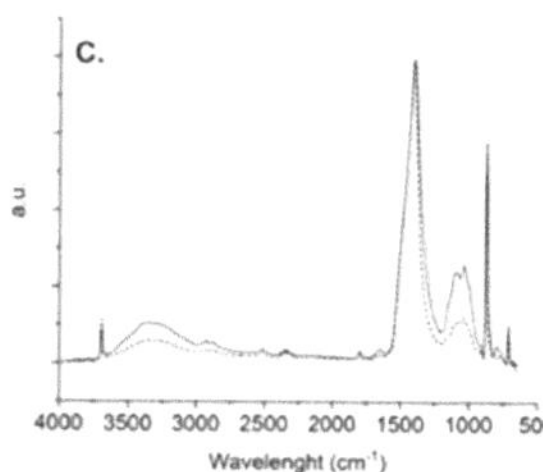

Figure 4. ATR spectra of not-aged (dot-line) and aged Proconsol® (**a**), Paraloid B72® (**b**), and Chitosan AgNPs (**c**).

Overall, the two spectra of the samples treated with Proconsol® do not show any chemical alterations due to degradation phenomena; the variations in the spectral intensity can be attributed, in this case, to a different pressure applied on the diamond cell by the investigated surface. It is unclear, however, due to the Proconsol® formulation, if, in this instance, the weak absorption at 3640 cm^{-1} is related to the product itself or to the calcium hydroxide present in the mock-ups' composition.

ParaloidB72® exhibits good stability over artificial ageing; its main absorption bands present in the two spectra remain basically unchanged and are found at about: 3400–4000 (broad, due to the hydroxyl groups stretching modes), 2988, 2953 (C-H stretching modes), 1725 (ester carbonyl group stretching mode), 1476, 1445 (CH$_3$ asymmetric bending mode), 1387, 1366 (CH$_3$ symmetric bending modes), 1234, 1168, 1140, 1023 (C-C=O-O stretching modes), 968 (C-C stretching mode), 859, 835, and 752 cm^{-1} (C-H rocking modes) [45,46]. Nevertheless, a progressive decrease in all of the main absorptions of the spectrum, and an increase in intensity in the C-H stretching and bending region can be noticed: the first phenomenon could be due to the initial stage of loss of monomers and small fragments formed as a result of the chain scissions, whereas the second one could suggest the beginning of an oxidation process [47–49].

Similarly to the Proconsol®-coated samples, those treated with the new chitosan–AgNPs coating present ATR spectra that reflect the composition of the underlying first two strata and do not display any spectral differences due to degradation, apart from the variations in the spectral intensity. Furthermore, it is possible to identify spectral features that are characteristic of glycerol, present in the formulation, in the region between 1200 and 900 cm^{-1}. Specifically, the signals at about 1039 and 1116 cm^{-1} are attributable to the C-O stretching vibration of the primary and secondary alcohol group, while the band at about 928 cm^{-1} and the small shoulder at 995 cm^{-1} are due to the C-H vibrations [50,51]. After ageing, the bands attributed to glycerol are still present; nevertheless, it is possible to observe an increase in the intensities of the absorptions between 2700 and 3600 cm^{-1}, and changes in band intensities ratios at lower wavenumbers (890 and 1200 cm^{-1} range), due to the curing of the protective layer and the evaporation of the glycerol [52,53]. In particular, the increase in the intensity of the broad band between 3000 and 3600 cm^{-1} is ascribable to the OH-O stretching vibration of the chitosan film; this band grows with the increase of the glycerol content, suggesting a larger number of chitosan-glycerol hydrogen bonds due to the presence of more OH groups provided by the glycerol molecules [52]. The variations in the intensities of the bands at 1036 and 1112 cm^{-1}, and the sharper shape of the spectrum profile in this region, suggest an interaction via hydrogen bonds between chitosan and glycerol OH groups, as reported by Leceta et al. (2015) [53]. There are no variations in the polymeric fraction of the formulation since the signal does not arise at 1378 cm^{-1} due to chitosan deacetylation, which allows for higher mobility in the polymer and favors a Maillard reaction. Moreover, if there was a structural change in the film as time progressed, a small peak at 986 cm^{-1} should appear [53]. Therefore, it is possible to state that the new formulation based on chitosan is chemically stable within this study and the ageing conditions adopted.

The morphological characterization is pivotal to determining the occurrence of visible degradation or alteration phenomena during the artificial ageing. Moreover, it allows for evaluation of whether the coating's application on the surface altered it or not, and, consequently, its appearance due to a different interaction with light.

The obtained results show that Proconsol® is transparent and very stable over the entire artificial ageing process, and that it does not change the mock-ups' surface morphology, both after its application and after its ageing (Figure S1). On the other hand, after the application of Paraloid B72®, the mock-ups immediately present a shiny/glossy aspect on the surface, which does not decrease after ageing (Figure 5). Lastly, concerning the application of the new product formulated with chitosan gel and silver nanoparticles, the coating seems to be quite stable over time and does not change the surface appearance of the mock-ups on which it was applied, while a slightly white patina is observed after ageing. This phenomenon could be ascribable to the coffee stain effect due to water deposition on the surface during ageing or could be due to salt transport during solvent evaporation, and requires further examination. No crack formation and/or increased roughness is registered due to the degradation of the chitosan polymeric film, induced by UV exposure and temperature fluctuations [54]. However, the morphological observations must be compared with colorimetric data (Figure S2).

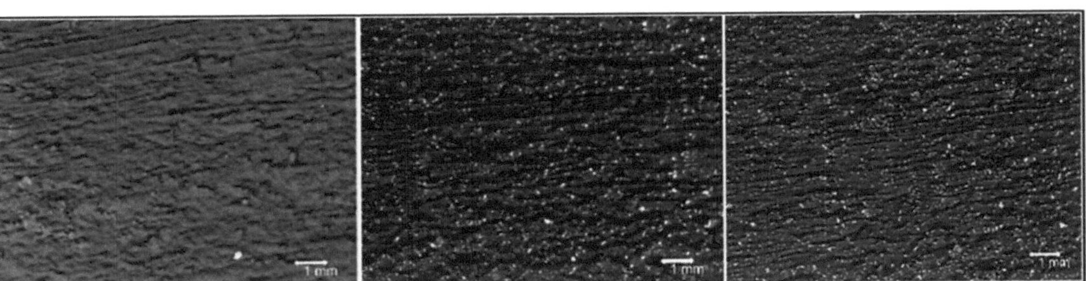

Figure 5. (From left to right) Stereomicroscopy images of untreated, treated with ParaloidB72, and aged with ParalodB72 surfaces.

The colorimetric tests are useful to determine the color variation and the gloss change between before and after the application of the three coatings and to verify if there is any variation after the ageing process. Figure 6 summarizes the ΔE variation of the samples coated with the three products, before (on the left section) and after (on the right section) the artificial ageing process. In the sample treated with Proconsol® (White), the ΔE presents a small variation, lower than 3, and its gloss variation (ΔGloss) is, on average, equal to 0. This means that the Proconsol® coating positively passes the colorimetric test, after the artificial ageing procedure as well. On the contrary, the ΔE variation graph of Paraloid B72® (dark Grey) shows that this product reaches the color limit of the human eye detection. This means that the color change is also visible by the naked eye. Nevertheless, it is worth comparing the ΔGloss between before and after applying Paraloid B72®: unlike the other products which have a Δgloss equal to 0, it presents an average gloss of 7 (Figure S3). This glossy characteristic is a main issue for the use of Paraloid B72®, since it radically changes the aesthetical aspect of mock-ups. Finally, the mock-ups coated with chitosan and silver NPs show good colorimetric results: as reported by Bussiere et al., the exposure to UV light could photo-oxidize the polymeric film, resulting in the yellowing of the treated surface, not registered here (light Grey) [54]. All treated samples are under the limit of detection of the human eye both after the application, and after the artificial ageing cycles.

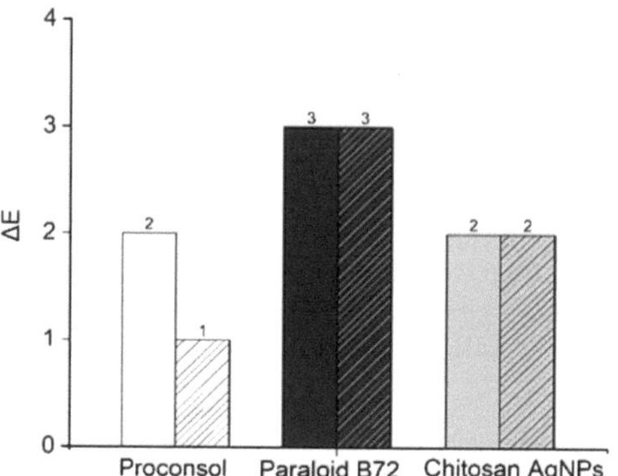

Figure 6. ΔE before (left) and after (right) ageing of the three coatings applied to mock-up surfaces.

Analyses of the contact angle are carried out to investigate the hydrophobicity of the different coatings. As can be seen in Table 3, Proconsol® does not show any sign of increasing the hydrophobicity of the material. This result is a critical aspect in the evaluation of the protective coating features, since water is one of the main degradation factors which affects Cultural Heritage materials exposed outdoors [3], and hydrophobicity is one of the main characteristics required of a superficial coating [55]. Contrarily, Paraloid B72® presents, in all the mock-ups, a higher contact angle, meaning that its application makes the surface more hydrophobic. Its hydrophobicity is also preserved after the artificial ageing process. Likewise, the chitosan gel with silver NPs, even if it has a lower contact angle (51 vs 75 on average), maintains good hydrophobic features over time (only a partially and small loss is observed −13%).

Table 3. Contact angle data of mock-ups coated with Proconsol®, Paraloid B72®, or Chitosan gel.

Contact Angle θ	After the Coating Application	After the Artificial Ageing
Proconsol®	0°	0°
Paraloid B72®	75°	71°
Chitosan AgNPs product	51°	44°

This characterization shows that Paraloid B72® is more effective in making the surface hydrophobic than the formulation presented in this study. Nevertheless, these results must be compared with the ones of the permeability to water vapor (δp) test. This test is crucial to understanding if the product that is being spread on the surface could lead to positive protective interactions or if it would result in a completely impermeable layer over the surface. Since Proconsol® does not present any hydrophobicity, for this test, only Paraloid B72® is taken as a reference product. In Figure 7, it is possible to notice that both tested products, freshly applied and aged, present a lower δp than the uncoated mock-up. However, after ageing, their water vapor permeability increases slightly. These analyses showed that the chitosan-based protective coating is more permeable to water vapor than Paraloid B72®.

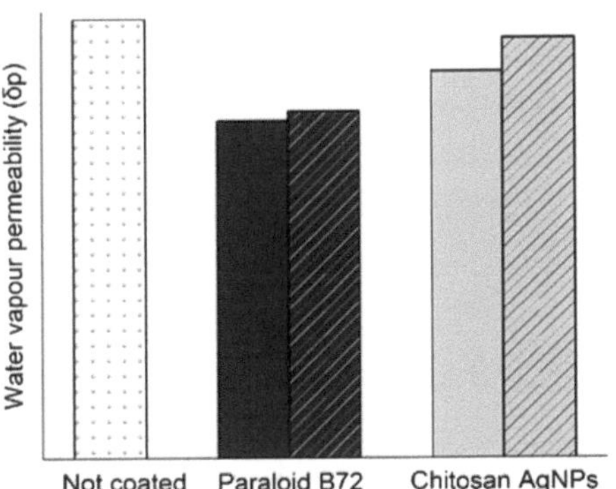

Figure 7. Water vapor permeability of treated mock-ups before (left) and after (right) ageing.

In general, it must be said that the results of all the characterizations proposed in this work do not have to be considered separately, and that only a complete view of the results can lead to appropriate conclusions, correlating the information obtained by the analyses.

3.3. Evaluation of the Coating Performances

To easily evaluate the performances of the investigated coatings, radar charts of six parameters are drawn. For each piece of information (compositional stability, morphology, color variance, gloss variance, hydrophobicity, and permeability to water vapor), a numerical value, on a scale between 0 and 3, is given, where 0 represents unsatisfying results and 3 represents well-performing outcomes. In the adopted evaluation method, assessing the quality of the overall performance, all the parameters are considered as equally important and are all assigned with the same unitary weight. Moreover, it is important to note that the numerical values are assigned based on a purely qualitative assessment by averaging the results obtained both after application and after aging, except for the compositional stability which is evaluated only after stressing the coatings in the climatic chamber. The evaluated parameters and their score values are listed in Table 4.

Table 4. Evaluation parameters and their score value for each coating (0 for unsatisfying results and 3 for well-performing outcomes).

Evaluation Parameters	Chitosan Gel	Proconsol®	Paraloid B72®
Compositional stability	2	3	2
Morphology	3	3	0
Color variance	2	2	1
Gloss variance	3	3	0
Hydrophobicity	2	0	3
Water vapor permeability (δp)	3	0 (n.a.)	2

Figure 8 shows the radar charts of the three products presented in this study: the bigger the area, the better-performing the product. Proconsol® presents very appealing performances in terms of morphology, color, gloss, and compositional stability; however, the doubts about the use of a $CaCO_3$-based material finally came up: the product, which is sold as a protective-consolidant material, does not present any hydrophobicity, hence the score attributed to the water vapor permeability is set to zero. Nevertheless, Proconsol® obtains a larger area than the still commonly-used Paraloid B72®, which excels in making the surface hydrophobic, presenting also passing permeability (δp) and chemical stability results; however, it fails in the aesthetic tests: color, gloss, and morphology variations, already visible before the artificial ageing, are far too evident to be ignored. On the other hand, the newly-proposed chitosan gel with AgNPs, even if does not show outstanding results, is high-performing in most of the evaluation parameters, especially because it simultaneously offers both aesthetic stability and a compromise between hydrophobicity and water vapor permeability.

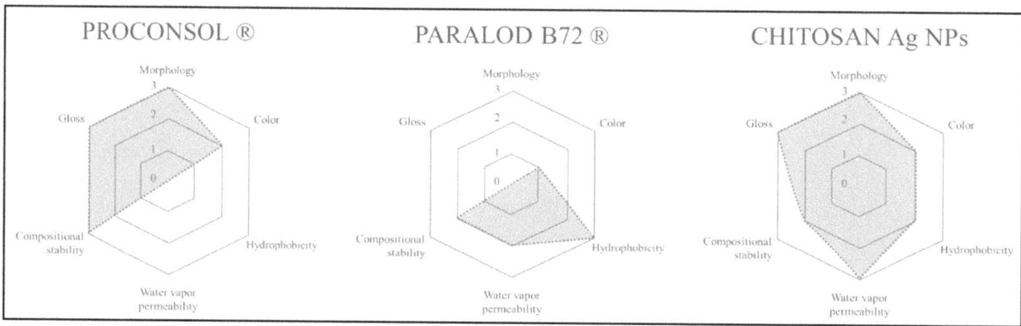

Figure 8. Radar charts used to evaluate the overall performance of the investigated coatings.

Thus, it can be said that the new formulation with chitosan and silver nanoparticles represents a valid alternative to the existing commercial products on the market today, proving to be effective and safe.

4. Conclusions

This work presented the formulation and characterization of a new product for outdoor frescoes protection, introducing the novelty of the synthesis method: nucleation and growth of silver nanoparticles in bio-polymer-based matrixes provides several chemical routes for organic-inorganic hybrid coating preparation. An enhanced control of particles growth seems to be provided by the stabilizing effect of bio-polymer-based gel introduced in the reaction with the double function of reducing, and capping and stabilizing agent. The controlled growth in shape and size of the silver nanoparticles is due to the azelaic and lactic acids solubilizing chitosan and obtaining the hosting gel matrix. Furthermore, the calcium hydroxide, as a precipitating agent, guarantees compatibility with the treated surface. Additional properties, such as a good transparency, hydrophobicity, intrinsic reversibility, and antimicrobial activity, are obtained by an adequate choice of the biopolymer and the inorganic nanoparticles type. Chitosan and silver nanoparticles emerge to produce an effective combination to obtain the desired properties for frescoes protection: the nanoparticles added to the coating base act on the surface roughness at the nanoscale, resulting in an increased hydrophobicity and consequent protection from water decay mechanisms.

The presented one-pot synthesis allows for obtaining a zero-waste formulation starting from non-toxic reagents, thus providing a new environment-friendly approach to the formulation of products designed for the conservation of Cultural Heritage. The one-pot method facilitates the optimization of reagents, avoiding product loss and assuring high homogeneity of the final product.

The application of the chitosan-based coating on fresco replicas was overall successful. Nanoparticles do not penetrate the surface layer of the *intonachino* and have been shown to endow the stratified system with hydrophobicity. This result strengthens the initial assumption that natural aging of frescoes can be strongly reduced by the application of hybrid coatings. The newly formulated coating shows good performances since it does not excessively alter surface morphology, has no gloss effects, and presents color variations lower than the human eye limit ($\Delta E = 2$). The interaction between water and the compound leads to an increase in surface hydrophobicity ($\theta = +51°$) without significantly altering the water vapor permeability (-12%). The artificial ageing does not affect the chemical stability of the synthesized product.

The comparison with the two traditional products for outdoor frescoes protection (ParaloidB72®and Proconsol®) shows the effectiveness of the Ag NPs-chitosan hybrid system in increasing the barrier properties of the treated samples, while keeping the external appearance unaltered. Proconsol does not confer hydrophobicity, leaving the surface sensitive to interaction with water; the ParaloidB72®, instead, creates a thicker shiny layer, limiting the water vapor permeability.

The encouraging results obtained represent the initial step toward new and extensive research on in situ frescoes protection. The next planned steps will be aimed to improve the overall performance of the coating, dealing with the optimization of the formulation to reduce the color variations and to further increase the hydrophobicity. Furthermore, although chitosan, azelaic acid, and AgNPs have documented antimicrobial properties, these compounds will be studied in relation to their interactions with microbiological colonization. Tests of natural ageing and of mechanical and thermal resistance will be conducted to obtain an overall evaluation of the performance of the formulation for subsequent commercial scale-up. Finally, the potential release rate of AgNPs into the environment, consequential to the leaching and/or degradation of the coating, will also be studied to complete the characterization of the material performances of the coating in outdoor environments.

Supplementary Materials: The following supporting information can be downloaded at: https://www.mdpi.com/article/10.3390/coatings13020277/s1, Figure S1: (From left to right) Stereomicroscopy images of untreated, treated with Proconsol® and aged with Proconsol® surfaces; Figure S2: (From left to right) Stereomicroscopy images of untreated, treated with Chitosan AgNPs and aged with Chitosan AgNPs surfaces; Figure S3: Δgloss before (left) and after (right) ageing of the 3 coatings applied on mock-up surfaces.

Author Contributions: Conceptualization, R.L., E.G., F.M. and M.S.; methodology, A.C. and R.L.; validation, E.G. and F.M.; formal analysis, A.C., R.L., A.D.L.P. and A.D.M.; investigation, A.C., R.L., G.F., A.D.L.P. and A.D.M.; resources, A.T. and M.S.; data curation, E.G. and F.M.; writing—original draft preparation, R.L. and A.C.; writing—review and editing, G.F. and A.T.; supervision, F.M. All authors have read and agreed to the published version of the manuscript.

Funding: This research received no external funding.

Institutional Review Board Statement: Not applicable.

Informed Consent Statement: Not applicable.

Data Availability Statement: Data is contained within the article or Supplementary Material.

Acknowledgments: The authors gratefully acknowledge Cimmino Calce s.r.l., Casoria Arpino, Naples, Italy for supplying Proconsol® tested as benchmark product. The authors are sincerely thankful to Alessia Artesani for the suggestions provided during the fulfillment of this research. Andrea Campostrini thanks for financing his Ph.D. the "PON Ricerca e Innovazione 2014–2020" and "Fondo per la promozione e lo sviluppo delle politiche del Programma Nazionale per la Ricerca (PNR) 2021–2027".

Conflicts of Interest: The authors declare no conflict of interest.

References

1.	Costantini, I.; Aramendia, J.; Tomasini, E.; Castro, K.; Manuel Madariaga, J.; Arana, G. Detection of Unexpected Copper Sulfate Decay Compounds on Late Gothic Mural Paintings: Assessing the Threat of Environmental Impact. *Microchem. J.* **2021**, *169*, 106542. [CrossRef]
2.	Pérez-Diez, S.; Pitarch Martí, A.; Giakoumaki, A.; Prieto-Taboada, N.; Fdez-Ortiz de Vallejuelo, S.; Martellone, A.; De Nigris, B.; Osanna, M.; Madariaga, J.M.; Maguregui, M. When Red Turns Black: Influence of the 79 AD Volcanic Eruption and Burial Environment on the Blackening/Darkening of Pompeian Cinnabar. *Anal. Chem.* **2021**, *93*, 15870–15877. [CrossRef] [PubMed]
3.	Sesana, E.; Gagnon, A.S.; Ciantelli, C.; Cassar, J.A.; Hughes, J.J. Climate Change Impacts on Cultural Heritage: A Literature Review. *Wiley Interdiscip. Rev. Clim. Chang.* **2021**, *12*, e710. [CrossRef]
4.	Veneranda, M.; Prieto-Taboada, N.; de Vallejuelo, S.F.O.; Maguregui, M.; Morillas, H.; Marcaida, I.; Castro, K.; Madariaga, J.M.; Osanna, M. Biodeterioration of Pompeian Mural Paintings: Fungal Colonization Favoured by the Presence of Volcanic Material Residues. *Environ. Sci. Pollut. Res.* **2017**, *24*, 19599–19608. [CrossRef] [PubMed]

5. Morena, S.; Bordese, F.; Caliano, E.; Freda, S.; De Feo, E.; Barba, S. Architectural Survey Techniques for Degradation Diagnostics. an Application for the Cultural Heritage. *Int. Arch. Photogramm. Remote Sens. Spat. Inf. Sci.—ISPRS Arch.* **2021**, *46*, 449–454. [CrossRef]

6. Unković, N.; Dimkić, I.; Stupar, M.; Stanković, S.; Vukojević, J.; Grbić, M.L. Biodegradative Potential of Fungal Isolates from Sacral Ambient: In Vitro Study as Risk Assessment Implication for the Conservation of Wall Paintings. *PLoS ONE* **2018**, *13*, e0190922. [CrossRef]

7. Mang, S.M.; Scrano, L.; Camele, I. Preliminary Studies on Fungal Contamination of Two Rupestrian Churches from Matera (Southern Italy). *Sustainability* **2020**, *12*, 6988. [CrossRef]

8. De Luca, D.; Caputo, P.; Perfetto, T.; Cennamo, P. Characterisation of Environmental Biofilms Colonising Wall Paintings of the Fornelle Cave in the Archaeological Site of Cales. *Int. J. Environ. Res. Public Health* **2021**, *18*, 8048. [CrossRef]

9. Randazzo, L.; Montana, G.; Alduina, R.; Quatrini, P.; Tsantini, E.; Salemi, B. Flos Tectorii Degradation of Mortars: An Example of Synergistic Action between Soluble Salts and Biodeteriogens. *J. Cult. Herit.* **2015**, *16*, 838–847. [CrossRef]

10. Pérez, M.C.; García-Diego, F.J.; Merello, P.; D'Antoni, P.; Fernández-Navajas, A.; Ribera I Lacomba, A.; Ferrazza, L.; Pérez-Miralles, J.; Baró, J.L.; Merce, P.; et al. Pinturas Murales de La Casa de Ariadna (Pompeya, Italia): Un Estudio Multidisciplinar de Su Estado Actual Enfocado a Una Futura Restauración y Conservación Preventiva. *Mater. Constr.* **2013**, *63*, 449–467. [CrossRef]

11. Wollner, J.L. Planning Preservation In Pompeii: Revising Wall Painting Conservation Method and Management. *Stud. Mediterr. Antiq. Class.* **2013**, *3*, 5.

12. D'Andrea, A.; DI Lillo, A.; Laino, A.; Pesaresi, P.M. Documenting Large Archaeological Sites, Managing Data, Planning Conservation and Maintenance: The Herculaneum Conservation Project Experience. *Int. Arch. Photogramm. Remote Sens. Spat. Inf. Sci.—ISPRS Arch.* **2019**, *42*, 359–364. [CrossRef]

13. Matteini, M.; Mazzeo, R.; Moles, A. *Chemistry for Restoration: Painting and Restoration Materials*; Nardini Editore: Florence, Italy, 2016; p. 343.

14. Available online: https://www.cimminocalce.com/ (accessed on 12 May 2022).

15. Vaz, M.F.; Pires, J.; Carvalho, A.P. Effect of the Impregnation Treatment with Paraloid B-72 on the Properties of Old Portuguese Ceramic Tiles. *J. Cult. Herit.* **2008**, *9*, 269–276. [CrossRef]

16. Commentary on the ICOM-CC Resolution on Terminology for Conservation. The Working Method Note about the Task Force. In Proceedings of the 15th Triennial Conference, New Delhi, India, 22–26 September 2008.

17. Brandi, C. *Il Restauro. Teoria e Pratica (1939–1986)*; Saagi Arte, Editori Riuniti: Rome, Italy, 2009.

18. Artesani, A.; Di Turo, F.; Zucchelli, M.; Traviglia, A. Recent Advances in Protective Coatings for Cultural. *Coatings* **2020**, *10*, 217. [CrossRef]

19. Andreotti, S.; Franzoni, E.; Fabbri, P.; Fabbri, P. Poly(Hydroxyalkanoate)s-Based Hydrophobic Coatings for the Protection of Stone in Cultural Heritage. *Materials* **2018**, *11*, 165. [CrossRef]

20. Ocak, Y.; Sofuoglu, A.; Tihminlioglu, F.; Böke, H. Protection of Marble Surfaces by Using Biodegradable Polymers as Coating Agent. *Prog. Org. Coatings* **2009**, *66*, 213–220. [CrossRef]

21. Infurna, G.; Cavallaro, G.; Lazzara, G.; Milioto, S.; Dintcheva, N.T. Bionanocomposite Films Containing Halloysite Nanotubes and Natural Antioxidants with Enhanced Performance and Durability as Promising Materials for Cultural Heritage Protection. *Polymers* **2020**, *12*, 1973. [CrossRef]

22. Ivanov, V.; Stabnikov, V. *Basic Concepts on Biopolymers and Biotechnological Admixtures for Eco-Efficient Construction Materials*; Elsevier Ltd.: Amsterdam, The Netherlands, 2016; ISBN 9780081002148.

23. Giuliani, C.; Pascucci, M.; Riccucci, C.; Messina, E.; Salzano de Luna, M.; Lavorgna, M.; Ingo, G.M.; Di Carlo, G. Chitosan-Based Coatings for Corrosion Protection of Copper-Based Alloys: A Promising More Sustainable Approach for Cultural Heritage Applications. *Prog. Org. Coatings* **2018**, *122*, 138–146. [CrossRef]

24. Silva da Conceição, D.K.; Nunes de Almeida, K.; Nhuch, E.; Raucci, M.G.; Santillo, C.; Salzano de Luna, M.; Ambrosio, L.; Lavorgna, M.; Giuliani, C.; Di Carlo, G.; et al. The Synergistic Effect of an Imidazolium Salt and Benzotriazole on the Protection of Bronze Surfaces with Chitosan-Based Coatings. *Herit. Sci.* **2020**, *8*, 40. [CrossRef]

25. Trang, T.T.C.; Takaomi, K. Chitosan and Its Biomass Composites in Application for Water Treatment. *Curr. Opin. Green Sustain. Chem.* **2021**, *29*, 100429. [CrossRef]

26. Goy, R.C.; De Britto, D.; Assis, O.B.G. A Review of the Antimicrobial Activity of Chitosan. *Polimeros* **2009**, *19*, 241–247. [CrossRef]

27. Sanpui, P.; Murugadoss, A.; Prasad, P.V.D.; Ghosh, S.S.; Chattopadhyay, A. The Antibacterial Properties of a Novel Chitosan-Ag-Nanoparticle Composite. *Int. J. Food Microbiol.* **2008**, *124*, 142–146. [CrossRef] [PubMed]

28. Wang, X.; Hu, Y.; Zhang, Z.; Zhang, B. The Application of Thymol-Loaded Chitosan Nanoparticles to Control the Biodeterioration of Cultural Heritage Sites. *J. Cult. Herit.* **2022**, *53*, 206–211. [CrossRef]

29. Indrie, L.; Bonet-Aracil, M.; Ilieş, D.C.; Albu, A.V.; Ilieş, G.; Herman, G.V.; Baias, Ş.; Costea, M. Heritage Ethnographic Objects—Antimicrobial Effects of Chitosan Treatment. *Ind. Text.* **2021**, *72*, 284–288. [CrossRef]

30. Kim, J.S.; Kuk, E.; Yu, K.N.; Kim, J.H.; Park, S.J.; Lee, H.J.; Kim, S.H.; Park, Y.K.; Park, Y.H.; Hwang, C.Y.; et al. Antimicrobial Effects of Silver Nanoparticles. *Nanomed. Nanotechnol. Biol. Med.* **2007**, *3*, 95–101. [CrossRef]

31. Wang, C.; Sun, M.; Wang, H.; Zhao, G. Cubic Halide Double Perovskite Nanocrystals with Anisotropic Free Excitons and Self-Trapped Exciton Photoluminescence. *J. Phys. Chem. Lett.* **2023**, *14*, 164–169. [CrossRef]

32. Wang, Y.; Wang, Y.; Aravind, I.; Cai, Z.; Shen, L.; Zhang, B.; Wang, B.; Chen, J.; Zhao, B.; Shi, H.; et al. In Situ Investigation of Ultrafast Dynamics of Hot Electron-Driven Photocatalysis in Plasmon-Resonant Grating Structures. *J. Am. Chem. Soc.* **2022**, *144*, 3517–3526. [CrossRef]
33. Hasebe, S.; Hagiwara, Y.; Komiya, J.; Ryu, M.; Fujisawa, H.; Morikawa, J.; Katayama, T.; Yamanaka, D.; Furube, A.; Sato, H.; et al. Photothermally Driven High-Speed Crystal Actuation and Its Simulation. *J. Am. Chem. Soc.* **2021**, *143*, 8866–8877. [CrossRef]
34. Mokrzycki, W.S.; Tatol, M. Colour Difference ΔE—A Survey Mokrzycki. *Mach. Graph. Vis.* **2011**, *20*, 383–411.
35. Cimino, D.; Lamuraglia, R.; Saccani, I.; Berzioli, M.; Izzo, F.C. Assessing the (In)Stability of Urban Art Paints: From Real Case Studies to Laboratory Investigations of Degradation Processes and Preservation Possibilities. *Heritage* **2022**, *5*, 581–609. [CrossRef]
36. Costantini, R.; Vanden Berghe, I.; Izzo, F.C. New Insights into the Fading Problems of Safflower Red Dyed Textiles through a HPLC-PDA and Colorimetric Study. *J. Cult. Herit.* **2019**, *38*, 37–45. [CrossRef]
37. Striova, J.; Camaiti, M.; Castellucci, E.M.; Sansonetti, A. Chemical, Morphological and Chromatic Behavior of Mural Paintings under Er:YAG Laser Irradiation. *Appl. Phys.* **2011**, *104*, 649–660. [CrossRef]
38. UNI EN 15802:2010. Available online: http://store.uni.com/catalogo/uni-en-15802-2010?josso_back_to=http://store.uni.com/josso-security-check.php&josso_cmd=login_optional&josso_partnerapp_host=store.uni.com (accessed on 12 May 2022).
39. UNI EN 15803:2010. Available online: http://store.uni.com/catalogo/uni-en-15803-2010 (accessed on 12 May 2022).
40. Boanić, D.K.; Trandafilović, L.V.; Luyt, A.S.; Djoković, V. "Green" Synthesis and Optical Properties of Silver-Chitosan Complexes and Nanocomposites. *React. Funct. Polym.* **2010**, *70*, 869–873. [CrossRef]
41. Nishimura, S.; Mott, D.; Takagaki, A.; Maenosono, S.; Ebitani, K. Role of Base in the Formation of Silver Nanoparticles Synthesized Using Sodium Acrylate as a Dual Reducing and Encapsulating Agent. *Phys. Chem. Chem. Phys.* **2011**, *13*, 9335–9343. [CrossRef]
42. Berlangieri, C.; Poggi, G.; Murgia, S.; Monduzzi, M.; Dei, L.; Carretti, E. Structural, Rheological and Dynamics Insights of Hydroxypropyl Guar Gel-like Systems. *Colloids Surfaces B Biointerfaces* **2018**, *168*, 178–186. [CrossRef]
43. Farhadi Khouzani, M.; Chevrier, D.M.; Güttlein, P.; Hauser, K.; Zhang, P.; Hedin, N.; Gebauer, D. Disordered Amorphous Calcium Carbonate from Direct Precipitation. *CrystEngComm* **2015**, *17*, 4842–4849. [CrossRef]
44. Sanjuan, B.; Girard, J. Review of kinetic data on carbonate mineral precipitation. *Tetrahedron* **1996**, *52*, 13837–13866.
45. Hussein Elsayed, N.; Najmi, A.H.; Mohamed, G.M.; Mohamed, A.A.; Al-Sayed, H.M.A. Preparation and Characterization of Paraloid B-72/TiO$_2$ Nanocomposite and Their Effect on the Properties of Polylactic Acid as Strawberry Coating Agents. *J. Food Saf.* **2020**, *40*, e12838. [CrossRef]
46. Ciccola, A.; Serafini, I.; Guiso, M.; Ripanti, F.; Domenici, F.; Sciubba, F.; Postorino, P.; Bianco, A. Spectroscopy for Contemporary Art: Discovering the Effect of Synthetic Organic Pigments on UVB Degradation of Acrylic Binder. *Polym. Degrad. Stab.* **2019**, *159*, 224–228. [CrossRef]
47. Pintus, V.; Wei, S.; Schreiner, M. UV Ageing Studies: Evaluation of Lightfastness Declarations of Commercial Acrylic Paints. *Anal. Bioanal. Chem.* **2012**, *402*, 1567–1584. [CrossRef]
48. Papliaka, Z.E.; Andrikopoulos, K.S.; Varella, E.A. Study of the Stability of a Series of Synthetic Colorants Applied with Styrene-Acrylic Copolymer, Widely Used in Contemporary Paintings, Concerning the Effects of Accelerated Ageing. *J. Cult. Herit.* **2010**, *11*, 381–391. [CrossRef]
49. Shanti, R.; Bella, F.; Salim, Y.S.; Chee, S.Y.; Ramesh, S.; Ramesh, K. Poly(Methyl Methacrylate-Co-Butyl Acrylate-Co-Acrylic Acid): Physico-Chemical Characterization and Targeted Dye Sensitized Solar Cell Application. *Mater. Des.* **2016**, *108*, 560–569. [CrossRef]
50. Hammoude, A.Y.; Obeida, S.M.; Abboushi, E.K.; Mahmou, A.M. FT-IR Spectroscopy for the Detection of Diethylene Glycol (DEG) Contaminant in Glycerin-Based Pharmaceutical Products and Food Supplements. *Acta Chim. Slov.* **2020**, *67*, 530–536. [CrossRef]
51. Joshi, R.; Joshi, R.; Amanah, H.Z.; Faqeerzada, M.A.; Jayapal, P.K.; Kim, G.; Baek, I.; Park, E.-S.; Masithoh, R.E.; Cho, B.-K. Quantitative Analysis of Glycerol Concentration in Red Wine Using Fourier Transform Infrared Spectroscopy and Chemometrics Analysis. *Korean J. Agric. Sci.* **2021**, *48*, 299–310.
52. Debandi, M.V.; Bernal, C.; Francois, N.J. Development of Biodegradable Films Based on Chitosan/Glycerol Blends Suitable for Biomedical Applications. *J. Tissue Sci. Eng.* **2016**, *7*, 3. [CrossRef]
53. Leceta, I.; Peñalba, M.; Arana, P.; Guerrero, P.; De La Caba, K. Ageing of Chitosan Films: Effect of Storage Time on Structure and Optical, Barrier and Mechanical Properties. *Eur. Polym. J.* **2015**, *66*, 170–179. [CrossRef]
54. Bussiere, P.O.; Gardette, J.L.; Rapp, G.; Masson, C.; Therias, S. New Insights into the Mechanism of Photodegradation of Chitosan. *Carbohydr. Polym.* **2021**, *259*, 117715. [CrossRef] [PubMed]
55. Pino, F.; Fermo, P.; La Russa, M.; Ruffolo, S.; Comite, V.; Baghdachi, J.; Pecchioni, E.; Fratini, F.; Cappelletti, G. Advanced Mortar Coatings for Cultural Heritage Protection. Durability towards Prolonged UV and Outdoor Exposure. *Environ. Sci. Pollut. Res.* **2017**, *24*, 12608–12617. [CrossRef]

Article

Removing Aged Polymer Coatings from Porous Stone Surfaces Using the Gel Cleaning Method

Maduka L. Weththimuni [1,2,*], Giacomo Fiocco [3,4], Alessandro Girella [1,2], Barbara Vigani [5], Donatella Sacchi [1], Silvia Rossi [5] and Maurizio Licchelli [1,2,*]

1. Department of Chemistry, University of Pavia, Via Taramelli 12, 27100 Pavia, Italy; alessandro.girella@unipv.it (A.G.); donatella.sacchi@unipv.it (D.S.)
2. Research Centre for Cultural Heritage (CISRiC), University of Pavia, Via Ferrata 1, 27100 Pavia, Italy
3. Arvedi Laboratory of Non-Invasive Diagnostics, Research Centre for Cultural Heritage (CISRiC), University of Pavia, Via Bell'Aspa, 26100 Cremona, Italy; giacomo.fiocco@unipv.it
4. Department of Musicology and Cultural Heritage, University of Pavia, Corso Garibaldi, 178, 26100 Cremona, Italy
5. Department of Drug Sciences, University of Pavia, Via Taramelli 14, 27100 Pavia, Italy; barbara.vigani@unipv.it (B.V.); silvia.rossi@unipv.it (S.R.)
* Correspondence: madukalankani.weththimuni@unipv.it (M.L.W.); maurizio.licchelli@unipv.it (M.L.)

Abstract: Acrylic polymers were extensively used in past restoration practices, usually as consolidants or protecting agents. Their removal is often required because polymer coatings can improve some decay processes of stone substrates and, after ageing, may generate undesirable materials on the surface of artifacts. Therefore, the removal of old polymer coating from the surface of artifacts has become a common operation in the conservation of cultural heritage. As with other cleaning operations, it is a delicate process that may irreversibly damage the artifacts if not correctly carried out. The main aim of this study was to determine the appropriate cleaning procedure for efficiently removing old acrylic polymers (e.g., Paraloid B-72) from the surface of historical buildings. For this purpose, a polymer was applied to two different porous stone substrates (bio-calcarenite and arenaria stone). The hydrogel cleaning approach was used for the present study, as preliminary results suggested that it is the most promising polymer-removing method. The considered hydrogel (based on a semi-interpenetrating polymer network involving poly(2-hydroxyethyl methacrylate) and polyvinylpyrrolidone) was prepared and characterized using different techniques in order to assess the gel's properties, including the gel content, equilibrium water content, retention capability, hardness, Young's modulus, and morphology. After that, the hydrogel was loaded with appropriate amounts of nano-structured emulsions (NSEs) containing a surfactant (EcoSuf[TM]), organic solvents, and H_2O, then applied onto the coated surfaces. Moreover, plain EcoSurf[TM] in a water emulsion (EcoSurf/H_2O) was also used to understand the polymer-removing behavior of the surfactant without any organic solvent. A comparative study was carried out on artificially aged and unaged polymer-coated samples to better understand the cleaning effectiveness of the considered emulsions for removing decayed polymer coatings. The experimental results showed that the NSE-loaded hydrogel cleaning method was more effective than other common cleaning procedures (e.g., cellulose pulp method). In fact, only one cleaning step was enough to remove the polymeric material from the stone surfaces without affecting their original properties.

Keywords: gel materials; characterization; polymer coatings; eco-friendly surfactant; nano-emulsions; cleaning process; SEM-EDS; micro FTIR-mapping

Citation: Weththimuni, M.L.; Fiocco, G.; Girella, A.; Vigani, B.; Sacchi, D.; Rossi, S.; Licchelli, M. Removing Aged Polymer Coatings from Porous Stone Surfaces Using the Gel Cleaning Method. *Coatings* **2024**, *14*, 482. https://doi.org/10.3390/coatings14040482

Academic Editor: Patricia Sanmartín

Received: 28 February 2024
Revised: 3 April 2024
Accepted: 11 April 2024
Published: 14 April 2024

1. Introduction

Cultural heritage includes artifacts, monuments, and buildings that have a diversity of values, including symbolic, historic, artistic, aesthetic, ethnological or anthropological, scientific, and social significance. It includes tangible heritage (movable, immobile, and underwater), intangible cultural heritage embedded in culture, and natural heritage artifacts, sites, or monuments. Hence, the conservation of cultural heritage, particularly of tangible elements, is a common operation to maintain their original structure against degradation processes [1,2]. Cleaning, i.e., removing undesired materials from the substrate surface, is an essential step of the conservation process. It represents a delicate operation because, if is not correctly performed, it may irreversibly damage the heritage [3]. The undesired materials to be removed from the surfaces of artifacts may include deposits of pollutants, grime or dirt, soil materials, and aged polymeric coatings (and related degraded materials) applied in previous conservative interventions. Synthetic polymers have been widely used in past restoration practices, usually for consolidation and/or protection purposes. The removal of aged polymeric and other undesired materials without damaging the substrates and preserving their original properties is a challenging task [3]. A traditional methodology involves the use of organic solvents and a cleaning intervention based on mechanical action or solubilization processes. This method has many drawbacks such as poor selectivity, health and environmental risk, uncontrolled penetration, alterations of the substrate's original properties, and re-deposition of the undesired materials inside the pores of the substrate [3–6]. In order to overcome these problems, water-based nano-structured materials have been proposed for removing old coatings [4,5,7–9]. For instance, nano-emulsions (NSEs) can be used as an alternative to non-confined organic solvents. They are water-based fluids containing small amounts of organic solvents dispersed as nano-droplets, which are stabilized by the addition of a surfactant. Due to the nanometric size of the solvents/surfactant droplets and their consequently large surface area, the cleaning effect can be significantly enhanced [4].

There are different methods to apply nano-structured emulsions for removing undesired coating and related materials: they can be placed into direct contact with the surfaces to be treated, or they can be supported on proper transporting mediums, e.g., cellulose pulp poultice or hydrogels. In a previous work, we thoroughly investigated and discussed the cleaning performances of the direct contact method as well as the cellulose pulp poultice method [10]. In the present work, we mainly focused on the hydrogel method. In the last decades, scientists have introduced physical gels in which a cross-linking process is obtained via non-covalent interactions, (i.e., xanthan gum, gellan gum, agar, chitosan, sodium alginate-calcium) and chemical gels based on covalently cross-linked polymers (e.g., polyacrylamide gels) [3,11–13]. The chemical gels had better performances than the physical gels due to their good mechanical properties, well-defined shape, and swelling properties [3,14–17].

The main aims of the present study were the synthesis, characterization, and application of a gel material that could load water-based cleaning systems, particularly nano-emulsions. Such a hydrogel would behave as a suitable transporting medium of NSEs for removing aged polymer coating from stone artifacts. For this purpose, a semi-interpenetrating polymer network involving poly(2-hydroxyethyl methacrylate) and polyvinylpyrrolidone was selected and synthesized following a method reported in the literature [18]. Poly(2-hydroxyethyl methacrylate), often crosslinked with N,N-methylenebisacrylamide (HEMA-MBA copolymer), has been used as main component of various hydrogels due to its good biocompatibility, and it has found different applications including drug delivery systems, porous sponges, and the manufacturing of soft contact lenses and artificial corneas [19–24]. Moreover, in combination with the highly hydrophilic polyvinyl pyrrolidone (PVP) [25], a HEMA-MBA copolymer was used to make hydrogels that displayed excellent properties and suitability for the preparation of cleaning tools for cultural heritage items. In particular, highly retentive hydrogel materials based on this polymeric matrix were prepared and used in the cleaning of different water-sensitive substrates [3]. Nevertheless, the effectiveness of the HEMA-MBA copolymer/PVP hydrogel in the removal of aged coatings from stone artifacts has been poorly explored [4].

The properties of the investigated hydrogel were studied in detail using different techniques to verify its suitability for cleaning applications, including its gel content, equilibrium water content, retention capability, hardness, Young's modulus, and morphology (via SEM observations). Moreover, the properties were compared with well-known agar gel. To our knowledge, a full characterization of this type of hydrogel has not been reported elsewhere.

The cleaning effectiveness of the NSE-loaded hydrogel was tested for the removal of an aged acrylic polymer (e.g., Paraloid B-72) that was widely employed in the past for restoration processes [26–34] from the surface of two different stone substrates (a bio-calcarenite and an arenaria stone). Cleaning tests using the same hydrogel loaded with an emulsion containing only the surfactant were also carried out for comparison. For the present study, Lecce stone (LS) was used as the bio-calcarenite, which has been highly employed for different buildings and artifacts in the south of Italy, especially in the Puglia region during the Baroque period [35–39]. Due to its high open porosity (>30%), and high content of calcite (~95), it has been facing many deterioration problems compared to other stone materials [1,35]. The considered sandstone is an arenaria (AS, mainly composed of SiO_2 and silicates) used as a building material (e.g., palaces and churches) mainly in the surroundings of Pavia (North Italy). The magnificent façade of the San Michele Maggiore church (XI century) and other representative buildings in Pavia were made using this Arenaria stone. It is a soft stone with a low surface cohesion and high open porosity (around 15%) that shows a heterogeneous morphology because of its chemical composition. Due to its heterogeneous structure and high open porosity, this sandstone is also affected by many decay problems [10,40].

Previous restoration interventions performed on LS and AS using synthetic polymer materials have badly affected the stone substrates (strong variation in the original surface properties, water capillarity and vapor permeability reduction, alterations in surface wettability) [5]. As the old coatings must be removed prior to any new possible conservation action, i.e., consolidation using a more appropriate material [1,35,41], investigations on suitable methods for the removal of the aged polymer materials are highly desired.

In order to investigate the decayed polymer cleaning process as in the real cases, polymer-treated specimens were exposed to artificial ageing cycles (up to 35 days) at a high temperature (inside an oven, 70 ± 2 °C), corresponding to more than 50 years of natural ageing of the polymer material [5,42]. The cleaning ability for the aged polymer samples (12, 25, and 35 days) of a hydrogel loaded with two different emulsions (different chemical proportions of surfactant, organic solvents, and water) was evaluated using different experimental techniques. Chromatic variations and the surface wettability were determined to evaluate the effects of the cleaning process on the original stone properties, while the surface morphology and cross-sectional properties were studied using an optical microscope and SEM (scanning electron microscopy). Moreover, the EDS (energy dispersive X-ray spectra) technique was used to obtain semi-quantitative analyses of the stone substrate before and after the cleaning process. Finally, the presence of possible residual coatings or any other organic residue on the stone surfaces after cleaning was examined using micro-FTIR (ATR mode) spectroscopy and mapping experiments.

The experimental results showed that the investigated cleaning tool based on the NSE-loaded hydrogel was able to efficiently remove the old polymer coatings from the surface of the very porous stones. Therefore, it can be considered as a promising alternative to other traditional cleaning methods (e.g., cellulose pulp, agar gel) for the conservation of stone cultural heritage items.

2. Materials and Experimental Methods

2.1. Chemicals and Lithotypes

The selected polymer material was Paraloid B-72 (100% acrylic resin, Bresciani s.r.l, Milan, Italy). Nano-emulsions were prepared using the surfactant (ECOSURF™ EH-6, $C_8H_{18}O \cdot (C_3H_6O)_x \cdot (C_2H_4O)_y$, non-ionic surfactant, Sigma Aldrich, St. Louis, MO, USA) and the organic solvents 2-butanol (purity 99% BuOH, $C_2H_5CH(OH)CH_3$, Fluka Chemicals, Buchs, Switzerland) and 2-butanone (purity 99%–101% $CH_3C(O)CH_2CH_3$, BDH Chemicals Ltd., Poole, UK). Ethyl acetate (ACS reagent, purity $\geq$ 99.5% $CH_3COOC_2H_5$, Sigma Aldrich, St. Louis, MO, USA) was used for preparing the polymer solution. The HEMA-MBA/PVP hydrogel was synthesized using different chemicals: 2-hydroxyethyl methacrylate (HEMA, $CH_2=C(CH_3)COOCH_2CH_2OH$, Sigma Aldrich, St. Louis, MO, USA), N,N-methylenebisacrylamide (MBA, $(H_2C=CHCONH)_2CH_2$, Sigma Aldrich, St. Louis, MO, USA), polyvinylpyrrolidone (PVP, $(C_6H_9NO)_n$, Sigma Aldrich, St. Louis, MO, USA), and α,α'-azobisisobutyronitrile (AIBN, $(CH_3)_2C(CN)N=NC(CH_3)_2CN$, purity 98%, Sigma Aldrich, St. Louis, MO, USA). Agar (Agarose, 3,6-Anhydro-α-L-galacto-β-D-galactan, Sigma Aldrich, St. Louis, MO, USA) was used to compare the properties of the synthesized hydrogel. Water was purified using a Millipore Organex system: $R \geq 18$ M cm (Burlington, MA, USA).

One month before the coating application, the stone samples were cleaned according to the standard method (UNI 10921 Protocol) [43]. In brief, Lecce stone (LS, supplied by Tarantino and Lotriglia, Nardò, Lecce, Italy), and Arenaria stone (AS, taken from a quarry in Monte Arzolo, Province of Pavia, North Italy) samples (5 cm $\times$ 5 cm $\times$ 1 cm and 5 cm $\times$ 5 cm $\times$ 2 cm) were smoothed using abrasive carbide paper (No: 180 mesh). After that, they were washed with deionized water, dried in an oven at 60 °C, and stored in a desiccator until they reached room temperature. The dry weights of the samples were recorded every day until they were constant [44].

2.2. Synthesis of the Hydrogel

The HEMA-MBA copolymer/PVP hydrogel was synthesized according to a modification of a method reported in the literature [18]. A resistant, transparent, and highly retentive semi-interpenetrating polymer network was created by embedding PVP into a poly HEMA hydrogel network. In a typical preparation, PVP and water (25.1 and 57.9% *w/w* of the final mixture, respectively) were accurately mixed in a round-bottom glass flask until a solution was obtained. Then, a mixture made of 2-hydroxyethyl methacrylate monomer (HEMA) and a N,N-methylenebisacrylamide cross-linker (MBA) was prepared (16.8 and 0.2% *w/w* of the final mixture, respectively) and added to the PVP aqueous solution. Nitrogen (N_2) gas was passed through the resulting solution for some minutes to remove the oxygen, then a radical initiator α,α-azoisobutyronitrile (10^{-2} mol with respect to the HEMA) was added into the reaction mixture. After that, the mixture was gently sonicated in an ultrasonic bath for 30 min to eliminate gas bubbles. In order to complete the polymerization reaction, the mixture was heated to 60 °C for 4 h under a nitrogen atmosphere. Then the resulting gel was separated and washed several times with distilled water. It was stored in water in a beaker, and the water was replaced every day (up to 7 days) to remove the unreacted materials.

2.3. Characterization of the Synthesized Hydrogel

Several experimental analyses were performed to study the properties of the synthesized hydrogel material. Moreover, some of the properties were compared with the agar physical gel.

2.3.1. Gel Content

The gel content (G) was calculated according to the following equation [18,45]:

$$G\ (\%) = (W_d / W_0) \times 100 \tag{1}$$

where W_d is the dry weight of the synthesized hydrogel and W_0 is the weight of the starting components (i.e., HEMA and PVP for the investigated material) in the initial reaction mixture.

2.3.2. Water Content and Release Capacity

The equilibrium water content was determined according to Equation (2) [18,45]:

$$(EWC) \quad EWC = [(W_w - W_d)/W_w] \times 100 \tag{2}$$

where W_w is the water swollen hydrogel (at least 7 days immersed in water) and W_d is the dry weight of the hydrogel as previously indicated.

The retention capability (RC; capacity of releasing water) was calculated considering a gently dried (vacuum dried at 25 °C for 1 h) fully swollen gel sample (about 12.5 cm^2 and 2 mm thickness gel film). It was kept on a filter paper (five sheets of Whatman$^®$) inside a covered petri dish. The sheets were weighed before applying the gel and after 30 min from the application. The retention capability was obtained based on the weight difference with respect to the surface area of the gel film, as indicated in Equation (3). The solvent-releasing capacity was calculated similarly.

$$RC = (W_{wf} - W_{df})/SA \tag{3}$$

Here, W_{wf} is the weight of the wet filter paper due to the releasing of water from the gel film, W_{df} is the dry weight of the filter paper, and SA is the surface area of the gel film.

2.3.3. Rheological Properties

The mechanical properties (compression strength and Young's modulus) of the HEMA-MBA/PVP hydrogel were measured, compared with physical gel (agar), and examined considering their possible relationship with the texture, the durability, and the usability of the gel materials. All the considered measurements were performed on the prepared gels without drying them. The considered gel samples were prepared in cylindrical molds (30 mm × 30 mm, diameter × height) and were subjected to a compression test using a TA.XT plus Texture Analyzer (Stable Micro Systems, Godalming, UK) equipped with a 5 kg load cell and a P/10 measuring system consisting of a cylindrical probe with a diameter of 10 mm [46–48]. The probe was lowered with a test speed equal to 1.00 mm/s in order to determine a 70% sample deformation. The following parameters were determined: (a) hardness; that is, the maximum compressive force per unit area required for sample destructuring; (b) Young's modulus (YM), calculated as the slope of the tangent at the first part of the compressive stress–strain curve. Stress–strain curves were also produced. Three replicates were carried out for each gel. The experimental values of the various types of measurements were subjected to statistical analysis, which was carried out using the statistical package Statgraphics 5.0 (Statistical Graphics Corporation, Rockville, MD, USA). In particular, a *t*-test was carried out to evaluate whether the difference between the mean of the two groups was statistically significant.

2.4. Preparation of Coated Stone Specimens

Polymer coatings were applied to the considered lithotypes in a similar way as previously reported [10]. In brief, commercially available Paraloid B-72 was dissolved in ethyl acetate to prepare a 10% (*w/w*) polymer solution, which was then applied to the LS and AS specimens. In particular, one half of each specimen surface (half of the available surface, 2.5 cm × 5 cm) was treated with the polymer solution (600 µL) using a small pipette, and the other part was kept without any treatment and used as a reference surface. Some stone specimens were fully coated and used for specific analyses. The treated samples (LSPB, and ASPB) underwent different ageing cycles in order to obtain artificially weathered coatings before testing the removing ability of the emulsion-loaded hydrogel as in the real cases. In the artificial ageing process, the coated specimens were divided into three groups, which were then stored in an oven (70 ± 2 °C) for 12, 25, and 35 days, respectively. Some coated samples were kept without exposing them to ageing cycles for comparison with the aged samples and were named as unaged LSPB and ASPB.

2.5. Preparation of Nano-Structured Emulsions and Application of the Emulsion-Loaded Gels

After completing several preliminary trials, two different emulsions were chosen with different compositions (v/v ratio): (i) EcoSurf/H_2O: H_2O 95.0%, surfactant (ECOSURF^{TM} EH-6, non-ionic surfactant) 5.0%; (ii) NSE: H_2O, 65.9%, surfactant, 3.5%, BuOH, 9.7%, and butanone 20.9%. The NSE was prepared according to the literature [4] and analyzed using dynamic light scattering (DLS) using a MALVERN ZS90 apparatus (Malvern Panalytical Private Limited, Malvern, UK) to assess the dimensions of the droplets in the emulsion [49]. The emulsion was well mixed, and a small portion (1 mL) was placed into the plastic stub. The measurements (time = 10 s) were performed at 25 °C. The analysis was repeated three times in order to obtain an accurate measurement.

The prepared emulsions were loaded in the synthesized hydrogel before being applied to the surfaces of the stones. Thin hydrogel films (2 mm thickness) were properly cut to fit the area of the treated specimen surface (about 12.5 cm^2). The correctly shaped hydrogel films were gently dried (vacuum at 25 °C for 1 h) and then immersed into the envisaged emulsion (NSE or Ecosurf/H_2O, 2.1 mL) for 24 h. Then, they were applied to the surface of the stone specimens to be cleaned for 15 min. After removing the gel film, the stone surface was gently cleaned using a wet cotton swab to remove any possible residues. This procedure was performed on three samples of each specimen group to collect accurate results.

The cleaning performances of the emulsion-loaded hydrogel were evaluated using different analyses, as reported in Section 2.6.

2.6. Instrumental Techniques

A Nicolet iN10 Thermo Fischer μFT-IR spectrometer was used to collect micro-FTIR spectra in attenuated total reflectance mode (ATR, germanium crystal, Thermo Fisher Scientific, Waltham, MA, USA) and to perform μ-FTIR mapping experiments on the cleaned stone surfaces. A total of 64 scans were collected for each spectrum.

Chromatic variations after ageing and after the cleaning processes were evaluated using a Konica Minolta CM-2600D spectrophotometer (Konica Minolta, Inc., Tokyo, Japan) considering the L*, a*, and b* coordinates of the CIELAB space and the global chromatic variations, expressed as ΔE* according to the UNI EN 15886 protocol [50]. In order to obtain accurate results, 15 different measurements (3 specimens for each kind of sample and 5 measurements on each specimen) were performed, and the average values were determined as recommended in the literature [44,51].

The contact angle measurements on the treated surfaces as well as after the ageing and cleaning processes were performed using a Lorentzen and Wettre instrument (Zurich, Switzerland) according to the UNI EN 15802 Protocol [52]. A total of 15 different measurements of each group of samples were used to calculate the average values, as previously reported [53].

The morphological features were observed using optical microscopy using a light polarized microscope Olympus BX51TF equipped with the Olympus TH4-200 lamp (Olympus Corporation, Tokyo, Japan) and using scanning electron microscopy (SEM) using a Tescan FE-SEM apparatus (MIRA XMU series, TESCAN, Brno, Czech Republic). The elemental composition of each material (semi-quantitative analysis) was determined using energy dispersive X-ray spectrometry (EDS) using a Bruker Quantax 200 instrument (Bruker, Billerica, MA, USA) combined with the SEM apparatus. The SEM-EDS instrument operated at both low and high vacuum settings and was located at the Arvedi Laboratory, CISRiC, University of Pavia, Pavia, Italy. The gel samples were examined after drying in a vacuum at room temperature (20 ± 2 °C).

3. Results and Discussion

3.1. Characterization of the Synthesized Hydrogel

A semi-interpenetrating hydrogel network was created by synthesizing the HEMA-MBA copolymer in the presence of linear PVP polymer chains, as reported in Scheme 1. The copolymer was synthesized via free radical polymerization with α,α'-azobisisobutyronitrile (AIBN) as a radical initiator. It is known that the HEMA monomer and MBA cross-linker react very fast in the presence of radical initiators due to their double bonds [54–58]. On the contrary, no reactivity is expected for PVP in these experimental conditions, and it only interacts via H-bonds contributing to the interpenetrated structure. The synthesized hydrogel was washed very well with water and stored for up to 7 days in water in order to remove any unreacted substances. It appeared as a soft and transparent gel material, whose properties were investigated using different techniques, as mentioned in the experimental section.

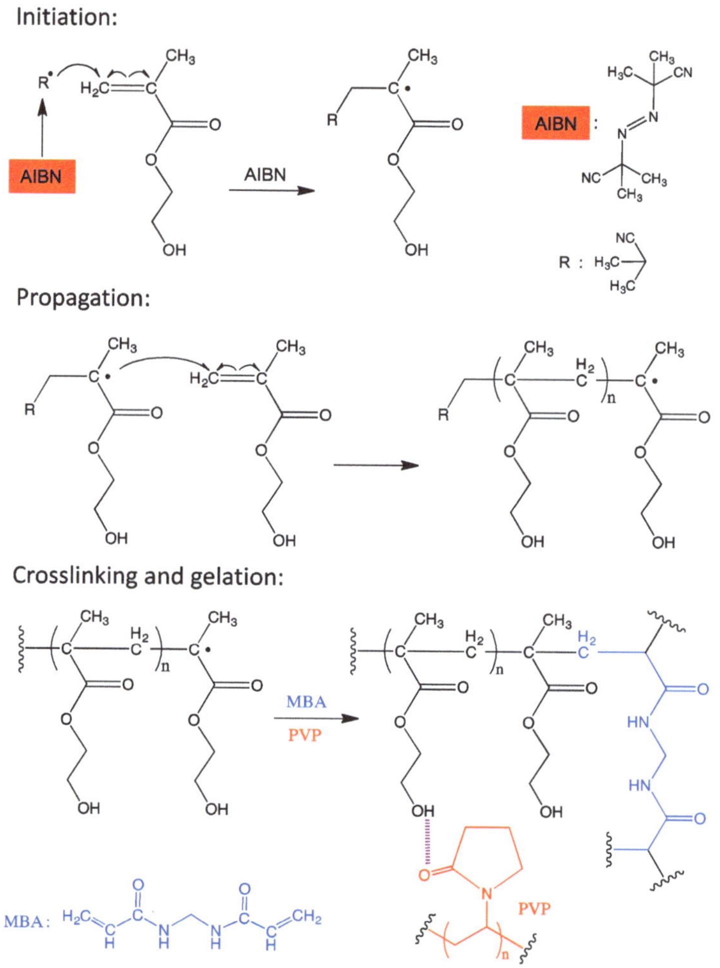

Initiation:

Propagation:

Crosslinking and gelation:

Scheme 1. Reaction steps of the synthesis of the HEMA/MBA/PVP hydrogel.

The gel content, the equilibrium water content, and the retention capability are summarized in Table 1 and compared with the corresponding properties of agar and other commonly used gels. Investigating the equilibrium water content (EWC) as well as the capacity of water release (RC) of the hydrogel is very important, especially when its application concerns the cleaning of water-sensitive substrates. The EWC and RC values determined for the HEMA-MBA/PVP hydrogel (81%, and 13 mg/cm^2, respectively) were similar to those already reported [3]. Both parameters were distinctly lower than the agar gel (used as a reference) and other common hydrogels (see Table 1). The lower values of water content and water release displayed by the HEMA-MBA/PVP material were more acceptable, considering its possible use for the cleaning of water-sensitive artefacts. In particular, the amount of released water for the agar gel was more than double with respect to the HEMA-MBA/PVP material. The RC values reported for Kelcogel or acrylamide (soft) were even larger (Table 1) [3].

Table 1. The physicochemical properties of the HEMA-MBA/PVP copolymer compared to other common hydrogels [3].

Samples	Gel Content (G, %)	Equilibrium Water Content (EWC, %)	Retention Capability (RC, Water Released, mg/cm^2)
HEMA-MBA/PVP [a]	76 ± 5	81 ± 2	13 ± 2
Agar [a]	-	95 ± 2	28 ± 2
Acrylamide (soft) [b]	88	97	56
Kelcogel [b]	-	97	33

[a] This work. [b] from ref. [3].

After being left to rest in cylindrical molds and stored at 4 °C, the HEMA-MBA/PVP and agar gels were subjected to a compression test. An evaluation of the gels' mechanical properties is useful to deeply investigate the gels' inner microstructures and to investigate their durability and usability. Figure 1a shows that the hardness of the HEMA-MBA/PVP gel, measured as its resistance to compression, was slightly higher than the agar gel, although the observed differences were poorly significant from a statistical point of view. On the contrary, the two materials displayed a strongly different behavior when their Young's modulus values were considered. In Figure 1b, the compressive strength vs. strain profiles of the two specimens are reported. It can be observed that the agar gels and HEMA-MBA/PVP hydrogel were characterized by completely different mechanical behaviors when they were subjected to a uniaxial compressive force: the former showed a fracture zone (peak stress) at low strain values [59], while the latter exhibited a delayed fracture only at the higher level of the considered strain range. Therefore, the agar gel appeared to be more brittle, quickly breaking down, with respect to the HEMA-MBA/PVP hydrogel, which instead showed a longer duration without easy de-structuring [6,59–61]. Based on these experimental data as well as the literature data, agar was not considered for the preparation of the emulsion-loaded hydrogels to be applied to the removal of the aged coatings on the stone substrates.

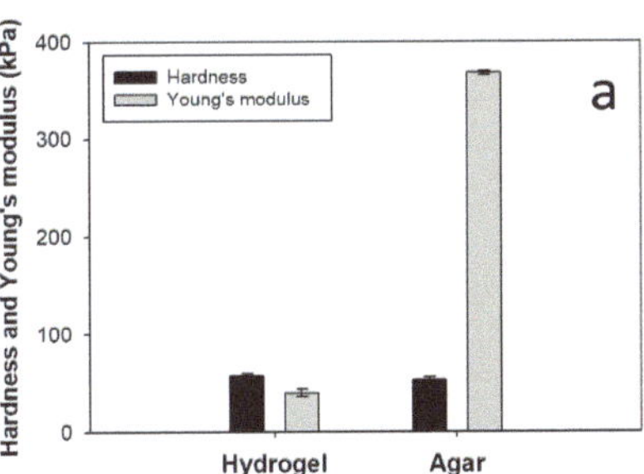

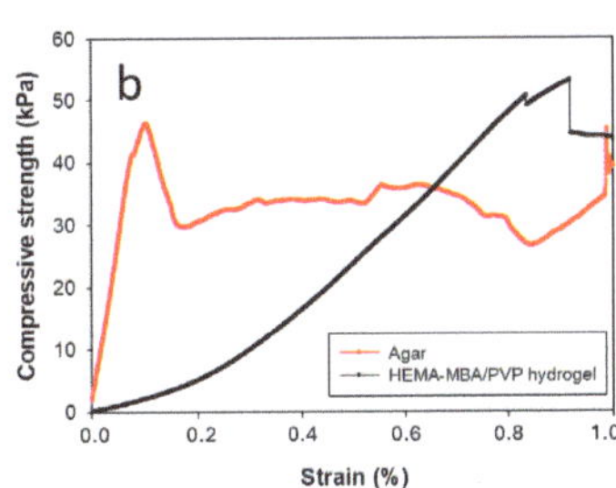

Figure 1. (**a**) The hardness and Young's modulus values of the HEMA-MBA/PVP hydrogel with respect to the agar gel (mean values $\pm$ S.E.; $n = 3$); (**b**) Stress–strain graphs of the considered gels.

The SEM micrographs of the HEMA-MBA/PVP material reported in Figure 2 show the porous structure at a microscopic level. The EDS analysis confirmed the presence of nitrogen due to the PVP component, in addition to carbon and oxygen, which are the most abundant elements, as expected for the organic nature of the polymer.

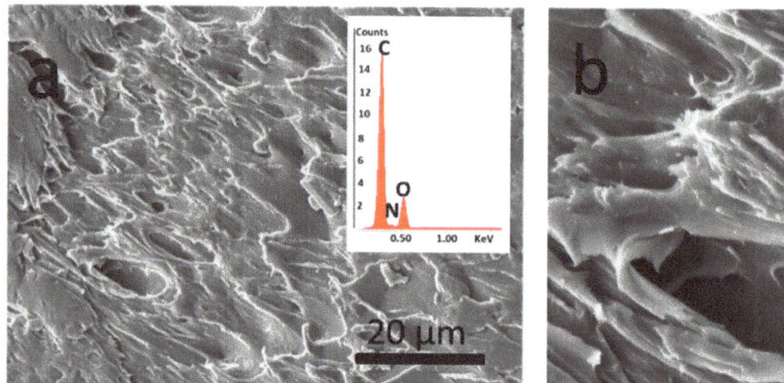

Figure 2. SEM images of the HEMA-MBA/PVP hydrogel with different magnifications (**a**,**b**). EDS spectrum in the inset.

3.2. Characterization of the Aged Polymer (Paraloid B-72) Coating

The polymer-coated stone specimens (LS and AS) were exposed to artificial ageing cycles, as explained in the experimental section, before investigating the cleaning ability of the gel-supported emulsions towards the artificially weathered polymer, which could be considered representative of naturally aged protective coatings on artifacts. The coated LS and AS specimens were artificially aged at different extents (0, 12, 25, and 35 days at T = 70 ± 2 °C) in order to correlate the cleaning effectiveness with the ageing degree of the polymer. The Paraloid B-72 ageing process was thoroughly investigated and reported in a previous paper [10]. In particular, the prolonged exposure to the artificial ageing conditions affected the physical appearance and the texture of the polymer coating, inducing color darkening (Figure 3), gloss decrease, and the partial loss of adhesion to the substrate. The effects of ageing were quite different on the two examined stones, most likely due to the distinct features of the original substrates. In any case, these property variations indicated that polymer decay, involving changes in its structure and chemical composition, occurred during the ageing process (both artificial and natural). The variations were particularly evident in the most aged samples. It should be noted that 35 days of ageing in an oven under the above-mentioned conditions could correspond to 50 years of natural ageing [5,42].

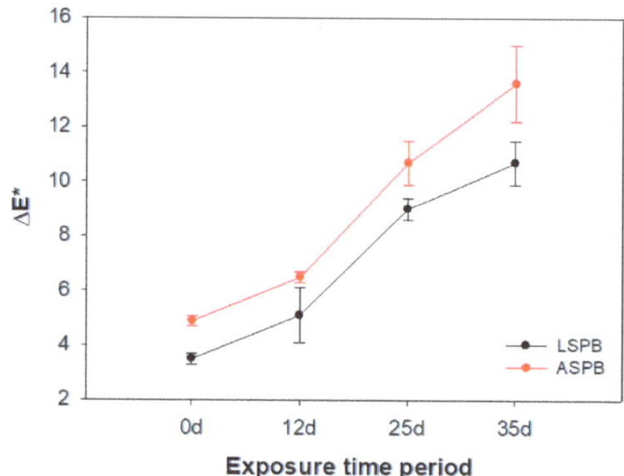

Figure 3. Overall chromatic changes (ΔE^*) observed on the coated stones before and after the ageing cycles (variations compared to uncoated/natural surfaces).

3.3. Cleaning Performances of the Loaded Hydrogels

Prior to loading them into the hydrogels, the nano-structured emulsions were analyzed using DLS in order to evaluate their droplet sizes. The analysis showed that 99% of the droplets contained in the NSEs had a size ranging between 90 and 120 nm (average size: 110.5 ± 6.3 nm, Figure S1). As reported in the literature, nano-sized droplets induce an improvement in the emulsion cleaning effectiveness [4]. Thin films of hydrogel loaded with emulsions (EcoSurf/H$_2$O and NSE) were used to test their ability to remove polymer coating from the surfaces of the LS and AS. The films (thickness of about 1 mm, area of about 12.5 cm^2) were applied to the coated stone specimens as described in Section 2.5. After removing the gel films from the surfaces, they were gently cleaned using a cotton swab to remove any possible residue. An initial naked-eye observation could determine the different behaviors of the two cleaning systems (Figure 4). In fact, the Paraloid B-72 coating seemed to be completely removed from both stones using the NSE-loaded hydrogel system, as can be clearly seen in Figure 4 (LSPB_NSE and ASPB_NSE). Meanwhile, in the other case, the polymer coating was still present on the stones' surfaces.

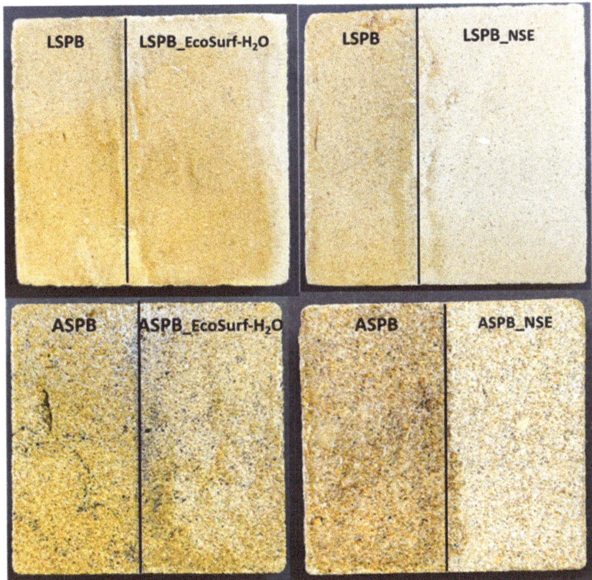

Figure 4. Coated LS and AS before and after exposure to the cleaning process using two different emulsions (EcoSurf/H$_2$O and NSE). For each specimen, the vertical bar demarcates the coated area (**left**) and the area treated with the emulsion-coated hydrogel (**right**).

To better characterize the surfaces of the LS and AS after the application of emulsions-loaded hydrogels, more investigations were performed using different experimental techniques, such as optical microscopy, SEM-EDS, chromatic and wettability measurements, and micro-FTIR (ATR mode).

3.3.1. Optical and Electron Microscopy

The morphological and micro-structural changes induced by the cleaning process on the coated stone specimens were examined using optical microscopy and SEM-EDS experiments. The OM images after the polymer removal process from the LS surface are reported in Figure 5 and Figure S2, while the corresponding images for the AS are reported in Figures S3 and S4. Images taken of the original stones and the coated stones before cleaning are also reported in both cases for comparison. After the application of the NSE-loaded hydrogel, the removal of the polymer coating from the Lecce stone surface appeared to be complete, regardless of the ageing degree (0–35 days ageing) of the considered specimens (Figure 5e,f and Figure S2e–h). On the contrary, residues of the polymer coating could be still observed on the surface of all the LS specimens (bot unaged and aged) after the application of the hydrogel loaded with the EcoSurf/H$_2$O system, indicating that the emulsion prepared without using any organic solvents displayed distinctly lower cleaning performances than the NSEs (Figure 5c,d and Figure S2a–d). Similar results were obtained after carrying out the cleaning procedure on the AS specimens. The Paraloid B-72 was exhaustively removed both from the unaged and aged specimens (Figures S3e,f and S4e–h) after the application of the NSE-loaded hydrogel, while EcoSurf/H$_2$O was poorly effective, even on the arenaria specimens (Figures S3c,d and S4a–d).

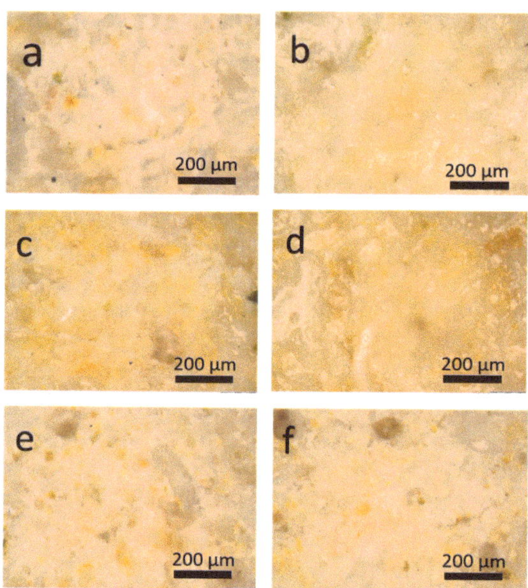

Figure 5. Optical microscope images of LS: (**a**) original/natural LS; (**b**) coated LSPB before cleaning; (**c**) unaged LSPB; (**d**) 35 d aged LSPB cleaned using EcoSurf/H$_2$O; (**e**) unaged LSPB; (**f**) 35 d aged LSPB cleaned using NSE.

The polymer removing process was further investigated via SEM-EDS analyses both on the surface and on the cross-section of the stone specimens. The results were used to understand the morphological, micro-structural, and compositional changes that occurred in both lithotypes due to the hydrogel cleaning process. The SEM-EDS experiments were performed on unaged and 35-day-old specimens, as well as on plain and coated-uncleaned stones for comparison. Observations of the aged specimens were particularly considered because they were closer to real cases in which naturally weathered polymer coatings need to be removed during conservative operations.

The microphotographs taken of the LS specimens and the corresponding EDS spectra are reported in Figure 6. The cleaning system EcoSurf/H$_2$O/hydrogel showed a very poor ability to remove the Paraloid B-72 coating (both unaged and 35-day aged) from the LS surface, confirming the results obtained via OM. In fact, the stone surface after the application of this hydrogel (Figure 6c,d) appeared to be quite similar to before the cleaning system was applied (Figure 6b). The polymer material can be clearly observed over the stone matrix. On the contrary, the NSE-based cleaning system was able to completely remove the polymer layer from the LS specimens, and the cleaned surface displayed the almost the similar morphological features to the original uncoated LS (Figure 6e,f). Moreover, no polymer residues could be observed on the stone surface (e.g., inside the pores).

The behavior of the emulsion-loaded hydrogel cleaning systems was also investigated via EDS measurements. The EDS spectra acquired for the uncleaned and cleaned LS specimens are reported in the insets of Figure 6. They show a drastic decrease in carbon content on the surfaces of the stone specimens that were cleaned using the NSE-containing hydrogel compared to the uncleaned surfaces due to the satisfactory removal of the organic coating. At the same time, an increase in the calcium content was observed on the cleaned surface because the main component of the LS (calcium carbonate) was no longer covered by the organic layer. Also, peaks corresponding to the minor components of LS (Mg, Al, Si, P) can be observed in the EDS spectra taken after the cleaning process (insets of Figure 6e,f), and the overall elemental composition was very similar to that of the plain LS (inset of Figure 6a). The EDS spectra taken after the application of the hydrogel loaded with EcoSurf/H_2O displayed an almost unaltered content of carbon and only a small increase in the calcium content with respect to the uncleaned specimen surface, suggesting that the polymer removal (if any) occurred only at a low extent.

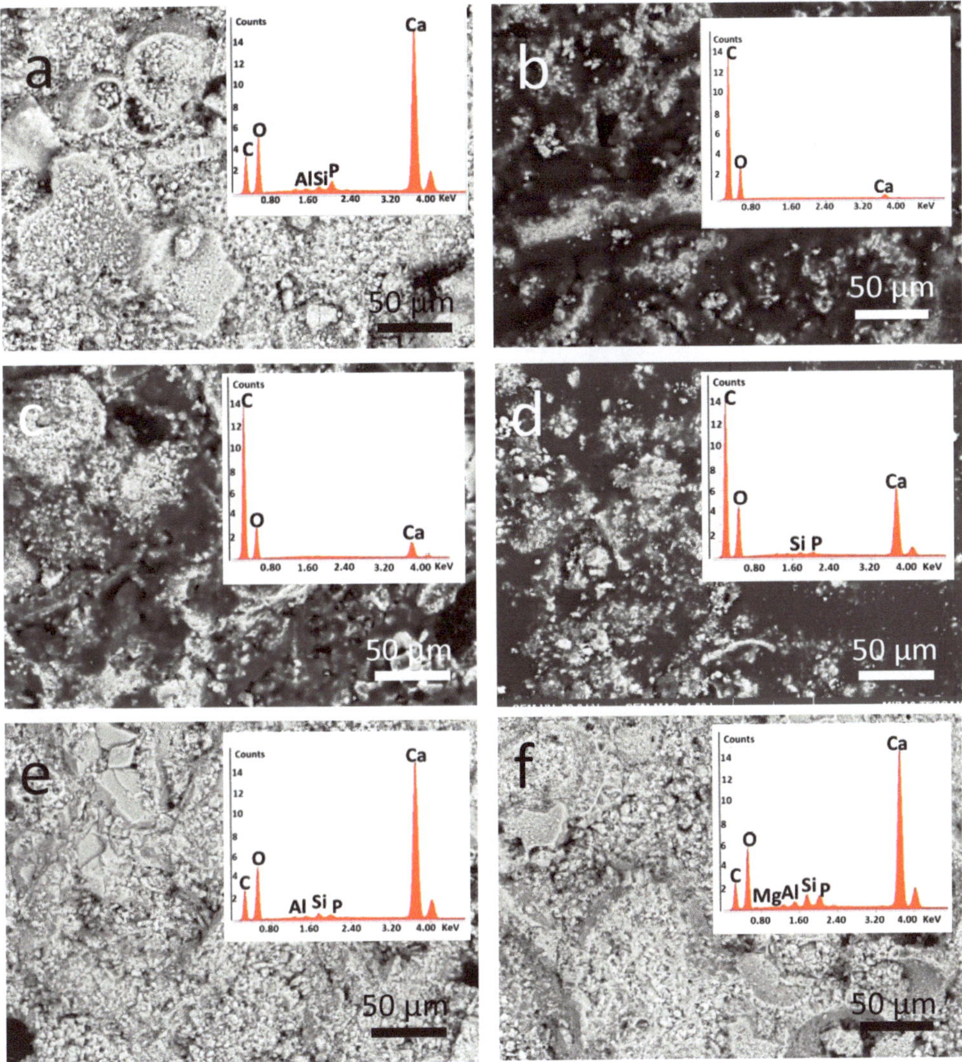

Figure 6. SEM images of the LS surfaces: (**a**) original/natural LS; (**b**) coated LSPB before cleaning; (**c**) unaged LSPB; (**d**) 35 d aged LSPB cleaned using the EcoSurf/H_2O hydrogel; (**e**) unaged LSPB; (**f**) 35 d aged LSPB cleaned using the NSE hydrogel. Corresponding EDS spectra are reported in the insets.

Similar results were obtained via the SEM-EDS experiments performed on the AS specimens. The homogeneous coating surfaces on the arenaria were not (or very poorly) affected by the cleaning procedure involving the EcoSurf/H_2O-hydrogel system, while the original appearance of the AS surface was completely recovered after the cleaning performed using the NSE-hydrogel system (Figure 7). In this case too, a drastic reduction in carbon content was observed after cleaning using NSEs, and the elemental composition was very similar to the original arenaria stone, indicating a satisfactory coating removal. The other cleaning system based on the EcoSurf/H_2O emulsion, on the contrary, did not induce significant variations in the elemental composition, confirming that the polymer remained on the surface.

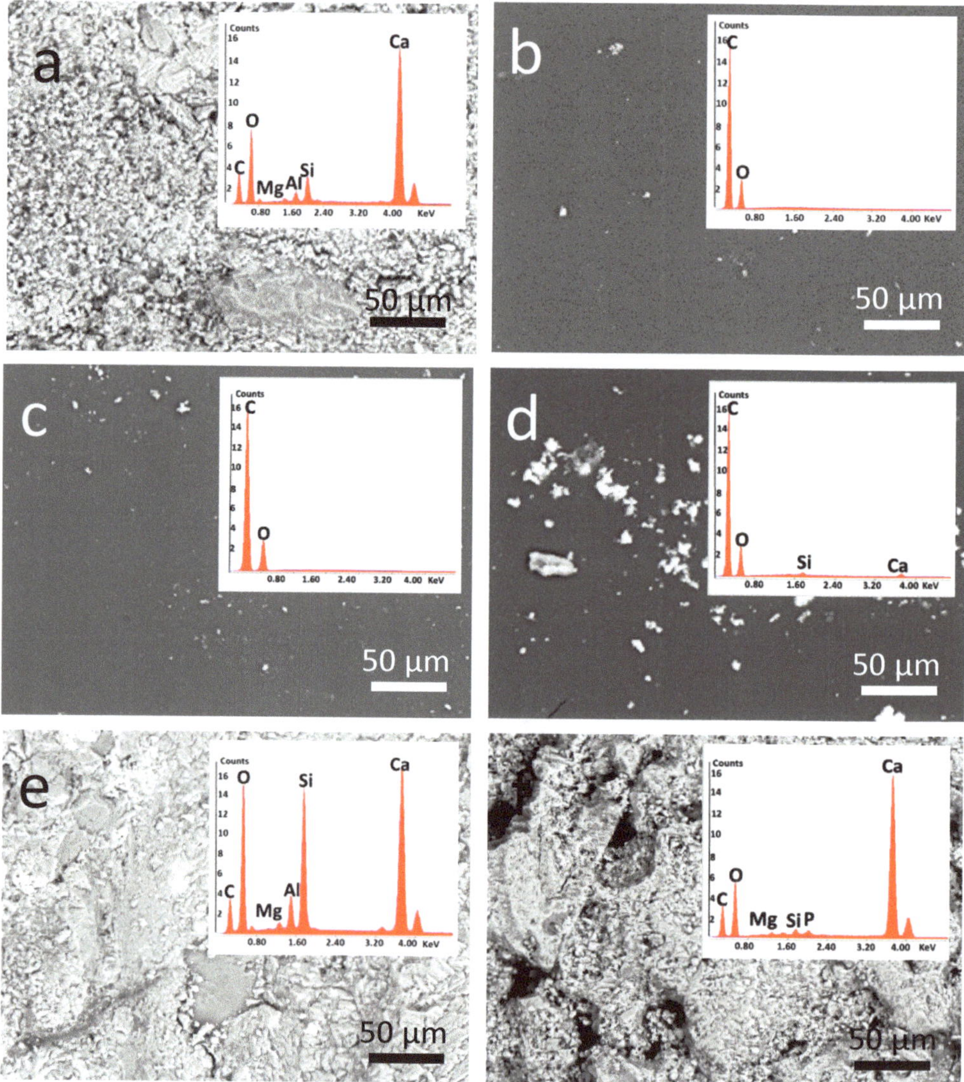

Figure 7. SEM images of AS surfaces: (**a**) original/natural AS; (**b**) coated ASPB before cleaning; (**c**) unaged ASPB; (**d**) 35 d aged ASPB cleaned using EcoSurf/H_2O; (**e**) unaged ASPB; (**f**) 35 d aged ASPB cleaned using NSEs. Corresponding EDS spectra are reported in the insets.

In addition to the qualitative observation of the EDS spectra, the results of the semi-quantitative analyses can also be considered to further explain the behavior of the cleaning systems. The C contents (weight %) are presented in Table 2, while Table S1 reports the results of the same EDS analyses expressed as carbon/calcium content ratios.

Table 2. The results of the EDS semi-quantitative analysis of both LS and AS (wt% of carbon): uncoated, coated, and after cleaning with two different emulsions (unaged and 35 d aged).

Stones	Uncoated	Coated		Cleaned—EcoSurf-H$_2$O		Cleaned—NSE	
		Unaged	35 d Aged	Unaged	35 d Aged	Unaged	35 d Aged
LS	11.3 ± 0.7	64.8 ± 0.9	46.8 ± 1.2	59.1 ± 0.7	43.7 ± 0.2	10.7 ± 0.5	10.2 ± 0.3
AS	10.5 ± 1.4	67.4 ± 0.7	67.4 ± 0.5	66.3 ± 0.2	65.2 ± 0.9	11.1 ± 0.2	11.7 ± 0.4

Furthermore, cross-section analyses were also performed to assess the effectiveness of the cleaning process carried out using the emulsion-loaded hydrogels. The SEM images taken of cross-sections of the coated LS and AS specimens (35 days-aged) before and after the cleaning process are reported in Figure 8 and Figure S4, respectively. Before the hydrogel application, the polymer layer was clearly visible on the outer stone surface (Figure 8a and Figure S5a), while the morphological features of the original stone substrates were completely recovered after the cleaning process (Figure 8b and Figure S5b,c).

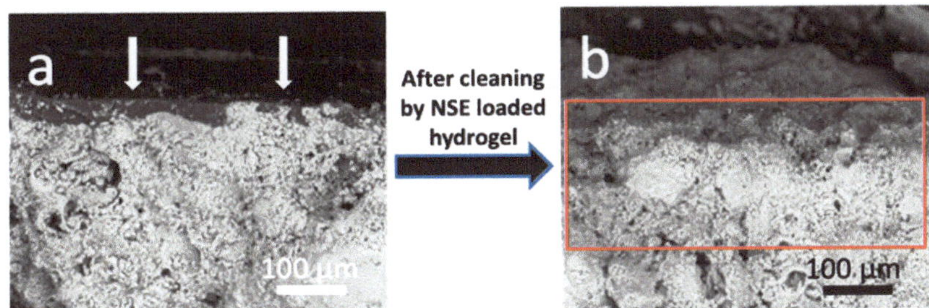

Figure 8. SEM images of the cross-section of 35 d aged LS before and after polymer removal using the NSE-loaded hydrogel: (**a**) before cleaning LSPB; (**b**) after cleaning LSPB (the red box indicates the cross-section area from the cleaned surface to a few micrometers depth).

These results further confirm the effectiveness of the NSE-loaded hydrogel as an appropriate cleaning tool for removing both the unaged and aged acrylic polymer coatings from the surfaces of the porous stone substrates.

3.3.2. Chromatic and Wettability Measurements

The chromatic variations induced by the investigated cleaning processes measured on both the LS and AS specimens and expressed as ΔE^* values (calculated with respect to the original stone materials) are reported in Figure 9. The variations in the single chromatic coordinates L^*, a^*, and b^* determined for the coated (35-days aged) and cleaned stone specimens are presented in Table S2. The presence of the aged polymer coating on the stone surface significantly affected the chromatic properties of both the LS ($\Delta E^* = 10.7 \pm 0.8$) and AS ($\Delta E^* = 13.6 \pm 1.4$). In particular, a color change towards yellow (positive Δb^* values compared to the untreated stones) and a decrease in surface brightness (negative ΔL^* values, Table S2) can be observed.

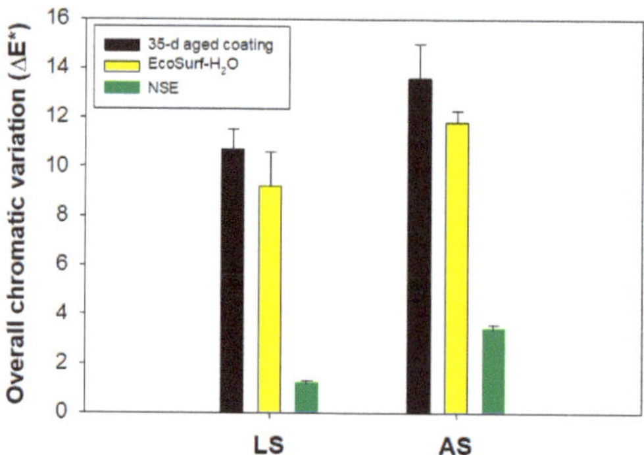

Figure 9. The overall chromatic variation (compared to natural stones) observed for the LS and AS surfaces coated with aged Paraloid B-72 coatings and after its removal using hydrogel loaded with two different emulsions (EcoSurf/H$_2$O and NSE).

After application of the EcoSurf/H$_2$O-loaded hydrogel, the chromatic properties of the treated stones were still very different from the original lithotypes and quite similar to those observed on the coated specimens. In fact, the ΔE* values underwent only a small decrease (by about 2 units for both stones), suggesting an unsatisfactory removal of the aged polymer layer from the stone surface. On the contrary, the chromatic properties observed after the application of the NSE-loaded hydrogel were closer to those of the original lithotypes. Consequently, the overall chromatic changes calculated for the cleaned LS and AS (1.2 $\pm$ 0.1 and 3.4 $\pm$ 0.2, respectively) corresponded to color variations which could be hardly detected by the naked eye (ΔE* values lower than 5) [62]. This indicates again that the NSE-based cleaning system efficiently removed the aged polymer coating from the stones' surfaces.

The wettability of the stones' surfaces was also assessed at the different stages of the cleaning process by carrying out contact angle (α) measurements (Table 3). The values of α were higher than 90° when the coated stones were considered, confirming the water-repellent behavior of the polymer coating. A decrease in the water-repellant character was observed after ageing due to the decay process undergone by the coating. The application of the EcoSurf/H$_2$O-loaded hydrogel induced only a moderate reduction in the α value measured on the LS surface (both unaged and aged specimens), while in the case of the AS, the decreased contact angle was very low and almost within the experimental error range. This strongly suggests that the coating removal (if any) was very unsatisfactory. The contact angle values could not be measured on the surface of the LS and AS after cleaning using the NSE-loaded hydrogel because water droplets were quickly absorbed by the stone matrix (Figure S6), as already observed in the case of untreated natural stones. This clearly indicates that the polymer coating was efficiently removed, restoring the original hydrophilic character of the stone surface.

Table 3. The contact angle measurements, α (°) of the coated stones before and after the ageing cycles (35 days) and after cleaning using the two different emulsions.

Stones	Contact Angle, α (°)					
	Coated		Cleaned—EcoSurf/H$_2$O		Cleaned—NSE	
	Unaged	35 d Aged	Unaged	35 d Aged	Unaged	35 d Aged
LS	95 $\pm$ 2	80 $\pm$ 3	78 $\pm$ 1	72 $\pm$ 3	n.d.	n.d.
AS	115 $\pm$ 5	95 $\pm$ 3	106 $\pm$ 3	90 $\pm$ 1	n.d.	n.d.

3.3.3. Micro-FTIR Mapping

The chemical composition of the stone surfaces after applying the polymer-removing tools (compared to the coated one, i.e., before the cleaning processes), was also investigated using the micro-FTIR technique, particularly by performing mapping experiments in the ATR mode. A picture of the examined LS specimen used for testing the NSE-loaded hydrogel, the related microscope image showing the examined area, the corresponding μ-FTIR map, and the related spectra are reported in Figure 10. The examined surface included both a coated area (left side of the specimens in Figure 10a,b) and a cleaned area (right side). The spectra were registered every 100 μm along an overall distance of around 0.5 mm so that measurements from 0 to 200 μm referred to the still coated surface and from 300 to 500 μm corresponded to the surface that underwent the cleaning treatment. The peaks observed at about 1750 and 1200 cm^{-1} in the spectra taken on coated stone can be ascribed to the acrylic polymer (C=O and C-O stretching, respectively). Bands corresponding to the main inorganic components of the LS (i.e., CaCO$_3$) were also present. When the area treated with the NSE-loaded hydrogel was examined (right side in Figure 10a,b), the representative peaks of the polymer and particularly the one ascribed to the C=O group (about 1750 cm^{-1}) [63–65] were no longer observed, and the main absorptions were only due to CaCO$_3$ (1425 and 873 cm^{-1}) [1,64,66–70], as expected for the cleaned surface. The false-color map (Figure 10c) graphically displays the results discussed above for the single FTIR spectra.

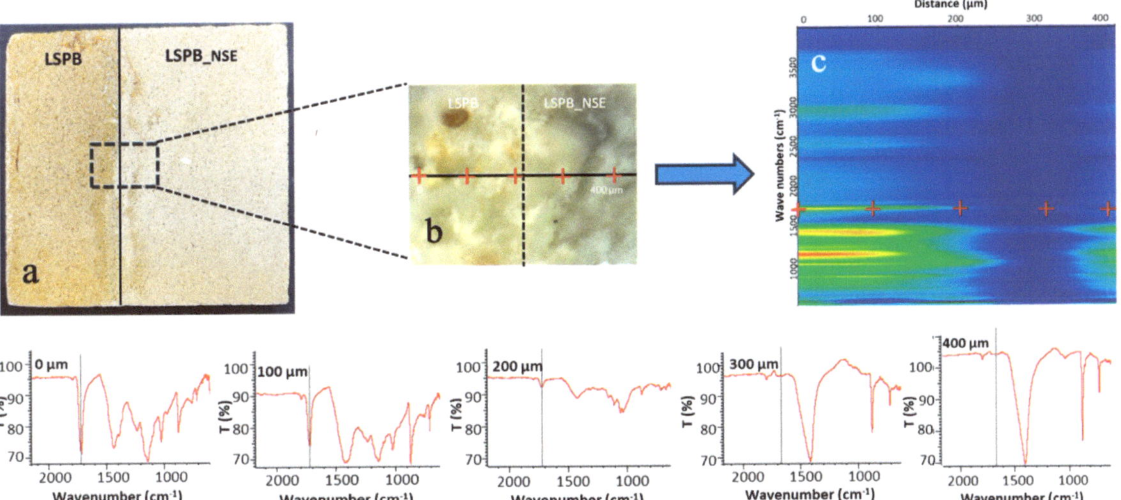

Figure 10. Micro-FTIR (ATR mode) mapping analysis performed on the 35 d coated LS surface before and after applying the NSE-loaded hydrogel. Upper side from the left: (**a**) the examined specimen with a vertical bar separating the coated surface on the left from the cleaned one on the right; (**b**) the examined area with the analysis points and their distances; (**c**) the resulting false-color μ-FTIR map. Lower side: FTIR spectra taken at different points in the examined area (from 0 to 400 μm distance); vertical bar indicates the position of the carbonyl stretching peak characteristic of the polyacrylate coating.

The same experiment was also performed on another LS specimen after the application of the EcoSurf/H$_2$O-loaded hydrogel (Figure S7). In this case, the strong peaks at about 1750 and 1200 cm^{-1}, ascribed to the polyacrylate polymer, were still detected on the treated area as well as on the coated area, confirming again the poor ability of this cleaning system to remove the coating from the stone surface.

The results of the micro-FTIR experiments also suggested that no residues of organic substances remained on the cleaned area after removing the polymer coating using the NSE-loaded hydrogel. In fact, the FTIR spectra did not show any absorption related to organic compounds (e.g., gel material, surfactant). In particular, no peaks in the carbonyl-stretching region (expected at about 1731 cm^{-1} for the HEMA-MBA/PVP copolymer) or in the C-H absorption region (both stretching and bending) were observed.

Similar experiments performed on the AS specimens did not provide useful results because their rough and irregular surfaces were not particularly suitable for the μ-FTIR-ATR technique. In fact, owing to its inhomogeneity, a satisfactory contact between the stone surface and germanium crystal could not be attained to the same extent across all the considered areas.

4. Conclusions

The present work was mainly focused on evaluating the effectiveness of the gel cleaning method for removing aged polymer coating (e.g., Paraloid B-72) from the surface of very porous stone substrates (Lecce stone, LS and arenaria stone, AS).

For this purpose, hydrogels based on the HEMA-MBA/PVP copolymer were prepared, and their properties were compared to those of agar gel, which is commonly used in the cleaning procedures that are currently applied to conservation of cultural heritage items. The considered hydrogel displayed a lower equilibrium water content (EWC) and retention capability (RC) than agar and other common hydrogels. This could be desirable when its application to water-sensitive artefacts is considered. Moreover, the investigated hydrogel was slightly harder (57.2 ± 2 kPa) than the physical gel of agar (52.3 ± 2 kPa), and at the same time, it showed lower Young's modulus values, indicating that it was more elastic than the reference gel.

To evaluate its applicability to the cleaning of stone surfaces, the HEMA-MBA/PVP hydrogel was loaded with a nano-structured emulsion (NSE) formed by the surfactant Eco-Surf™ EH-6 and limited amounts of organic solvents (2-butanol and butanone). A hydrogel loaded only with EcoSurf™ EH-6 and water was also prepared and tested for comparison.

Stone specimens (LS and AS) which had been previously coated with Paraloid B-72 were artificially aged in an oven to simulate the natural ageing (and consequent degradation) of the polymer coatings. After application of the emulsion-loaded gels on the stone specimens, the effectiveness of the polymer removal was assessed using different experimental techniques. The results showed that the NSE-loaded hydrogel allowed for satisfactory removal of the polymer, regardless of the aging time. This result was particularly demonstrated by the SEM-EDS and FTIR-mapping experiments as well as the colorimetric and wettability measurements. On the contrary, the hydrogel loaded only with EcoSurf/H$_2$O was not effective in coating removal, as the polymer was still present on the surface of both stones after the application of this cleaning system. The presence of emulsified organic solvents, although in limited amounts, seems to be essential in providing a good performance of the gel cleaning system.

In conclusion, the cleaning method based on the HEMA-MBA/PVP hydrogel and nano-structured emulsion can be considered as a promising cleaning tool that allows for the complete removal of the polymer coating from the surface of even porous stone substrates. Based on this study, this method is more effective than other common cleaning practices (e.g., cellulose pulp) because aged polymer removal is obtained after just one application without affecting the original properties of the stone substrate.

Supplementary Materials: The following supporting information can be downloaded at: https://www.mdpi.com/article/10.3390/coatings14040482/s1, Figure S1. Dynamic light scattering (DLS) analysis of prepared NSE. Figure S2. Optical microscope images of unaged and aged LS after polymer removal using emulsion-loaded hydrogels: (a) unaged LSPB; (b) 12 d LSPB; (c) 25 d LSPB; (d) 35 d LSPB using EcoSurf/H$_2$O; (e) unaged LSPB; (f) 12 d LSPB; (g) 25 d LSPB; (h) 35 d LSPB using NSE. Figure S3. Optical microscope images of AS: (a) original/natural AS; (b) coated ASPB before cleaning; (c) unaged ASPB; and (d) 35 d ASPB cleaned using EcoSurf/H$_2$O; (e) unaged ASPB; (f) 35 d ASPB cleaned using NSE. Figure S4. Optical microscope images of unaged and aged AS after polymer removal using emulsion-loaded hydrogels: (a) unaged ASPB; (b) 12 d ASPB; (c) 25 d ASPB; (d) 35 d ASPB using EcoSurf/H_2O; (e) unaged ASPB; (f) 12 d ASPB; (g) 25 d ASPB; (h) 35 d ASPB using NSE. Table S1. The results of the EDS semi-quantitative analysis of both LS and AS (carbon/calcium wt%): uncoated, coated, and after cleaning with two different emulsions (unaged and 35 days aged). Figure S5. SEM images of the cross-sections of 35 d aged AS before and after polymer removing using NSE-loaded hydrogels: (a) before cleaning ASPB; (b) after cleaning ASPB; (c) at a higher magnification of (b). Table S2. Chromatic coordinates of coated (Paraloid B-72) stone specimens after being exposed to 35 days of ageing cycles and after being removed by hydrogels loaded with two different emulsions (EcoSurf/H_2O and NSE): variations refer to the uncoated/natural stone surfaces. Figure S6. Deposition of water droplets on coated stone specimens (LSPB and ASPB) and on the same specimens after cleaning using NSE-loaded hydrogels. Figure S7. Micro-FTIR (ATR mode) mapping analysis performed on 35 d coated LS surface before and after applying EcoSurf/H_2O-loaded hydrogels. Upper side from the left: (a) the examined specimen with a vertical bar separating the coated surface on the left from the cleaned one on the right; (b) the examined area with the considered points and their distances; (c) the resulting false-color m-FTIR map. Lower side: FTIR spectra taken at the different points in the examined area (from 0 to 400 mm distance); vertical bar indicates the position of the carbonyl stretching peak characteristic of the polyacrylate coating.

Author Contributions: Conceptualization, M.L.W.; Methodology, M.L.W. and B.V.; Validation, S.R. and M.L.; Formal analysis, G.F., A.G., B.V., D.S. and S.R.; Investigation, M.L.W. and M.L.; Resources, M.L.; Data curation, M.L.W., B.V., S.R. and M.L.; Writing—original draft, M.L.W.; Writing—review & editing, M.L.W. and M.L.; Visualization, M.L.; Supervision, M.L. All authors have read and agreed to the published version of the manuscript.

Funding: The authors acknowledge support from the Ministero dell'Università e della Ricerca (MUR) and the University of Pavia through the program "Dipartimenti di Eccellenza 2023–2027".

Institutional Review Board Statement: Not applicable.

Informed Consent Statement: Not applicable.

Data Availability Statement: Data are contained within the article and Supplementary Materials.

Acknowledgments: The authors gratefully acknowledge Carlo Mangano, Department of Chemistry, University of Pavia, for his kind support during the experimental work in the laboratory, and Lavinia Doveri, Department of Chemistry, University of Pavia, for carrying out the DLS measurements.

Conflicts of Interest: The authors declare no conflict of interest.

References

1. Licchelli, M.; Malagodi, M.; Weththimuni, M.; Zanchi, C. Nanoparticles for conservation of bio-calcarenite stone. *Appl. Phys. A* **2014**, *114*, 673–683. [CrossRef]
2. Weththimuni, M.L.; Crivelli, F.; Galimberti, C.; Malagodi, M.; Licchelli, M. Evaluation of commercial consolidating agents on very porous biocalcarenite. *Int. J. Conserv. Sci.* **2020**, *11*, 251–260.
3. Domingues, J.A.L.; Bonelli, N.; Giorgi, R.; Fratini, E.; Gorel, F.; Baglioni, P. Innovative Hydrogel Based on Semi-Intrepenetrating p(HEMA)/PVP Networks for the Cleaning of Water-Sensitive Cultural Heritage Artifacts. *Langmuir* **2013**, *29*, 2746–2755. [CrossRef] [PubMed]
4. Baglioni, M.; Alterini, M.; Chelazzi, D.; Giorgi, R.; Baglioni, P. Removing Polymeric Coatings with Nanostructured Fluids: Influence of Substrate, Nature of the Film, and Application Methodology. *Front. Mater.* **2019**, *6*, 311. [CrossRef]
5. Brajer, I.; Rouzic, M.F.; Shashoua, Y.; Taube, M.; Chelazzi, D.; Baglioni, M.; Giorgi, R.; Baglioni, P. The Removal of Aged Acrylic Coatings from Wall Paintings using Microemulsions. In Proceedings of the ICOM-CC, 17th Triennial Conference 2014, Melbourne, Australia, 15–19 September 2014.
6. Casini, A.; Chelazzi, D.; Baglioni, P. Advanced methodologies for the cleaning of works of art. *Sci. China Technol. Sci.* **2023**, *66*, 2162–2182. [CrossRef]
7. Rosciardi, V.; Bandelli, D.; Bassu, G.; Casu, I.; Baglioni, P. Highly biocidal poly(vinyl alcohol)-hydantoin/starch hybrid gels: A "Trojan Horse" for Bacillus subtilis. *J. Colloid Interface Sci.* **2024**, *657*, 788–798. [CrossRef] [PubMed]
8. Kamoun, E.A.; Kenawy, E.R.S.; Chen, X. A Review on Polymeric Hydrogel Membranes for Wound Dressing Applications: PVA-Based Hydrogel Dressings. *J. Adv. Res.* **2017**, *8*, 217–233. [CrossRef] [PubMed]
9. Giorgi, R.; Baglioni, M.; Baglioni, P. Nanofluids and chemical highly retentive hydrogels for controlled and selective removal of overpainting and undesired graffiti from street art. *Anal. Bioanal. Chem.* **2017**, *409*, 3707–3712. [CrossRef] [PubMed]

10. Weththimuni, M.L.; Girella, A.; Ferretti, M.; Sacchi, D.; Licchelli, M. Nanostructured Emulsions as Smart Cleaning Materials for Removing Aged Polymer Coatings from Stone Substrates. *Sustainability* **2023**, *15*, 8117. [CrossRef]

11. Richteringa, W.; Saunders, B.R. Gel architectures and their complexity. *Soft Matter* **2014**, *10*, 3695–3702. [CrossRef]

12. Lee, C.; Volpi, F.; Fiocco, G.; Weththimuni, M.L.; Licchelli, M.; Malagodi, M. Preliminary Cleaning Approach with Alginate and Konjac Glucomannan Polysaccharide Gel for the Surfaces of East Asian and Western String Musical Instruments. *Materials* **2022**, *15*, 1100. [CrossRef] [PubMed]

13. Thakur, S.; Arotiba, O.A. Synthesis, swelling and adsorption studies of a responsive sodium alginate–poly(acrylic acid) superabsorbent hydrogel. *Polym. Bull.* **2018**, *75*, 4587–4606. [CrossRef]

14. Djabourov, M. Book Chapter; Chapter 1: Gels, Series: New Developments in NMR, NMR and MRI of Gels, Edited by Yves De Deener, *ISBN: 978-1-78801-317-8*, The Royal Society of Chemistry 2020, Published by the Royal Society of Chemistry. Available online: https://books.rsc.org/books/edited-volume/756/chapter/475969/Gels (accessed on 13 April 2024).

15. Rebers, L.; Reichsöllner, R.; Regett, S.; Tovar, G.E.M.; Borchers, K.; Baudis, S.; Southan, A. Diferentiation of physical and chemical cross-linking in gelatin methacryloyl hydrogels. *Sci. Rep.* **2021**, *11*, 3256. [CrossRef] [PubMed]

16. Shibayama, M.; Tsujimoto, M.; Ikkai, F. Static Inhomogeneities in Physical Gels: Comparison of Temperature-Induced and Concentration-Induced Sol-Gel Transition. *Macromolecules* **2000**, *33*, 7868–7876. [CrossRef]

17. Baglioni, M.; Bartoletti, A.; Bozee, L.; Chelazzi, D.; Giorgi, R.; Odlyha, M.; Pianorsi, D.; Poggi, G.; Baglioni, P. Nanomaterials for the cleaning and pH adjustment of vegetable-tanned leather. *Appl. Phys. A* **2016**, *122*, 114–124. [CrossRef]

18. Domingues, J.; Bonelli, N.; Giorgi, R.; Baglioni, P. Chemical semi-IPN hydrogels for the removal of adhesives from canvas paintings. *Appl. Phys. A* **2014**, *114*, 705–710. [CrossRef]

19. Hernandez-Martinez, A.R. Poly(2-Hydroxyethyl methacrylate-co-N,Ndimethylacrylamide)-Coated Quartz Crystal Microbalance Sensor: Membrane Characterization and Proof of Concept. *Gels* **2021**, *7*, 151. [CrossRef] [PubMed]

20. Huaman, M.A.L.; Vega-Chacòn, J.; Quispe, R.I.H.; Negròn, A.C.V. Synthesis and swelling behaviors of poly(2-hydroxyethyl methacrylate-co-itaconic acid) and poly(2-hydroxyethylmethacrylate-co-sodium itaconate) hydrogels as potential drug carriers. *Results Chem.* **2023**, *5*, 100917. [CrossRef]

21. Buemi, L.P.; Petruzzellis, M.L.; Chelazzi, D.; Baglioni, M.; Mastrangelo, R.; Giorgi, R.; Baglioni, P. Twin-chain polymer networks loaded with nanostructured fluids for the selective removal of a non-original varnish from Picasso's "L'Atelier" at the Peggy Guggenheim Collection, Venice. *Herit. Sci.* **2020**, *8*, 77. [CrossRef]

22. Baglioni, M.; Poggi, G.; Chelazzi, D.; Baglioni, P. Advanced Materials in Cultural Heritage Conservation. *Molecules* **2021**, *26*, 3967. [CrossRef]

23. Tighe, B.J. Hydrogels as Contact Lens Materials. *Hydrogels Med. Pharm.* **1987**, *3*, 53–82.

24. Mack, E.J.; Okano, T.; Kim, S.W. Biomedical Applications of Poly(2-Hydroxyethyl Methacrylate) and its Copolymers. *Hydrogels Med. Pharm.* **1987**, *2*, 65–93.

25. Haaf, F.; Sanner, A.; Straub, F. Polymers of N-Vinylpyrrolidone: Synthesis, Characterization and Uses. *Polym. J.* **1985**, *17*, 143–152. [CrossRef]

26. Carretti, E.; Dei, L. Physicochemical characterization of acrylic polymeric resins coating porous materials of artistic interest. *Prog. Org. Coat* **2004**, *49*, 282–289. [CrossRef]

27. Alessandrini, G.; Aglietto, M.; Castelvetro, V.; Ciardelli, F.; Peruzzi, R.; Toniolo, L. Comparative evaluation of fluorinated and unfluorinated acrylic copolymers as water-repellent coating materials for stone. *J. Appl. Polym. Sci.* **2000**, *76*, 962–977. [CrossRef]

28. Toniolo, L.; Poli, T.; Castelvetro, V.; Manariti, A.; Chiantore, O.; Lazzari, M. Tailoring new fluorinated acrylic copolymers as protective coatings for marble. *J. Cult. Herit.* **2002**, *3*, 309–316. [CrossRef]

29. Fardia, T.; Pintus, V.; Kampasakalia, E.; Pavlidou, E.; Schreiner, M.; Kyriacou, G. Analytical characterization of artist's paint systems based on emulsion polymers and synthetic organic pigments. *J. Anal. Appl. Pyrolysis* **2018**, *135*, 231–241. [CrossRef]

30. Jablonski, E.; Hayes, J.; Learner, T.; Golden, M. Conservation concerns for acrylic emulsion paints. *Rev. Conserv.* **2003**, *4*, 3–12. [CrossRef]

31. Learner, T.J.S.; Patricia Smithen, J.W.K.; Michael, R.S. (Eds.) Modern Paints Uncovered. In Proceedings of the Modern Paints Uncovered Symposium Organized by the Getty Conservation Institute, Tate and the National Gallery of Art, Tate Modern, London, UK, 16–19 May 2006; Getty Conservation Institute: Los Angeles, CA, USA, 2007.

32. Stoye, D.; Freitag, W. (Eds.) *Paints, Coatings and Solvents*, 2nd ed.; WILEY–VCH: Weinheim, Germany, 1998; pp. 125–129, ISBN 9783527611867.

33. Pintus, V.; Wei, S.; Schreiner, M. UV ageing studies: Evaluation of lightfastness declarations of commercial acrylic paints. *Anal. Bioanal. Chem.* **2012**, *402*, 1567–1584. [CrossRef]

34. Silva, M.F.; Doménech–Carbó, M.T.; Osete–Cortina, L. Characterization of additives of PVAc and acrylic waterborne dispersions and paints by analytical pyrolysis–GC–MS and pyrolysis–silylation–GC–MS. *J. Anal. Appl. Pyrol* **2015**, *113*, 606–620. [CrossRef]

35. Weththimuni, M.L.; Licchelli, M.; Malagodi, M.; Rovella, N.; La Russa, M. Consolidation of Bio-Calcarenite Stone by Treatment Based on Diammonium Hydrogenphosphate and Calcium Hydroxide Nanoparticles. *Measurement* **2018**, *127*, 396–405. [CrossRef]

36. Bergamonti, L.; Potenza, M.; Scigliuzzo, F.; Meli, S.; Casoli, A.; Lottici, P.P.; Graiff, C. Hydrophobic and Photocatalytic Treatment for the Conservation of Painted Lecce stone in Outdoor Conditions: A New Cleaning Approach. *Appl. Sci.* **2024**, *14*, 1261. [CrossRef]

37. Ghio, F.; Stefanelli, E.M.; Ampolo, E. The Flight of Saint Mary Magdalene—A Case Study of the Dismantling, Repositioning and Restoration of a Votive Aedicule and Wall Painting in Nardò, Lecce, Italy. *Heritage* **2023**, *6*, 3429–3447. [CrossRef]

38. Esposito, D.; Vitarelli, F.; Vita, L.; D'Onofrio, M. Conoscenza e progetto. Un caso di studio. Santa Maria di Cerrate. In *RPR, Rilievo, Progetto, Riuso*; Maggioli: Santarcangelo di Romagna, Italy, 2017; pp. 285–298, ISBN 9788891624833.

39. De Pascalis, D.G.; Leucci, G.; De Giorgi, L.; Giuri, F.; Scardozzi, G. The Basilica of Santa Caterina d'Alessandria in Galatina (Lecce, Italy): NDT surveys for the conservation project. In Proceedings of the 2019 IMEKO TC-4 International Conference on Metrology for Archaeology and Cultural Heritage, Florence, Italy, 4–6 December 2019; pp. 369–371, ISBN 978-92-990084-5-4.

40. Riganti, V.; Perotti, A.; Fiumara, A.; Veniale, F.; Zezza, U. Applicazione di tecniche strumentali al controllo del degrado delle pietre nei monumenti: Il caso della Basilica di S. Michele in Pavia. *Atti. Soc. Ital. Sci. Nat. Mus. Civ. Stor. Nat. Milano* **1981**, *122*, 109–138.

41. Baglioni, P.; Chelazzi, D.; Giorgi, R.; Poggi, G. Colloid and Materials Science for the Conservation of Cultural Heritage: Cleaning, Consolidation, and Deacidification. *Langmuir* **2013**, *29*, 5110–5122. [CrossRef] [PubMed]

42. Shashoua, Y.R. Resins in the conservation of three-dimensional works of art. In *Plastics and Resin Compositions*; Simpson, W.G., Ed.; The Royal Society of Chemistry: London, UK, 1995; pp. 294–328.

43. *UNI 10921:2001*; Beni Culturali—Materiali Lapidei Naturali Ed Artificiali—Prodotti Idrorepellenti—Applicazione Su Provini e Determinazione in Laboratorio Delle Loro Caratteristiche. UNI Ente Italiano di Normazione: Milan, Italy, 2001. Available online: https://www.biblio.units.it/SebinaOpac/resource/beni-culturali-materiali-lapidei-naturali-ed-artificiali-prodotti-idrorepellenti-applicazione-su-pro/TSA1388731?locale=eng (accessed on 28 February 2024).

44. Weththimuni, M.L.; Ben Chobba, M.; Sacchi, D.; Messaoud, M.; Licchelli, M. Durable Polymer Coatings: A Comparative Study of PDMS-Based Nanocomposites as Protective Coatings for Stone Materials. *Chemistry* **2022**, *4*, 60–76. [CrossRef]

45. Kuo, S.M.; Chang, S.J.; Wang, Y.J. Properties of PVA-AA Cross-linked HEMA-based Hydrogels. *J. Polym. Res.* **1999**, *6*, 191–196.

46. Vigani, B.; Valentino, C.; Sandri, G.; Carla, M.C.; Ferrari, F.; Rossi, S. Spermidine Crosslinked Gellan Gum-Based "Hydrogel Nanofibers" as Potential Tool for the Treatment of Nervous Tissue Injuries: A Formulation Study. *Int. J. Nanomed.* **2022**, *17*, 3421–3439. [CrossRef]

47. Hurler, J.; Engesland, A.; Kermany, P.B.; Škalko-Basnet, N. Improved Texture Analysis for Hydrogel Characterization: Gel Cohesiveness, Adhesiveness, and Hardness. *J. Appl. Poly. Sci.* **2012**, *125*, 180–188. [CrossRef]

48. Hackla, E.V.; Khutoryanskiy, V.V.; Ermolinaa, I. Hydrogels based on copolymers of 2-hydroxyethylmethacrylate and 2-hydroxyethylacrylate as a delivery system for proteins: Interactions with lysozyme. *J. Appl. Poly. Sci.* **2017**, *134*, 44768. [CrossRef]

49. Smułek, W.; Grząbka-Zasadzi'nska, A.; Kilian, A.; Ciesielczyk, F.; Borysiak, S.; Baranowska, H.M.; Walkowiak, K.; Kaczorek, E.; Jarzębski, M. Design of vitamin-loaded emulsions in agar hydrogel matrix dispersed with plant surfactants. *Food Biosci.* **2023**, *53*, 102559. [CrossRef]

50. *UNI EN 15886:2010*; Conservazione dei Beni Culturali, Metodi di Prova, Misura del Colore Delle Superfici. UNI Ente Italiano di Normazione: Milan, Italy, 2010.

51. Ben Chobba, M.; Weththimuni, M.L.; Messaoud, M.; Sacchi, D.; Bouaziz, J.; De Leo, F.; Urzi, C.; Licchelli, M. Multifunctional and Durable Coatings for Stone Protection Based on Gd-Doped Nanocomposites. *Sustainability* **2021**, *13*, 11033. [CrossRef]

52. *UNI EN 15802:2010*; Conservazione dei Beni Culturali—Metodi di Prova—Determinazione Dell'angolo di Contatto Statico. UNI: Milan, Italy, 2010.

53. Weththimuni, M.L.; Milanese, C.; Licchelli, M.; Malagodi, M. Improving the Protective Properties of Shellac-Based Varnishes by Functionalized Nanoparticles. *Coatings* **2021**, *11*, 419. [CrossRef]

54. Tamburini, G.; Canevali, C.; Ferrario, S.; Bianchi, A.; Sansonetti, A.; Simonutti, R. Optimized Semi-Interpenetrated p(HEMA)/PVP Hydrogels for Artistic Surface Cleaning. *Materials* **2022**, *15*, 6739. [CrossRef] [PubMed]

55. Oyarce, E.; Pizarro, G.; Del, C.; Oyarzún, D.P.; Zúñiga, C.; Sánchez, J. Hydrogels based on 2-hydroxyethyl methacrylate: Synthesis, characterization and hydration capacity. *J. Chil. Chem. Soc.* **2020**, *65*, 4682–4685. [CrossRef]

56. Bashir, S.; Hina, M.; Iqbal, J.; Rajpar, A.H.; Mujtaba, M.A.; Alghamdi, N.A.; Wageh, S.; Ramesh, K.; Ramesh, S. Fundamental Concepts of Hydrogels: Synthesis, Properties, and Their Applications. *Polymers* **2020**, *12*, 2702. [CrossRef] [PubMed]

57. Kou, J.H.; Amidon, G.L.; Lee, P.I. pH-Dependent Swelling and Solute Diffusion Characteristics of Poly (Hydroxyethyl Methacrylate–CO–Methacrylie Acid) Hydrogels. *Pharm. Res* **1988**, *5*, 592–597. [CrossRef] [PubMed]

58. Sikdar, P.; Uddin, M.; Dip, T.M.; Islam, S.; Hoque, S.; Dhar, A.K.; Wu, S. Recent advances in the synthesis of smart hydrogels. *Mater. Adv.* **2021**, *2*, 4532–4573. [CrossRef]

59. Lee, C.; Di Turo, F.; Vigani, B.; Weththimuni, M.L.; Rossi, S.; Beltram, F.; Pingue, P.; Licchelli, M.; Malagodi, M.; Fiocco, G.; et al. Biopolymer Gels as a Cleaning System for Different Featured Wooden Surfaces. *Polymers* **2023**, *15*, 36. [CrossRef]

60. Passaretti, A.; Cuvillier, L.; Sciutto, G.; Guilminot, E.; Joseph, E. Biologically Derived Gels for the Cleaning of Historical and Artistic Metal Heritage. *Appl. Sci.* **2021**, *11*, 3405. [CrossRef]

61. Giordano, A.; Caruso, M.R.; Lazzara, G. New tool for sustainable treatments: Agar spray—Research and practice. *Herit. Sci.* **2022**, *10*, 123. [CrossRef]

62. Weththimuni, M.; Ben Chobba, M.; Tredici, I.; Licchelli, M. ZrO_2-Doped ZnO-PDMS Nanocomposites as Protective Coatings for the Stone Materials. *Acta IMEKO* **2022**, *11*, 5. [CrossRef]

63. Smith, B.C. Infrared Spectroscopy of Polymers X: Polyacrylates. *Spectroscopy* **2023**, *38*, 10–14. [CrossRef]

64. Carretti, E.; Chelazzi, D.; Rocchigiani, G.; Baglioni, P.; Poggi, G.; Dei, L. Interactions between Nanostructured Calcium Hydroxide and Acrylate Copolymers: Implications in Cultural Heritage Conservation. *Langmuir* **2013**, *29*, 9881–9890. [CrossRef]
65. Spathis, P.; Karagiannidou, E.; Magoula, A.E. Influence of Titanium Dioxide Pigments on the Photodegradation of Paraloid Acrylic Resin. *Stud. Conserv.* **2003**, *48*, 57–64. [CrossRef]
66. Hwidi, R.S.; Izhar, T.N.T.; Saad, F.N.M. Characterization of Limestone as Raw Material to Hydrated Lime. *E3S Web Conf.* **2018**, *34*, 02042. [CrossRef]
67. Cizer, Ö.; Rodriguez-Navarro, C.; Ruiz-Agudo, E.; Elsen, J.; Gemert, D.V.; Balen, K.V. Phase and morphology evolution of calcium carbonate precipitated by carbonation of hydrated lime. *J. Mater. Sci.* **2012**, *47*, 6151–6165. [CrossRef]
68. Taylor, D.R.; Crowther, R.S.; Cozart, J.C.; Sharrock, P.; Wu, J.; Soloway, R.D. Calcium carbonate in cholesterol gallstones: Polymorphism, distribution, and hypotheses about pathogenesis. *Hepathology* **1995**, *22*, 488–496.
69. Kalbus, G.E.; Kalbus, L.H. Use of infrared spectrophotometry in the analysis of limestone. *J. Chem. Educ.* **1966**, *43*, 314–318. [CrossRef]
70. Gunasekarana, S.; Anbalagan, G. Spectroscopic characterization of natural calcite minerals. *Spectrochim. Acta. A* **2007**, *68*, 656–664. [CrossRef]

Article

Agar Foam: Properties and Cleaning Effectiveness on Gypsum Surfaces

Paulina Guzmán García Lascurain [1,*], **Sara Goidanich** [1], **Francesco Briatico Vangosa** [1], **Marilena Anzani** [2], **Alfiero Rabbolini** [2], **Antonio Sansonetti** [3] and **Lucia Toniolo** [1]

1 Department of Chemistry, Materials and Chemical Engineering "Giulio Natta", Politecnico di Milano, 20133 Milan, Italy
2 Aconerre Conservation Studio, 20154 Milan, Italy
3 Institute of Heritage Science, National Research Council (CNR-ISPC), 20125 Milan, Italy
* Correspondence: paulina.guzman@polimi.it

Abstract: In the past decade, the usage of soft materials, like gels, has allowed for a better control of the water release process into the substrate for cleaning interventions. Agar—a natural polysaccharide harvested from algae—has been used to perform cleaning of stone materials, gypsum works, and paintings with remarkably positive results. Agar presents the great advantage of being cheap, easily available, fast to produce and not toxic, allowing for more sustainable conservation works. More recently, a new type of agar fluid, agar foam, promises further control of the water release and ease of application on delicate surfaces. In the present study, this new type of agar, CO_2 and N_2O foams, has been characterized and compared with the conventional sol/gel agar system. Moreover, the cleaning effectiveness of the agar foam was tested both in laboratory conditions and in two case studies: a historical gypsum from the porch framing of the Abbey of Nonantola, and the 20th century gypsum cast of the Pietà Rondanini by Michelangelo, located in the Sforza Castle in Milan. The obtained results show that foaming changes the *sol-gel* transition temperature of the agar gel as well as incrementing its dissipative behavior. When freshly applied, the foams flow with a reduced velocity, thus allowing a better control and ease of application. Once gelified, they act as a soft *solid-like* material, as shown by their rheological properties. Moreover, it was found that CO_2 foam slightly reduces the water release to the surface, while maintaining the moldability and ease of application. The study allows for the conclusion that agar foam offers an interesting alternative for delicate surfaces, with a non-coherent mineral deposit, and with complex geometries that often represent a challenge for the conventional agar applications

Keywords: agar gel; agar foam; heritage cleaning; soft-matter cleaning; gel rheology

Citation: Guzmán García Lascurain, P.; Goidanich, S.; Briatico Vangosa, F.; Anzani, M.; Rabbolini, A.; Sansonetti, A.; Toniolo, L. Agar Foam: Properties and Cleaning Effectiveness on Gypsum Surfaces. *Coatings* **2023**, *13*, 615. https://doi.org/10.3390/coatings13030615

Academic Editor: Peter Rodič

Received: 14 February 2023
Revised: 8 March 2023
Accepted: 9 March 2023
Published: 14 March 2023

1. Introduction

Cleaning works of art is a complicated task that requires an understanding of both the surface to be cleaned and the materials to be used for cleaning. Today, the use of wet cleaning methods with water, solvents, or, if needed, specific active ingredients, are established as a standard methodology in conservation [1]. However, there are three main inconveniences arising from the use of wet cleaning methodologies: firstly, not all of the surfaces are water/solvent resistant; for example, in paints, the interaction with water or solvent might lead to swelling and/or leaching of organic components [2]. Secondly, these types of methods often require a swab to place and remove the cleaning agent, thus inducing some mechanical damage on the surface. Finally, some solvents can present health- and environmental-related problems due to their toxicity. The former is an inconvenience from the health point of view, and also introduces complications regarding the logistics of the intervention.

To maintain further control of the cleaning agent release and perform gentler interventions, scientists and restorers have focused on the use of soft matter cleaning methodologies [3–6]. These methods allow for the safe transportation of organic solvents to be released only at the targeted surfaces. Selective release is a great advantage of soft materials compared to ordinary cleaning methods because (a) their selective action allows for a reduction in the amount of solvents and surfactant molecules used, and (b) the entrapment of solvents by reduction of the rate of evaporation has a big impact on the reduced toxicity and safety for the operators.

Gels, in particular, offer the main advantage of controlling water release and reducing the mechanical action to the surface. Given their rheological properties, gels can ensure good contact with the surface and the reduction of residues after the cleaning process [7]. Moreover, gels can be designed to target specific substances to be removed from the surface [4,8,9], for example, the removal of soiling particles [6] and varnishes [10]. Additionally, these soft materials allow specific functionalization to ensure the complete removal of the gel [11].

Although lab-synthesized soft materials for cleaning offer many advantages, some of them (or their by-products) have one or a few inconveniences, namely: are hazardous, produce waste that is difficult to dispose, or require precursors that are harmful for the environment. Additionally, these materials might require higher user-safety considerations. Furthermore, some of these materials are complex to synthesize and/or expensive. Therefore, it is necessary to promote the usage of sustainable, safe, and economic cleaning methodologies that match the restoration's objectives and needs.

Among the natural soft materials, agar *gel* has been successfully implemented as a *green*, cheap, and operator-safe methodology for the cleaning of diverse surfaces including historic buildings, sculptures, and paintings [12–18]. Agar *gel* is appreciated by many restorers due to its capacity to control the water release onto the surface of interest. Given its reduced price, agar is often used in cleaning campaigns that require large quantities of material. Furthermore, agar minimizes the remaining residues over the surface after its removal, and it can be combined with surfactants that act as hydrophobic and lipophilic to remove specific compounds (for example, it is quite common among conservators, to include the commercial product Tween20®, a nonionic surfactant of the polysorbate family) [19,20]. Additionally, it can be used together with laser cleaning, improving color saturation of the surface, temperature control, and cleaning efficiency [21]. Some of the limitations of agar *gel* include its usage on extremely water sensitive materials, since they cannot tolerate the water released from the agar. Moreover, agar preferably should be prepared the day of the intervention to avoid the development of microorganisms.

The exact details of the cleaning mechanism of agar *gel* are not completely known, and have mainly been studied on stone porous materials [20,22,23]. However, it is known that an absorption–diffusion mechanism of the free water takes place through capillarity moving from the agar *gel* to the substrate. The former is followed by the reabsorption of water to the agar *gel* matrix [24]. However, in this second step, the free water returns to the *gel* containing the substances removed from the substrate. This process could be driven by counter diffusion of ions into the *gel* matrix (slow rate) or by counter absorption due to the minor capillary radius (fast rate). The water release (and absorption) capacity of the agar *gel* decreases with increasing agar concentration, since water mobility decreases with higher density of the polymeric network. In turn, agar *gel* at 1% concentration behaves as a free water source, whereas at 5%, it offers the highest control on water transport [20]. Concentration values above 5% are normally not used.

Finally, the agar *gel* is a thermoreversible material; this property makes agar a versatile method for application into surfaces. Specifically, agar will maintain a random coil formation at high temperatures (above 35–40 °C), this is the *sol* state. When the temperature of the agar is below the abovementioned range, the polysaccharide chains will reorganize into a double helix structure held together by hydrogen bonds and Van der Waals forces, forming a gel [20,25]. Furthermore, agar can be used on a variety of surfaces with diverse shapes and orientations. The three main forms of application are: *sol* (its fluid form), on preformed *gel* blocks, and *unstructured* (obtained through milling the rigid gel blocks), each with its own advantages depending on the specifications of the object to be cleaned [20,26]. For example, the *sol* form is mainly used in complex surface morphologies that can tolerate a slight adhesion of the *gel* and temperatures around 40 °C. Moreover, it is evident from the literature [20] that this form contains the highest amount of free water, thus allowing for a higher efficacy of cleaning. On the other hand, both *unstructured* and rigid *gel* blocks are suitable for delicate surfaces (a surface with intrinsic or induced matrix decohesion) thanks to their lower adhesion given by the reduced mobility of the macromolecules in the double helix conformation. Additionally, unstructured agar is sometimes used in the presence of nonpolar solvents, according to the specific needs of the restorers. The drawback of both (rigid gel and *unstructured* agar) is related to the lower soiling removal.

In recent years, a new form of application of agar has been proposed and is rapidly becoming popular among conservators: agar foam generated by insufflating gasses in the sol phase (agar foam). The latter appears to have some advantages over the traditional agar applications, displaying a gentler action on very delicate surfaces and improving the ease of application. Moreover, the use of thermal isolating foaming chamber allows for the agar to be retained at stable temperature for longer. However, up to now, there have only been practical notes regarding the usage of agar foams, and no in-depth studies about their properties and behavior have been conducted. Consequently, the aim of the present work is to study the differences in the material properties and cleaning effectiveness amongst three agar systems: agar *sol-gel*, agar CO_2 foam, and agar N_2O foam. This was done through the study of the rheological properties and water release of these materials, together with the study of the cleaning effectiveness in laboratory conditions, and real case studies: a historical gypsum cast from the internal porch framing of the Abbey of Nonantola, and the 20th century gypsum cast of the Pietà Rondanini by Michelangelo, located in the Sforza Castle in Milan.

2. Materials and Methods

2.1. Preparation of the Agar

The agar used in the present work is AgarArt from C.T.S. (M.W: 100,000–150,000 g/mol). This material is a fine ivory-color powder, obtained from the algae family *Rodoficee*, species *Gelidium* and *Gracilaria*. According to the datasheet, AgarArt in solution has a gelification temperature between 38–42 °C and a sol temperature between 85–90 °C. For each of the experiments, batches of 1000 mL of AgarArt *sol* are prepared in order to ensure that all of the agar used (foams, *sol*, and *gel*) would come from the same batch. In fact, AgarArt is a natural material for which chemical and physical properties can slightly change between batches; hence, this control is fundamental to reduce the intrinsic error in the experimental work.

The precursor for the agar foams is the agar *sol*. The former is prepared in a concentration of 4 wt% by dissolving agar powder in water, followed by microwave heating at 700 W for 7 to 7:15 min until a fast bubbling is observed. During cooking, manual stirring is performed on intervals of 3 min to promote good mixing. The resulting agar *sol* is left to gelify, and the same heating procedure is repeated a second time before introducing the *sol* to the foaming chamber. For the testing and the cleaning procedures, the reference agar *sol/gel* is that prepared as reported above.

2.2. Foaming Chamber

Each batch of agar foam is prepared with 400 mL of agar *sol*, poured into the foaming chamber (iSi Thermo Whipper) at 60 °C. To pressurize the chamber, 8.4 g of gas (CO_2 or N_2O from iSi) are inserted once the agar *sol* is inside. Finally, the system is left to stabilize for 3–10 min before the agar foam is used. When the foam is expelled from the foaming chamber, it has a temperature above the gelifying point of the agar *sol*; therefore, it can flow and be spread. With time, the agar foam gelifies and appears similar to the consolidated agar *gel*.

The thermal characterization of the foaming chamber revealed that the iSi Thermo Whipper is capable of maintaining the agar *sol* in its fluid form (above the gelifying point) for at least three hours (Figure S1). The system is characterized by a thermal loss rate of 0.050 °C/min when the ambient temperature is 25 °C and 0.055 °C/min when the ambient temperature is 18 °C.

2.3. Substrates

2.3.1. Water Release Samples

Gypsum samples sized $5 \times 5 \times 5$ cm where stabilized in an oven at 40 °C for 10 days, until constant weight was obtained. After this, the samples were stored in a desiccator with Silica gel Rubin (Fluka Analytical) before they were used. Gypsum was selected as an appropriate material to assess the water release due to its porosity that allows water penetration; moreover, gypsum is one of the materials often cleaned with agar because it can be easily harmed when cleaned with conventional systems.

2.3.2. Cleaning Effectiveness

Artificially soiled gypsum samples were used to perform the cleaning efficacy evaluation in laboratory conditions. The samples were covered with a soiling made of water with 4 wt% of Carboxymethyl Cellulose Sodium Salt (M.$\overline{\text{W}}$: 220,000 g/mol), 11 wt% of gypsum powder, and 1 wt% of carbon black. One layer of soiling was spread on the gypsum surface and left to set for one month.

The analyses of two real case studies are considered as well in the present work: a historical gypsum from the Abbey of Nonantola, and a gypsum cast of the Pietà Rondanini of Michelangelo, located in the Castello Sforzesco, Milano (Figure 1).

(a)   (b)

Figure 1. Historical surfaces used for the real case study cleaning analysis (**a**) cast element from the internal porch framing of the Abbey of Nonantola and (**b**) Cast of the Pietà Rondanini.

The soiling present in the gypsum cast from the internal porch framing of the Abbey of Nonantola was composed by a layer ranging from 27 and 63 µm made of silicates, quartz, and feldspars, as revealed from Micro-FTIR analysis. As regards the soiling layer in the cast of the Pietà, it was formed by organic residues of an intentional treatment with beeswax and silicon-based detaching agents as observed in the Micro-FTIR analysis. In both cases, the analysis was performed using a Micro-FTIR ThermoFisher Nicolet iN-10-MX with a diamond macro ATR system.

2.4. Agar Systems Characterization

2.4.1. Microscopy Observations

Both foams were observed using a Leica DM6 microscope with an objective of 5×. The images were captured within the first 2 min after the foam was expelled from the foaming chamber (temperature ~50 °C) to ensure that the foam remained as similar to the initial conditions.

2.4.2. Density Measurement

The density of all systems was characterized through gravimetry using a Sartorius Acculab ATL-224-1 scale. Measures were carried out in already gelified samples with the same volume, all obtained from the same batch of agar preparation. A total of 10 samples for each system were used.

2.4.3. Flow Characterization

During cleaning procedures, it is fundamental that the applied agar be maintained in the area of interest. The studied agar systems are often applied at temperatures above the gel point; thus, they behaves as a fluid. To understand the effect that foaming has in the flow of the agar *sol*, the three systems at 50 °C (agar *sol*, agar foam CO_2 and agar foam N_2O) were recorded while flowing along a gypsum surface. The former was fixed with a given degree of inclination. The working angles of inclination were 0°, 15°, 45°, and 90°. All moving profiles were recorded with a camera Nikon D610 mounted on a Manfrotto stabilizing arm, and the moving profile was analyzed with Tracker software.

2.4.4. Rheological Measurements

Rheological experiments in an oscillatory regime were performed to determine the gelification temperature and the mechanical response of the material. An Anton Paar MCR 502 rheometer equipped with a Peltier plate and hood system was used. The testing configuration was a plate–plate with a 50 mm diameter. For each of the tests, three measurements were taken to ensure reproducibility, and the average value was obtained from the three repetitions.

To measure the gel temperature, the materials were subjected to an oscillatory test at a constant frequency of 1 Hz and shear strain amplitude of 0.01%. The thermal history applied consisted of a 10-min isothermal step at 60 °C, followed by a cooling ramp from 60 to 25 °C at 1 °C/min, and a final isothermal step at 25 °C. The agar samples were loaded into the rheometer in their fluid form at 60 °C to avoid any thermal shock. The response in terms of the conservative and dissipative components of the dynamic modulus, (also referred to as storage and loss moduli, G' and G'' respectively) were obtained.

To determine the linear viscoelastic region (LVER) of the gel-agar foam (meaning the agar foam after this has gelified), a strain amplitude sweep test (AS) was performed. Following the protocol used by M. Bertasa, et al. [27], the agar samples for each system were prepared as cylinders of 2–3.5 mm thickness and 50 mm diameter, and left to gelify for one hour before performing the test. The AS was performed at a constant frequency of 1 Hz, temperature of 20 °C, and a normal force of 0.15 N to ensure good contact between the sample and the rheometer's plates. The strain amplitude was varied in the range of 0.0001–1%. The LVER limit was determined as the maximum strain amplitude for which G' and G'' remained constant.

The frequency response of the shear modulus was measured. For these tests a frequency sweep (FS) protocol was used in the range 0.1–100 Hz, with a constant strain amplitude value in the LVER. Sample preparation, testing temperature and normal force parameters were the same as AS test.

Finally, to understand the stability of the viscoelastic properties of the systems over time, a time sweep test (TS) was performed. In this case, a constant strain amplitude within the LVER and a constant frequency of 1 Hz were applied. The complex modulus was measured over a one-hour period, which is the average application time in the restoration practice. The temperature and normal force values were the same as the previous test protocols; the samples were prepared in the same manner as the AS and FW tests.

2.4.5. Evaluation of the Water Release

The phenomena of water release for all agar systems was analyzed based on the change in weight of the agar samples after 30 min in contact with the gypsum surface prepared as previously described (Section 2.3.1). The amount of weight loss by the agar was quantified taking into consideration the evaporation of water to the atmosphere.

The measurements were performed for all four agar systems: agar *sol*, preformed agar *gel*, agar foam with CO_2, and agar foam with N_2O. For each of these agar systems, six gypsum specimens were used. The reported result is the average for each system. Both agar and gypsum specimens were weighted individually before having any contact. Then, 15 g of each agar system were placed on the gypsum specimen's top surface for 30 min. This selected time of application is sufficiently fast to avoid water evaporation from the gypsum samples to the environment and compatible with that of agar application in real restoration works.

It is important to specify that the agar *sol* and both foams were applied at 50 °C. Since these systems are fluid at 50 °C, parafilm frameworks were used to maintain the agar systems on the desired surface of the gypsum sample.

After the 30 min, the agar was removed, and both the agar and the gypsum specimens were weighted separately. This allowed for a characterization of the water released onto the gypsum block and of the evaporated water from the agar system. The water released by the agar was calculated as the weight loss percentage of the applied agar system towards the gypsum surface.

2.5. Cleaning Effectiveness Evaluation

The effectiveness of cleaning of the agar foams was compared with that of the agar *sol*. This effectiveness was evaluated using colorimetric measurements before and after cleaning. Colors coordinates were measured with a Konica Minolta Spectrophotometer CM-600d in CIELab color space with the corresponding L^*, a^* and b^* color coordinates. Three agar systems (4 wt% of agar): agar *sol*, agar CO_2 foam, and agar N_2O foam, were applied at 50 °C, and left to clean for 30 min. The colorimetric coordinates were processed to obtain the color change after cleaning: $\Delta E = \sqrt{\Delta L^* + \Delta a^* + \Delta b^*}$, where, a higher value of ΔE indicated a higher cleaning effectiveness.

3. Results and Discussion

Agar foam is applied in the same manner as agar *sol*, meaning while it is still fluid, allowing the operator to spread the foam along the surface to be cleaned. However, unlike agar *sol*, which appears as a transparent film when applied, both agar foams have a whitish color due to the high light scattering of bubbles, as can be observed in Figure 2. Over the gelification time, and due to the bubble coalescence, the amount of bubbles is reduced, and progressively, the agar foam turns transparent.

Even though both foams appear similar to the naked eye, when observed under the microscope (Figure 3), it becomes clear that the bubble density inside each foam is different. Both foams are formed by bubbles that range between 70–420 μm in diameter; however, the amount of bubbles per unit volume is higher for CO_2 than for N_2O foam. This is consistent with the density measurements obtained for the agar *gel* (1.067 ± 0.035 g/cm^3), agar CO_2 foam (0.483 ± 0.167 g/cm^3), and the agar N_2O foam (0.760 ± 0.062 g/cm^3), with CO_2 foam being the least dense material. Therefore, the amount of agar inside the CO_2 foam is the lowest for the same volume of material.

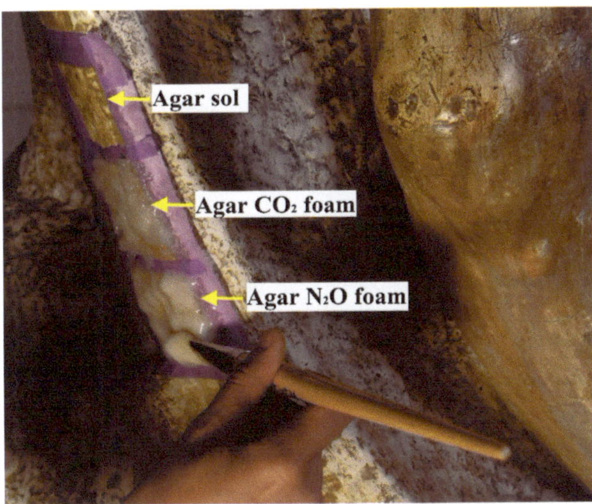

Figure 2. Application of the Agar foam to the gypsum cast of the Pietà Rondanini.

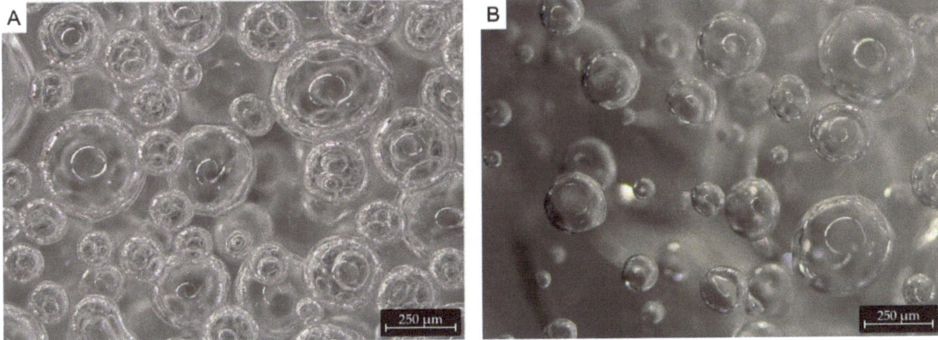

Figure 3. Stereomicroscopy ($5\times$ magn). Agar foam 2 min. after preparation (**A**) CO_2, and (**B**) N_2O.

3.1. Flow Characterization

From the practical point of view, when the agar is applied to a surface in its fluid state, it is important that the material remains as much as possible in the area of interest without dripping or flowing. If the surfaces to be cleaned are oriented with a certain inclination with respect to the horizontal, naturally, the agar will tend to flow. Furthermore, the absorption–diffusion mechanism through which the cleaning takes place is favored by a good adhesion between the agar and the surface. On this basis, it is hypothesized that the reduction of flow will also have an influence on the cleaning efficacy, since the material remains in the same position of contact for longer effective time.

In order to characterize the flow behavior of the three fluid systems (agar *sol*, agar CO_2, and N_2O foam), an inclined plane experiment was set. As can be observed in Figure 4, the front velocity of the three agar systems without any inclination is practically null; actually, the only motion is the layer's flattening under the material's weight. For all the following degrees of inclination, as expected from the density values, a clear dominance of velocity is observed for the agar *sol*. The faster flow is maintained until its matrix is gelified, hindering further movement.

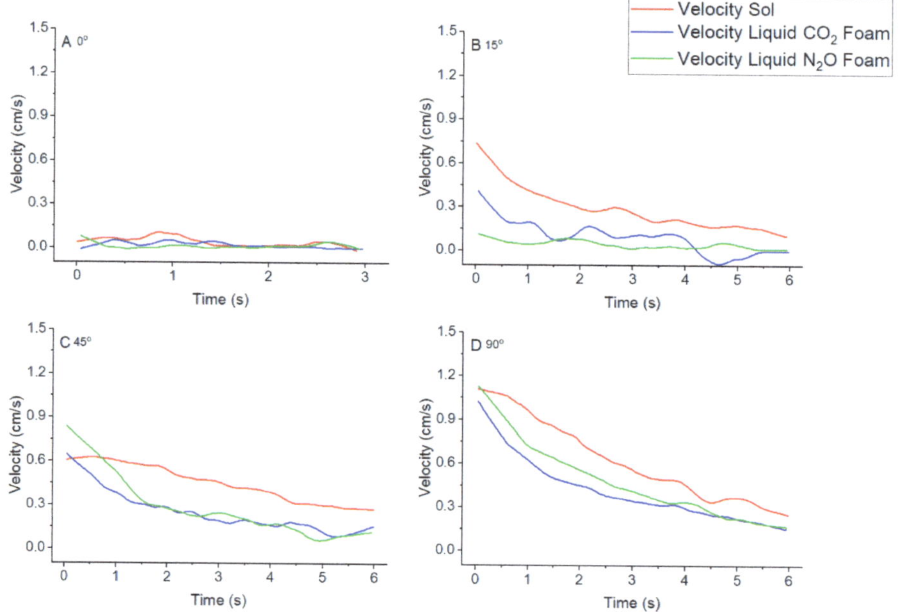

Figure 4. Behavior of the velocity front of the three fluid agar systems deposited onto a gypsum surface at different degrees of inclination (**A**) 0°, (**B**) 15°, (**C**) 45°, (**D**) 90°.

In contrast with agar *sol*, the agar foam systems show a lower flow velocity for all angles. In particular, it is observed that the movement is never totally impeded; however, the velocity decrease over time is faster, regardless of its onset (Figure 4C,D). It is possible to observe that for the limit case (90°), after one-minute motion, the velocity of the systems was reduced by 42.3% for CO_2 and 22.5% for N_2O foams, and only 9.5% for the *sol*.

One possible explanation of the flow reduction observed in the foams is that the gas bubbles inside the *sol* have a different behavior and flow differently than the *sol*. Bubbles apparently interfere with one another, slowing down the *sol* flow. Actually, these bubbles collide with each other, acting as a sort of *entanglement* to the free movement.

It is expected that for an increasing amount of bubbles, the velocity movement will be further impeded. This behavior is observed for both 45° and 90°, where the N_2O foam, which is denser (that is, with a lower amount of bubbles), has also a lower capacity of flow-speed reduction than the CO_2 foam. Slightly different is the case of an inclination of 15°; this is most probably due to the fact that at such shallow angles, the flowing effect is not so strong as to dominate over the initial velocity after pouring the samples onto the substrate.

The flowing behavior of the three systems confirms that both foams have the capacity to control the flowing movement of the agar in its fluid state. As stated before, this represents an advantage from the application point of view, since the reduction of the flow will also ease the application and reduce undesired dripping. The agar foam allows for a gentler approach towards the artworks: since the volumes of both foams are larger than that of the agar *sol*, when the foams are applied the layer formed by the foams is initially thicker than the one formed by the *sol*. Once applied, and while the material is still fluid, it is spread using a tool such as a painting brush. When using agar foams, the action of the spreading tool will naturally occur further away from the surface, acting as a sort of protective system toward the artwork surface.

3.2. Rheological Properties

The present rheological characterization of the agar systems was done through the analysis of both the storage (G′) and loss (G″) moduli. In general, the rheological behavior of a viscoelastic material can be characterized as *solid-like* behavior when G′ > G″, since the material will store more elastic energy than that dissipated, whereas the dominance of G″ over G′ depicts a more viscous or *liquid-like* behavior [28].

In a polymeric solution, such as the agar one, a *gel* will be formed when its constituents start to change from the random coil conformation to a double helix, which ultimately provides a network to form the *gel* [29]. This process is also referred to as *sol-gel* transition [20,25]. In systems such as the agar, the onset of the conformation transformation ($T_{sol\text{-}gel}$) is defined as the temperature at which G′ equals G″ during a cooling step [30].

This change in molecular conformation continues to progress over time until a plateau is reached and no further change in G′ and G″ is observed (T_{gel}). From this point on, there is a complete solidification of the material and this is no longer changing its structure. According to the datasheet of AgarArt, the gelation temperature of the *sol* occurs between 38–42 °C. Moreover, in the literature, it is reported that T_{gel} for agar *sols* oscillates between 33 and 39 °C, depending on increasing agar concentration [31]. However, these values also have an important dependency on the cooling rate, where a slower process will lead to a higher value of T_{gel}. The origin of the raw material for the agar preparation also affects the gelification temperature (T_{gel}) [32]. Considering all the possible parameters, the general accepted range for T_{gel} is 28–50 °C.

With the aim of understanding if foaming would have any effect on the *sol-gel* transition temperature ($T_{sol\text{-}gel}$) and the onset of the plateau region (T_{gel}), a temperature sweep experiment (TS) was performed starting from 60 °C. In this test, the effect of temperature on both the conservative component (G′) and the loss component (G″) of the complex modulus behavior is observed for the agar *sol* and the two foams (CO_2, and N_2O) (Figure 5). It was found that for the agar *sol*, the *sol-gel* temperature can be observed at around 58 °C, whereas for the foams, the *sol-gel* transition is never observed, because G′ > G″ at all tested temperatures, indicating that they already behave in a *solid-like* manner at 60 °C.

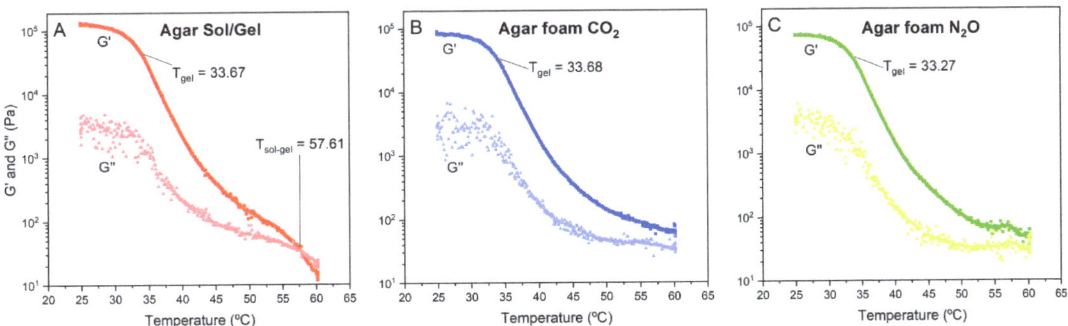

Figure 5. Temperature ramp analysis for the three systems (**A**) agar sol/gel, (**B**) CO_2 foam, and (**C**) N_2O foam. The *sol-gel* transition temperature ($T_{sol\text{-}gel}$) is marked for graph (**A**) and depicted as the crossover point between G′ and G″. The values of the gelification temperature (start of the plateau region) are marked as T_{gel} for all graphs.

The lack of a *sol-gel* transition point for both foams below 60 °C is an interesting result. It seems to indicate that even at a temperature where the agar molecules are non-structured, the material behaves as a solid, as would be expected for a material that has formed a three-dimensional network. It is hypothesized that the absence of $T_{sol\text{-}gel}$ is a consequence of the formation of the agar network at a higher temperature, possibly due to the local increment in pressure inside the foaming thermal chamber.

On the other hand, T_{gel} was obtained by calculating the first derivative of the curves and using this criterion to determine the inflection point. Despite the differences in the *sol-gel* transition temperature between agar *sol* and the two foams, the gelification temperature (start of the plateau behavior) is observed in all three systems around 34 °C, in agreement with the datasheet (38–42 °C) and the value reported on the literature (ca. 30 °C). The present result confirms that foaming affects the conformation transformation process, but only in terms of the kinetics, and not in the temperature at which the gelification is completed. Moreover, a decrease of 20 kPa in the magnitude of the storage modulus is observed for the foams when all three systems are completely gelified.

The shear strain response was measured in order to determine the linear visco-elastic region (LVER) for the three systems. This is considered as the region in which both G' and G'' are parallel and strain-independent. It was found that the response of the three agar systems to oscillatory strain is independent from shear strain amplitude up to a value of $10^{-1}\%$, which is taken as the limit of the LVER.

Within the LVER, frequency sweep tests reported in Figure 6 show a very limited frequency dependence of G' for all investigated systems. This, and the dominance of G' over G'', confirms the solid-like behavior expected for both the gel and the foams. At the mid frequency, the storage modulus, G', is 14.27 ± 1.24 kPa in the case of *sol*, 14.56 ± 2.75 kPa for CO_2 and 16.7 ± 8 kPa for N_2O foams, respectively. The values of all three systems are similar, confirming that foaming has little influence on the storage modulus of the material. It is worth observing that the value of the storage modulus (G') of all three systems is a low value, typical of soft, easily deformable materials [33,34].

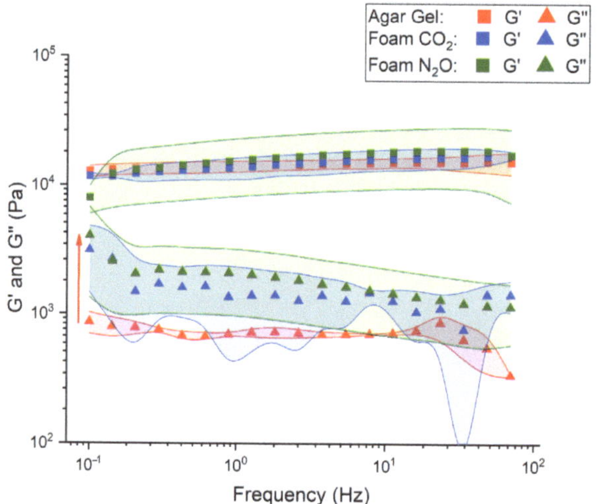

Figure 6. Viscoelastic modulus analyzed in frequency sweep test performed in the Linear Viscoelastic Region.

Unlike the behavior of G', the loss modulus (G'') for the foams is higher than that of the *gel* in the investigated frequency range. This indicates a higher dissipation ability of the foams, which may originate from both a contribution of gas entrapped in bubbles and a higher mobility of the chains, and which might result in a better adhesion to the surface.

The time stability of the agar systems was studied by means of a time sweep test at constant amplitude and frequency over one hour. The results shown in Figure 7 show a strong invariance of both G' and G'' for the three systems over the tested period. The initial variations of the systems are related to the immediate relaxation response at the onset of the experiment. The time response of the agar *gel* was as expected from the results obtained by Bertasa, et al. [27]; however, it is important to highlight that foaming does not affect this performance.

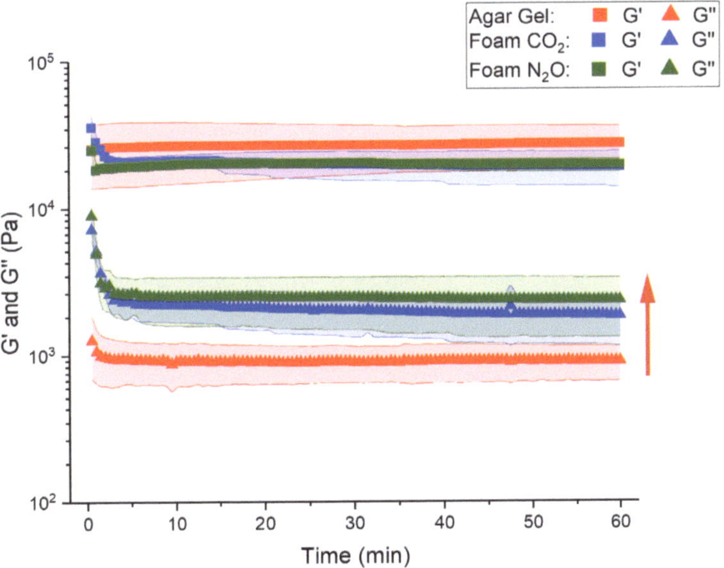

Figure 7. Viscoelastic modulus analyzed on a time sweep test.

Time stability of the systems is a desired characteristic for cleaning methodologies in the restoration framework. This becomes even more important for the mechanism of detachment of the cleaning agent. Agar is commonly used in periods of 30 min to 1 h, and its removal is performed by peeling of the layer with a movement that resembles the detachment of a thin membrane from a substrate. If the viscoelastic properties changed throughout the time of application, it could be possible that part of the material remained attached to the surface. Therefore, having a material that behaves in the same constant and predictable manner for the time of application ensures reliability of usage.

3.3. Water Release

Water release differences amongst agar systems can give insight into the cleaning process and efficacy. It is known that the amount of water released by the agar depends inversely on the concentration of the polysaccharide. The water contained by the agar gel is divided into free water (freezable) and bounded water (non-freezable). As the agar concentration increases, so does the amount of non-freezable water (which is bounded to the polymer network), and thus, the water release is reduced [20]. In this sense, agar at low concentrations, such as 1%, acts as a free water source (c.a. 90% freezable water), whereas agar at 3% concentration has only around 80% of freezable water. However, in practice, there is a limitation: the consistency of the agar and speed of gelation increase with concentration. Therefore, a compromise between freezable water and the practical consistency of the material has to be made.

The behavior of the water release for each of the agar systems showed that at constant agar concentration, foaming slightly reduces the amount of water that is released to the surface. Nevertheless, agar *gel*, *sol*, and N$_2$O foam show very similar water release values (Table 1). CO$_2$ agar foam releases a slightly lower amount of water (8.5%, as reported in Table 1). Moreover, these results are consistent with what is reported in the literature for agar *sol* and *gel*, in which, the former releases the highest amount of water [15,24].

Table 1. Percentage water release for the four different agar systems.

Systems	Water Release (%)
Agar *gel*	9.44 ± 1.18
Agar *sol*	9.63 ± 0.98
Agar CO_2 foam	8.49 ± 0.58
Agar N_2O foam	9.28 ± 1.06

The observed reduction of water release can be explained through the reduction in density due to foaming, where agar CO_2 foam, being the least dense material, releases the lowest amount of water. A foam drainage effect might also be considered, where a material with lower density corresponds to a lower amount of freezable water close to the surface [35].

As previously discussed in the present work, foaming has a relevant effect on the control of practical consistency of the agar, making it easier to handle. From the point of view of the application, the use of foam implies a better control when working on delicate surfaces. Moreover, CO_2 foaming slightly influences the amount of water that is released to a surface.

3.4. Cleaning Effectiveness Evaluation

Three cleaning tests were analyzed: one reference laboratory test performed on artificially soiled gypsum specimens and two case studies with different surface conditions. All the cleaning tests were performed with the reference agar *sol/gel* and the two foams in the same conditions (30-min cleaning). Cleaning effectiveness was evaluated through color measurements aimed to assess the total color difference achieved in the different tests (Figure 8).

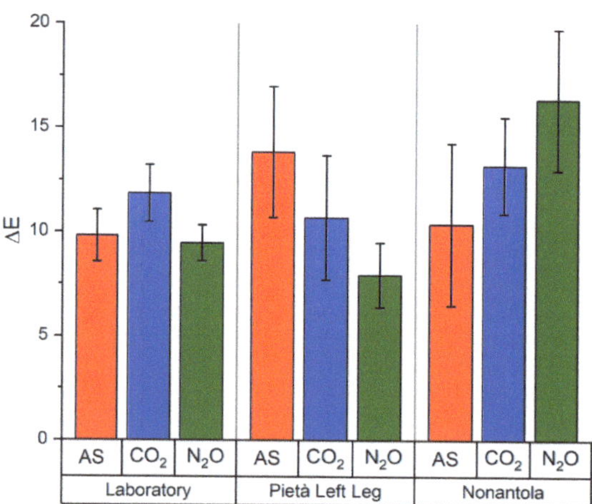

Figure 8. Cleaning performance based on ΔE^* for the different cleaning systems (agar sol/gel—AS; agar CO_2 foam—CO_2; and agar N_2O foam—N_2O) in laboratory conditions and with two case studies.

The laboratory cleaning experiments highlight that the three systems perform similarly in terms of their cleaning effectiveness. When observed more in depth, both agar *sol/gel* and N_2O foam have the same behavior, in accordance with the water release testing. On the other hand, CO_2 foam shows a slightly higher ΔE, despite the lower water release.

In the case of the cast of the Pietà Rondanini (Figure 8), it is possible to observe that the agar *sol/gel* has the highest effect on ΔE, followed closely by the CO_2, and finally, by the N_2O foam. This was clearly evident upon visual observation on site (Figure 9). On the other hand, the cleaning behavior on the historical gypsum cast of the porch framing of the Nonantola Abbey shows the complete reversed order of ΔE, with the N_2O foam providing the highest cleaning effect. This difference between the results of the laboratory testing and the real case studies highlights the relevance of the nature and characteristics of the layers to be removed. The cast of the Pietà Rondanini and the historical gypsum cast from the porch framing of the Nonantola Abbey have very different surface layers. Micro-FTIR investigations (Figure 10b) allowed for a characterization of the surface patination over the Pietà cast, showing organic residues of an intentional treatment with beeswax (red regions in the Figure 10b) and silicon-based detaching agents (1260, 1020, 870, and 800 cm^{-1}). As regards the Nonantola Abbey historical gypsum cast (Figure 10a), micro-FITR revealed that the surface deposit is composed of inorganic minerals coming from the environment: mainly feldspars, aluminosilicates (1355, 667 and 597 cm^{-1}), and quartz. Therefore, the adhesion and solubility of these two different surface mixtures is quite different.

Figure 9. Image after cleaning procedure with agar sol, agar CO_2 foam and agar N_2O foam on the cast of the Pietà Rondanini.

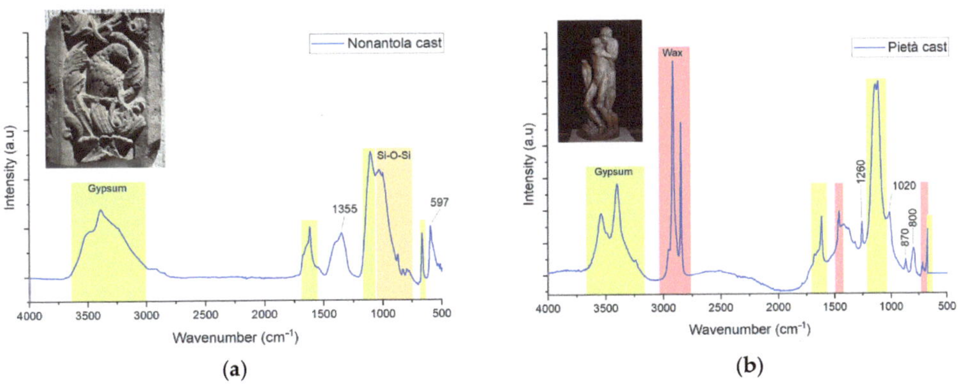

Figure 10. FTIR spectra of (a) atmospheric soiling on gypsum from cast of Nonantola Abbey and (b) wax over gypsum of the cast of the Pietà Rondanini.

The most coherent and adherent surface layer present on the Pietà cast is better removed by the agar *sol/gel*. While the less coherent deposit present in the Nonantola Abbey gypsum cast is better cleaned by N_2O foam. This correlates with the rheological properties, and specifically with the difference in G″ values reported in Figure 6, where it is demonstrated that the agar N_2O foam shows the highest *viscous* behavior.

It is worth mentioning that in real-case applications, it is important to control the cleaning process as much as possible. The obtained results showed that thanks to the different properties of the three agar systems, it is possible to tune the cleaning effects according to the nature and level of soiling. Restorers have a great sensibility in selecting specific methodology according to the surface and the desired cleaning level. Despite the common chemical nature of the three systems, the rheological properties modified by the foaming process thus allow for a differentiation of the cleaning effectiveness.

4. Conclusions

The main objective of this work was to characterize and compare the material properties and cleaning effectiveness of three agar systems: *sol/gel*, CO_2 foam, and N_2O foam.

As expected, microscopy and density measurements showed that foaming reduces the density with respect to the conventional agar. In particular, CO_2 foaming produces a material with a higher amount of bubbles and a lower density, whereas N_2O provides a material with a moderate density reduction.

By means of the rheological test, it was possible to confirm that the gelification temperature remained unmodified for all three agar systems (33 °C), while the onset of *sol-gel* transition increased for the foams. Moreover, it is noteworthy to point out that the foams, when applied fresh, flow with a reduced velocity, thus allowing for better control and ease of application. Once gelified, the foams act as a soft *solid-like* material, as shown by rheological studies. Finally, agar foams, as the reference agar *sol/gel*, show a stable behavior for the typical time of application (30–60 min).

As concerns the water release, the laboratory test showed that the agar N_2O foam and the *sol/gel* behave similarly, while CO_2 foam releases a slightly lower amount of water.

Finally, the cleaning effectiveness was evaluated by colorimetric measurements on both artificially soiled gypsum samples and historical gypsum case studies. Results showed that in laboratory cleaning experiments, the three systems perform similarly. However, in real cases, the behavior of foams highly depends on the nature of the surface and soiling. For instance, the presence of organic materials, such as waxes, strongly affects the interaction with water, and therefore, the effectiveness of the cleaning. The availability of the three agar systems allows restorers to tune the desired cleaning effect.

To summarize, although the selection of a cleaning methodology depends on the specific case at hand, agar foam is a good cleaning system that is ideal for delicate surfaces with geometries and orientations that represent a challenge for conventional cleaning methods. Agar foam is more likely to remain in the application area and allow the water diffusion to remain active for longer. From the practical point of view, when the restorers apply the agar foam they use a lower amount of active material than with the conventional agar methodologies. The former has a direct impact in the cost of the intervention and the supply of material required. Additionally, the use of the thermally insulating chamber permits the maintainance of the precursor agar *sol* at an adequate temperature for longer periods of time. In this sense, the ability to control better the application temperature allows for the use of a larger spectrum of active ingredients.

Supplementary Materials: The following supporting information can be downloaded at: https://www.mdpi.com/article/10.3390/coatings13030615/s1, Figure S1: Temperatures of the agar gel inside the ThermoWhipper for initial temperature of 80 °C.

Author Contributions: Conceptualization, S.G. and L.T.; methodology, S.G., A.S. and F.B.V.; investigation, P.G.G.L. and F.B.V.; resources, M.A., A.R. and A.S.; data curation, P.G.G.L. and S.G.; writing—original draft preparation, P.G.G.L.; writing—review and editing, P.G.G.L., S.G., F.B.V., M.A., A.R., A.S. and L.T.; supervision, S.G. and L.T.; project administration, S.G.; funding acquisition, L.T. All authors have read and agreed to the published version of the manuscript.

Funding: The present work was financed by Mexican Consejo Nacional de Ciencia y Tecnología (CONACyT) scholarship number: 2019-000034-01EXTF-00140, and the silver merit-based scholarship given by Politecnico di Milano. The elaboration and writing phase was supported by PhD Fellowship DOT1316197 (IMTR) CUP D45F21003710001 (Green) financed by Italian PON "Ricerca e Innovazione" 2014–2020 funding program.

Institutional Review Board Statement: Not applicable.

Informed Consent Statement: Not applicable.

Data Availability Statement: The data that support the findings of this study are available from the coordinator of the project, Sara Goidanich (sara.goidanich@polimi.it), upon reasonable request.

Acknowledgments: The authors would like to thank Davide Moscatelli and Edoardo Terreni for providing the Carboxymethyl Cellulose Sodium Salt employed in the present work.

Conflicts of Interest: The authors declare no conflict of interest.

References

1. *UNI EN 17138:2019*; Conservation of Cultural Heritage—Methods and Materials for Cleaning Porous Inorganic Materials. Ente Nazionale Italiano di Unificazione: Milano, Italy, 2019.
2. Ormsby, B.; Learner, T. The Effects of Wet Surface Cleaning Treatments on Acrylic Emulsion Artists' Paints—A Review of Recent Scientific Research. *Stud. Conserv.* **2009**, *54*, 29. [CrossRef]
3. Baglioni, M.; Giorgi, R.; Berti, D.; Baglioni, P. Smart Cleaning of Cultural Heritage: A New Challenge for Soft Nanoscience. *Nanoscale* **2012**, *4*, 42. [CrossRef]
4. Segel, K.; Brajer, I.; Taube, M.; de Fonjaudran, C.M.; Baglioni, M.; Chelazzi, D.; Giorgi, R.; Baglioni, P. Removing Ingrained Soiling from Medieval Lime-Based Wall Paintings Using Nanorestore Gel®Peggy 6 in Combination with Aqueous Cleaning Liquids. *Stud. Conserv.* **2020**, *65*, 284. [CrossRef]
5. Domingues, J.; Bonelli, N.; Giorgi, R.; Baglioni, P. Chemical Semi-IPN Hydrogels for the Removal of Adhesives from Canvas Paintings. *Appl. Phys. A* **2014**, *114*, 705. [CrossRef]
6. Bartoletti, A.; Barker, R.; Chelazzi, D.; Bonelli, N.; Baglioni, P.; Lee, J.; Angelova, L.; Ormsby, B. Reviving WHAAM! A Comparative Evaluation of Cleaning Systems for the Conservation Treatment of Roy Lichtenstein's Iconic Painting. *Herit. Sci.* **2020**, *8*, 1.
7. Baglioni, P.; Baglioni, M.; Bonelli, N.; Chelazzi, D.; Giorgi, R. Smart Soft Nanomaterials for Cleaning. In *Nanotechnologies and Nanomaterials for Diagnostic, Conservation and Restoration of Cultural Heritage*; Lazzara, G., Fakhrullin, R.F., Eds.; Elsevier: Amsterdam, The Netherlands, 2018; pp. 171–204.
8. Baglioni, P.; Berti, D.; Bonini, M.; Carretti, E.; Dei, L.; Fratini, E.; Giorgi, R. Microemulsions, and Gels for the Conservation of Cultural Heritage. *Adv. Colloid Interface Sci.* **2014**, *205*, 361. [CrossRef] [PubMed]
9. Carretti, E.; Grassi, S.; Cossalter, M.; Natali, I.; Caminati, G.; Weiss, R.G.; Baglioni, P.; Dei, L. Poly (Vinyl Alcohol)-Borate Hydro/Cosolvent Gels: Viscoelastic Properties, Solubilizing Power, and Application to Art Conservation. *Langmuir* **2009**, *25*, 8656. [CrossRef]
10. Buemi, L.P.; Petruzzellis, M.L.; Chelazzi, D.; Baglioni, M.; Mastrangelo, R.; Giorgi, R.; Baglioni, P. Chain Polymer Networks Loaded with Nanostructured Fluids for the Selective Removal of a Non-Original Varnish from Picasso's "L'Atelier" at the Peggy Guggenheim Collection, Venice. *Herit. Sci.* **2020**, *8*, 77. [CrossRef]
11. Bonini, M.; Lenz, S.; Giorgi, R.; Baglioni, P. Nanomagnetic Sponges for the Cleaning of Works of Art. *Langmuir* **2007**, *23*, 8681. [CrossRef]
12. Bertasa, M.; Canevali, C.; Sansonetti, A.; Lazzari, M.; Malandrino, M.; Simonutti, R.; Scalarone, D. An In-Depth Study on the Agar Gel Effectiveness for Built Heritage Cleaning. *J. Cult. Herit.* **2021**, *47*, 12. [CrossRef]
13. Gulotta, D.; Saviello, D.; Gherardi, F.; Toniolo, L.; Anzani, M.; Rabbolini, A.; Goidanich, S. Setup of a Sustainable Indoor Cleaning Methodology for the Sculpted Stone Surfaces of the Duomo of Milan. *Herit. Sci.* **2014**, *2*, 6. [CrossRef]
14. Diamond, O.; Barkovic, M.; Cross, M.; Ormsby, B. The Role of Agar Gel in Treating Water Stains on Acrylic Paintings: Case Study of Composition, 1963 by Justin Knowles. *J. Am. Inst. Conserv.* **2019**, *58*, 144. [CrossRef]
15. Giordano, A.; Caruso, M.R.; Lazzara, G. New Tool for Sustainable Treatments: Agar Spray—Research and Practice. *Herit. Sci.* **2022**, *10*, 1. [CrossRef]
16. Senserrich-Espuñs, R.; Anzani, M.; Rabbolini, A.; Font-Pagès, L. Deconstructed Agar Gels in the Restoration of the Wall Paintings in the Chapel of Sant Miquel, in Monestir Reial de Santa Maria de Pedralbes, Barcelona. In Proceedings of the Gels in the Conservation; Angelova, L.V., Bronwyn, O., Eds.; Archetype Pubblications Ltd.: London, UK, 2017; pp. 148–151.
17. Sansonetti, A.; Casati, M.; Striova, J.; Canevali, C.; Anzani, M.; Rabbolini, A. A Cleaning Method Based on the Use of Agar Gels: New Tests and Perspectives. In Proceedings of the 12th International Congress on the Deterioration and Conservation of Stone, New York, NY, USA, 21–25 October 2012.

18. Gulotta, D.; Anzani, M.; Petiti, C.; Rabbolini, A.; Cucciniello, O.; Goidanich, S. Un Approccio Integrato per La Conoscenza e Conservazione in Funzione Della Mostra Alla Galleria d'Arte Moderna: 100 Anni. Scultura a Milano 1815–1915. In Proceedings of the "Lo Stato Dell'Arte 15" XV Congresso Nazionale IGIIC, Bari, Italy, 12–14 October 2017; Nardini editore. pp. 209–216.

19. Anzani, M. Utilizzo Dei Gel Rigidi Di Agar per La Pulitura Delle Superfici Lapidee. In *Tecniche Di Restauro*; Musso, S.F., Ed.; UTET: Belluno, Italy, 2013.

20. Sansonetti, A.; Bertasa, M.; Canevali, C.; Rabbolini, A.; Anzani, M.; Scalarone, D. A Review in Using Agar Gels for Cleaning Art Surfaces. *J. Cult. Herit.* **2020**, *44*, 285–296. [CrossRef]

21. Striova, J.; Anzani, M.; Borgioli, L.; Brunetto, A.; Rabbolini, A.; Roati, G.; Sansonetti, A. A Novel Approach:Laser Cleaning Intermediated by Agar Gel. In Proceedings of the LACONA IX: 9th Conference on Lasers in the Conservation of Artworks, London, UK, 7–10 September 2011.

22. Canevali, C.; Fasoli, M.; Bertasa, M.; Botteon, A.; Colombo, A.; di Tullio, V.; Capitani, D.; Proietti, N.; Scalarone, D.; Sansonetti, A. A Multi-Analytical Approach for the Study of Copper Stain Removal by Agar Gels. *Microchem. J.* **2016**, *129*, 249. [CrossRef]

23. Volk, A.; van den Berg, K.J. Agar—A New Tool for the Surface Cleaning of Water Sensitive Oil Paint? In *Issues in Contemporary Oil Paint*; van den Berg, K.J., Burnstock, A., de Keijzer, M., Krueger, J., Learner, T., Tagle, A., Heydenreich, G., Eds.; Springer International Publishing: New York City, NY, USA, 2014; pp. 389–406.

24. Bertasa, M.; Poli, T.; Riedo, C.; di Tullio, V.; Capitani, D.; Proietti, N.; Canevali, C.; Sansonetti, A.; Scalarone, D. A Study of Non-Bounded/Bounded Water and Water Mobility in Different Agar Gels. *Microchem. J.* **2018**, *139*, 306. [CrossRef]

25. Arnott, S.; Fulmer, A.; Scott, W.E.; Dea, I.C.M.; Moorhouse, R.; Rees, D.A. The Agarose Double Helix and Its Function in Agarose Gel Structure. *J. Mol. Biol.* **1974**, *90*, 269. [CrossRef]

26. Anzani, M.; Rabbolini, A. Il Restauro Del Materiale Lapideo Con i Gel Di Agar Agar: Il Gesso "La Bella Pallanza" e Il Marmo "Busto Di Donna" Di Paolo Troubetzkoy. In Proceedings of theVII Congresso Nazionale IGIIC—Lo Stato Dell'Arte, Naples, Italy, 8–10 October 2009. Atti del VII Congresso Nazionale IGIIC Napoli.

27. Bertasa, M.; Dodero, A.; Alloisio, M.; Vicini, S.; Riedo, C.; Sansonetti, A.; Scalarone, D.; Castellano, M. Agar Gel Strength: A Correlation Study between Chemical Composition and Rheological Properties. *Eur. Polym. J.* **2020**, *123*, 109442. [CrossRef]

28. Stokes, J.R.; Frith, W.J. Rheology of Gelling and Yielding Soft Matter Systems. *Soft Matter* **2008**, *4*, 1133. [CrossRef]

29. Mohammed, Z.H.; Hember, M.W.N.; Richardson, R.K.; Morris, E.R. Kinetic and Equilibrium Processes in the Formation and Melting of Agarose Gels. *Carbohydr. Polym.* **1998**, *36*, 15. [CrossRef]

30. Zhao, Y.; Cao, Y.; Yang, Y.; Wu, C. Rheological Study of the Sol−Gel Transition of Hybrid Gels. *Macromolecules* **2003**, *36*, 855. [CrossRef]

31. Labropoulos, K.C.; Rangarajan, S.; Niesz, D.E.; Danforth, S.C. Dynamic Rheology of Agar Gel Based Aqueous Binders. *J. Am. Ceram. Soc.* **2001**, *84*, 1217. [CrossRef]

32. Labropoulos, K.C.; Niesz, D.E.; Danforth, S.C.; Kevrekidis, P.G. Dynamic Rheology of Agar Gels: Theory and Experiments. Part I. Development of a Rheological Model. *Carbohydr. Polym.* **2002**, *50*, 393. [CrossRef]

33. Ross-Murphy, S. Rheological Characterisation of Gels 1. *J. Texture Stud.* **1995**, *26*, 391. [CrossRef]

34. Wyss, H.M. *Rheology of Soft Materials. Fluids, Colloids and Soft Materials: An Introduction to Soft Matter Physics*; John Wiley & Sons: Hoboken, NJ, USA, 2016; p. 149.

35. Cox, S.J.; Weaire, D.; Hutzler, S.; Murphy, J.; Phelan, R.; Verbist, G. Applications and Generalizations of the Foam Drainage Equation. *Proc. R. Soc. London. Ser. A Math. Phys. Eng. Sci.* **2000**, *456*, 2441. [CrossRef]

MDPI AG
Grosspeteranlage 5
4052 Basel
Switzerland
Tel.: +41 61 683 77 34

Coatings Editorial Office
E-mail: coatings@mdpi.com
www.mdpi.com/journal/coatings